THEORY OF THE INHOMOGENEOUS ELECTRON GAS

PHYSICS OF SOLIDS AND LIQUIDS

SUPERIONIC CONDUCTORS
Edited by Gerald D. Mahan and Walter L. Roth

HIGHLY CONDUCTING ONE-DIMENSIONAL SOLIDS
Edited by Jozef T. Devreese, Roger P. Evrard, and Victor E. van Doren

ELECTRON SPECTROSCOPY OF CRYSTALS
V. V. Nemoshkalenko and V. G. Aleshin

MANY-PARTICLE PHYSICS
Gerald D. Mahan

THE PHYSICS OF ACTINIDE COMPOUNDS
Paul Erdös and John M. Robinson

THEORY OF THE INHOMOGENEOUS ELECTRON GAS
Edited by S. Lundqvist and N. H. March

THEORY OF THE INHOMOGENEOUS ELECTRON GAS

Edited by

S. Lundqvist
Chalmers University of Technology
Göteborg, Sweden

and

N. H. March
University of Oxford
Oxford, England

Springer Science+Business Media, LLC

Library of Congress Cataloging in Publication Data

Main entry under title:

Theory of the inhomogeneous electron gas.

(Physics of solids and liquids)
Includes bibliographical references and index.
1. Electron gas. 2. Thomas-Fermi theory. I. Lundqvist, Stig, 1925–
II. March, Norman H. (Norman Henry), 1927– . III. Series.
QC175.16.E6T46 1983 533'.7 83-2128

DOI 10.1007/978-1-4899-0415-7

CONTRIBUTORS

W. KOHN • Institute for Theoretical Physics, University of California, Santa Barbara, California 93108

N. D. LANG • IBM Thomas J. Watson Research Center, Yorktown Heights, New York 10598

S. LUNDQVIST • Chalmers University of Technology, S-41296 Göteborg, Sweden

N. H. MARCH • Theoretical Chemistry Department, University of Oxford, 1 South Parks Road, Oxford OX1 3TG, England

P. VASHISHTA • Solid State Science Division, Argonne National Laboratory, Argonne, Illinois 60439

U. VON BARTH • Institute for Theoretical Physics, University of Lund, Sweden

A. R. WILLIAMS • IBM Thomas J. Watson Research Center, Yorktown Heights, New York 10598

FOREWORD

The theory of the inhomogeneous electron gas had its origin in the Thomas–Fermi statistical theory, which is discussed in the first chapter of this book. This already leads to significant physical results for the binding energies of atomic ions, though because it leaves out shell structure the results of such a theory cannot reflect the richness of the Periodic Table. Therefore, for a long time, the earlier method proposed by Hartree, in which each electron is assigned its own personal wave function and energy, dominated atomic theory.

The extension of the Hartree theory by Fock, to include exchange, had its parallel in the density description when Dirac showed how to incorporate exchange in the Thomas–Fermi theory. Considerably later, in 1951, Slater, in an important paper, showed how a result similar to but not identical with that of Dirac followed as a simplification of the Hartree–Fock method.

It was Gombás and other workers who recognized that one could also incorporate electron correlation consistently into the Thomas–Fermi–Dirac theory by using uniform electron gas relations locally, and progress had been made along all these avenues by the 1950s.

In 1964, the Thomas–Fermi theory was formally completed when Hohenberg and Kohn showed that the ground-state energy of an inhomogeneous electron gas was a unique functional of the electron density, just as had been assumed in the Thomas–Fermi–Dirac theory and its extension to include electron correlation. Of course, the calculation of the functional is equivalent to solving the many-electron problem exactly, and needless to say this is presently not possible. Nevertheless, the last fifteen years or so have seen a great deal of progress, both in the foundations of the theory and in its applications. In this book, both the theoretical foundations and its numerical consequences for atoms, molecules, and solids are fully treated.

The outline of the book is such that the fundamentals are largely set out in Chapters 1–3, with Chapter 3 stressing especially dynamical aspects of the theory of the inhomogeneous electron gas. These three chapters should be sufficient for the reader to learn the basic theory. Chapters 4 and 5 should then be valuable for the reader who wishes to learn about the applications, and also, because of the extensive references given, to the research worker who wishes to apply the theory to atoms, molecules, and solids, including surface problems.

The background needed for profitable study of the book is a basic course in quantum mechanics, plus a working knowledge of statistical mechanics such as would be provided in an introductory course. Basic solid state theory such as is presented in the book by Ashcroft and Mermin is also taken for granted. The book is written in a manner in which, for the most part, the reader will understand the basic physical foundations even if he is not trained in advanced many-body techniques.

S. Lundqvist
N. H. March

CONTENTS

3. Density Oscillations in Nonuniform Systems

S. Lundqvist

5. Density Functional Approach to the Electronic Structure of Metal Surfaces and Metal–Adsorbate Systems

N. D. Lang

CHAPTER 1

ORIGINS—THE THOMAS–FERMI THEORY

N. H. MARCH

1. Introduction

This book is concerned entirely with the description of electronic systems, whether atoms, molecules, perfect or defect solids, in terms of the electron density. Much of the ensuing discussion will be concerned with the ground state, in which the number of electrons per unit volume at $\mathbf{r}$ will be denoted by $n(\mathbf{r})$. The objective of the theory of the inhomogeneous electron gas is to describe the properties of the ground state by this electron density, as well as to calculate this quantity $n(\mathbf{r})$.

It is important to stress here that, unlike a many-electron wave function, the electron density is an observable; for example the X-ray scattering from an atomic or molecular gas gives rather direct information about the spatial distribution of electrons $n(\mathbf{r})$. More directly, in a perfect crystal in which $n(\mathbf{r})$ is periodic, the Fourier components of $n(\mathbf{r})$ are essentially given by the intensity of X-ray scattering at the Bragg reflections. Thus, any many-electron theory which is set up for the density $n(\mathbf{r})$ can be confronted with scattering experiments and it has obvious merit to pose the theory of the inhomogeneous electron gas in terms of such an observable $n(\mathbf{r})$.

The origins of the theory of the inhomogeneous electron gas are to be found in the statistical theory of Thomas[1] and Fermi.[2] In this theory, one starts from relations appropriate to the homogeneous electron gas and then applies these locally in the inhomogeneous charge cloud such as exists in

N. H. MARCH • Theoretical Chemistry Department, University of Oxford, 1 South Parks Road, Oxford OX1 3TG, England.

atoms, molecules, or solids. Such a local assumption naturally implies approximation, which will be best when the electron density $n(\mathbf{r})$, or the potential generating it, varies by but a small fraction of itself over a characteristic electron wavelength.

In spite of the fact that such a condition is not well obeyed in some regions of the electron density in atoms and molecules (for example near nuclei, at which the potential becomes singular), there is a regime of validity of the Thomas–Fermi theory, namely, the statistical limit of many electrons, and furthermore it already embodies many of the essential features of modern density functional theory, which is the theory of the inhomogeneous electron gas. Therefore, in this first chapter we shall develop fully the Thomas–Fermi theory and its main consequences when applied to atoms and molecules. Some discussion of material under high pressure, and of defects in metals and semiconductors, will also be included.

We restrict ourselves, until the later part of Section 10, to static ground-state properties. Furthermore, the discussion mainly focuses on nonrelativistic Thomas–Fermi theory, though in the final section, Section 12, a brief review of the relativistic generalization of the theory is given. The dynamical generalization of the Thomas–Fermi theory is treated by Lundqvist in Chapter 3.

2. Electron Density–Potential Relation

We shall begin by studying N electrons moving independently in a common potential energy $V(\mathbf{r})$. If we solve the Schrödinger equation

$$\nabla^2 \psi_i + \frac{8\pi^2 m}{h^2}\Big[\epsilon_i - V(\mathbf{r})\Big]\psi_i = 0 \tag{1}$$

for the electronic wave functions $\psi_i(\mathbf{r})$ with corresponding eigenvalues ϵ_i, the electronic density is evidently given by

$$n(\mathbf{r}) = \sum \psi_i^*(\mathbf{r})\,\psi_i(\mathbf{r}) \tag{2}$$

where the summation is taken over the occupied levels ϵ_i and the wave functions are assumed to be normalized. For an atom, where we can assume a central field $V(\mathbf{r}) = V(r)$, we can readily solve equation (1) for the wave functions for a given $V(r)$, but troubles arise for the following reasons: (a) $V(r)$ must incorporate some average account of the electron–electron interactions. This obviously will involve knowledge of the density $n(r)$ in equation (2) and hence of the wave functions one is trying to find. (b) When N is very large, the computations involved in constructing the density via the wave functions are substantial. Actually, the situation is not too bad in atoms where $N \sim 100$ at most, but becomes serious in really large molecules, say.

Therefore, a method in which the density $n(\mathbf{r})$ can be obtained directly from $V(\mathbf{r})$, by-passing the wave functions, is very attractive, and this is the achievement of the Thomas–Fermi theory. Of course, the choice of $V(\mathbf{r})$ remains and much attention will be paid to that question in this Volume.

Use of Uniform Electron Gas Relations Locally

We begin by treating the electronic cloud as an electron gas, which, however, in atoms, molecules, and solids is not uniform. This is then the beginning of a general treatment of the inhomogeneous electron gas.

Since the Thomas–Fermi theory uses uniform electron gas relations locally, let us consider first a uniform electron gas, having N electrons contained in a volume $\mathcal{V}$, with mean density $n_0 = N/\mathcal{V}$. In the ground state, all the states in momentum space will be filled out to a maximum momentum $p_{\text{Fermi}} = p_f$, and all higher momentum states will be unoccupied. Thus, the volume of occupied momentum space is $\frac{4}{3}\pi p_f^3$ and the volume of occupied phase space is $\frac{4}{3}\pi p_f^3\mathcal{V}$.

As a direct consequence of the uncertainty principle, each cell in phase space of volume h^3 can hold two electrons with opposed spins in the ground state and hence we can write that the total number of electrons N is given by

$$N = \frac{2.4\pi}{3}\frac{p_f^3}{h^3}\mathcal{V} \tag{3}$$

where the right-hand side is twice the number of occupied phase space cells of volume h^3, the factor of 2 coming from the double occupancy in accord with the Pauli exclusion principle. Hence we have the basic equation relating electron density n_0 to the Fermi momentum p_f in a uniform electron gas as

$$n_0 = \frac{8\pi}{3h^3}p_f^3 \tag{4}$$

As previously mentioned, the next step is to assume that in the inhomogeneous electron cloud existing in, say, an atom, we can immediately employ equation (4) at position $\mathbf{r}$ to obtain the number of electrons per unit volume $n(\mathbf{r})$ in the ground state as

$$n(\mathbf{r}) = \frac{8\pi}{3h^3} p_f^3(\mathbf{r}) \tag{5}$$

where evidently the maximum momentum $p_f(\mathbf{r})$ has now to be specified at each position vector $\mathbf{r}$.

The second important step is to write down the classical energy equation for the fastest electron as

$$\mu = \frac{p_f^2(\mathbf{r})}{2m} + V(\mathbf{r}) \tag{6}$$

where the electrons are assumed as in equation (1) to move in a common potential energy $V(\mathbf{r})$. While the right-hand side of this energy equation is built up as a sum of kinetic and potential energies which individually depend on the position $\mathbf{r}$ in the charge cloud, we must stress that the left-hand side is constant throughout the entire electronic distribution. For if it varied from point to point, then this energy could adjust by electronic redistribution, and thereby lower the total energy of the assembly. We have used the symbol μ to denote the maximum energy in equation (6), anticipating the result we shall prove below that it is indeed the chemical potential of the electronic system.

We can now eliminate $p_f(\mathbf{r})$ from equation (6) in favor of $n(\mathbf{r})$ in equation (5) when we obtain

$$\mu = \frac{1}{2m}\left(\frac{3h^3}{8\pi}\right)^{2/3} [n(\mathbf{r})]^{2/3} + V(\mathbf{r}) \tag{7}$$

which is the basic relation between electron density $n(\mathbf{r})$ and potential energy $V(\mathbf{r})$ of the Thomas–Fermi theory.

It is an important pillar of the theory that this equation (7) can also be obtained as the Euler equation of a minimum energy principle. This point is so central for the theory of the inhomogeneous electron gas that we turn immediately to construct this variational principle from simple physical arguments.

3. Minimum Energy Principle and Chemical Potential

We can again use free electron relations locally to construct the kinetic energy density of the inhomogeneous electron gas in the Thomas–Fermi theory. Thus, let us write first the probability, $I_{\mathbf{r}}(p)dp$ say, of an electron at position **r** in the charge cloud having a momentum of magnitude between p and $p + dp$. From the phase space argument of Section 2.1, this is evidently given by

$$\begin{aligned} I_{\mathbf{r}}(p)\,dp &= \frac{4\pi p^2\,dp}{\frac{4}{3}\pi p_f^3(\mathbf{r})}, \qquad p < p_f(\mathbf{r}) \\ &= 0 \qquad \text{otherwise} \end{aligned} \tag{8}$$

Therefore, since there are $n(\mathbf{r})$ electrons per unit volume at **r**, we can write for the kinetic energy per unit volume, t say,

$$t = \int_0^{p_f(\mathbf{r})} n(\mathbf{r}) \frac{p^2}{2m} \frac{3p^2}{p_f^3(\mathbf{r})}\,dp \tag{9}$$

Using equation (5) for the electron density $n(\mathbf{r})$ we find immediately the result

$$t = c_k[n(\mathbf{r})]^{5/3}, \qquad c_k = \frac{3h^2}{10m}\left(\frac{3}{8\pi}\right)^{2/3} \tag{10}$$

where again, after performing the integration in equation (9) we have used equation (5) to write the kinetic energy density t solely in terms of the electron density $n(\mathbf{r})$ in the inhomogeneous electron gas.

Thus, by integrating through the entire charge cloud, we have the total kinetic energy T of the inhomogeneous electron gas as

$$T = c_k \int [n(\mathbf{r})]^{5/3}\,d\mathbf{r} \tag{11}$$

which shows that in this theory the kinetic energy is determined explicitly by the density $n(\mathbf{r})$ through equation (11).

Turning to the potential energy, we can write this as, first of all, the interaction of the electron density $n(\mathbf{r})$ with the potential energy of the nuclear framework, which we denote by $V_N(\mathbf{r})$, giving an electron–nuclear potential energy U_{en} as

$$U_{en} = \int n(\mathbf{r}) V_N(\mathbf{r}) d\mathbf{r} \tag{12}$$

and secondly the term from the electron–electron interactions. Thus, provided we work only with the classical Coulomb interaction energy of the electronic charge cloud (that is we neglect for the moment quantum mechanical contributions to the electron–electron potential energy, arising from exchange and correlation effects—see below), then we can write the total potential energy U as

$$U = U_{en} + U_{ee} \tag{13}$$

where U_{ee} has the explicit form

$$U_{ee} = \frac{1}{2} e^2 \int \frac{n(\mathbf{r})n(\mathbf{r}')}{|\mathbf{r} - \mathbf{r}'|} \, d\mathbf{r} \, d\mathbf{r}' \tag{14}$$

the factor of 1/2 being introduced to avoid counting interactions twice over. Of course, in the multicenter problems of molecules and solids we have also the nuclear–nuclear interaction energy U_{nn} but since this does not depend on the electron density, but only on the nuclear charges and the geometry of the nuclear framework, we shall not need to carry it along at present.

Therefore, combining equations (11) and (13), we have the total electronic energy as

$$E_{\text{el}} = c_k \int [n(\mathbf{r})]^{5/3} \, d\mathbf{r} + \int n(\mathbf{r}) V_N(\mathbf{r}) \, d\mathbf{r} + \frac{1}{2} e^2 \int \frac{n(\mathbf{r})n(\mathbf{r}')}{|\mathbf{r} - \mathbf{r}'|} \, d\mathbf{r} \, d\mathbf{r}' \tag{15}$$

This energy expression evidently depends only on the electron density and the given nuclear potential energy V_N. We now insist that it be minimized with respect to the electron density $n(\mathbf{r})$, subject only to the condition of normalization, namely, that

$$\int n(\mathbf{r}) \, d\mathbf{r} = N \tag{16}$$

N being the total number of electrons in the system under consideration. The standard method of taking care of the subsidiary condition (16) is to make the variational principle read

$$\delta(E - \mu N) = 0 \tag{17}$$

where μ is the Lagrange multiplier. Carrying out the variation by the standard procedures we find

$$\begin{aligned} \mu &= \tfrac{5}{3} c_k [n(\mathbf{r})]^{2/3} + V_N(\mathbf{r}) + e^2 \int \frac{n(\mathbf{r}')}{|\mathbf{r} - \mathbf{r}'|} \, d\mathbf{r}' \\ &= \tfrac{5}{3} c_k [n(\mathbf{r})]^{2/3} + V_N(\mathbf{r}) + V_e(\mathbf{r}) \end{aligned} \tag{18}$$

where in the second line of equation (18) we have written $V_e(\mathbf{r})$ for the potential energy created by the electron cloud $n(\mathbf{r})$. By substituting for c_k from equation (10) it follows immediately that this equation (18) is equivalent to tbe density–potential relation (7), provided we take $V(\mathbf{r})$ in that latter equation as the nuclear potential $V_N(\mathbf{r})$ plus the potential energy $V_e(\mathbf{r})$ created by the electron cloud.

That μ is the chemical potential is now clear if we write from equation (17) that

$$\mu = \frac{\partial E}{\partial N} \tag{19}$$

This equation (19) will be widely useful in applications of the Thomas–Fermi theory, to which we turn immediately.

4. Properties of Atoms and Ions

If we solve equation (7) for $n(\mathbf{r})$ we find immediately that the electron density is given by

$$\begin{aligned} n(\mathbf{r}) &= \frac{8\pi}{3h^3}(2m)^{3/2}[\mu - V(\mathbf{r})]^{3/2} \qquad \text{for } \mu \geqq V(\mathbf{r}) \\ &= 0 \qquad \text{otherwise} \end{aligned} \tag{20}$$

The reason the density is zero for $\mu < V(\mathbf{r})$ is easy to see. The theory used the classical energy equation (6), and therefore, though it crucially involves Planck's constant, it has semiclassical character. There is no leaking out of electron density into classically forbidden regions, the condition $\mu < V(\mathbf{r})$ implying negative kinetic energy.

We have already seen that $V(\mathbf{r})$ must be constructed using electrostatics from equation (18). An equivalent way to do this is to insist that n and V are related by Poisson's equation

$$\nabla^2(\mu - V) = 4\pi e^2 n(\mathbf{r}) = \frac{32\pi^2 e^3}{3h^3}(2m)^{3/2}(\mu - V)^{3/2} \tag{21}$$

where we have used (i) the constancy of μ and (ii) equation (20) for $n(\mathbf{r})$.

One of the elegant features of the above statistical theory applied to atoms and ions is that one can scale the solutions. To see this let us introduce the dimensionless function $\phi(x)$ through

$$\mu - V(r) = \frac{Ze^2}{r}\phi(x) \tag{22}$$

where x is also dimensionless and defined by

$$r = bx, \qquad b = \frac{1}{4}\left[\frac{9\pi^2}{2Z}\right]^{1/3} a_0 = 0.88534Z^{-1/3}\, a_0 \tag{23}$$

Here Z is evidently the atomic number, while a_0 is the Bohr radius $h^2/4\pi^2me^2$. Then it is a straightforward matter to show that $\phi(x)$ satisfies the dimensionless Thomas–Fermi equation

$$\frac{d^2\phi}{dx^2} = \frac{\phi^{3/2}}{x^{1/2}} \tag{24}$$

We have evidently from the definition (22) that as r tends to zero, $\phi(x)$ tends to 1. The small x expansion of $\phi(x)$, due to Baker,(3) has the form

$$\phi(x) = 1 + a_2x + a_3x^{3/2} + \cdots \tag{25}$$

the coefficients up to a_{11} being recorded in Table 1.

There is, furthermore, an exact solution of equation (24), namely, $144/x^3$, as is readily verified. A solution tending to zero as $144/x^3$ as x tends to infinity has been given by Coulson and March(4) in the form of the series

$$\phi(x) = \frac{144}{x^3}\left(1 - \frac{F_1}{x^c} + \frac{F_2}{x^{2c}} + \frac{F_3}{x^{3c}} + \cdots\right) \tag{26}$$

where $c = (73^{1/2} - 7)/2$. The relation of the coefficients F_n to F_1 is recorded in Table 2. It turns out that for only one choice of F_1, namely, $F_1 = 13.271$,

TABLE 1. Coefficients in Baker's Series Solution (25) of Dimensionless Thomas–Fermi Equation (24)

$a_3 = \frac{4}{3}$	$a_6 = \frac{1}{3}$	$a_9 = \frac{2}{27} - \frac{1}{252}a_2^3$
$a_4 = 0$	$a_7 = \frac{3}{70}a_2^2$	$a_{10} = \frac{a_2^2}{175}$
$a_5 = \frac{2a_2}{5}$	$a_8 = \frac{2a_2}{15}$	$a_{11} = \frac{31}{1485}a_2 + \frac{a_2^4}{1056}$

TABLE 2. Coefficients in Asymptotic Expansion (26) for Large x of Dimensionless Thomas–Fermi Equation (24)

Coefficient	Value	Coefficient	Value	Coefficient	Value
F_2	$0.62570F_1^2$	F_5	$0.05508F_1^5$	F_8	$0.00256F_1^8$
F_3	$0.31339F_1^3$	F_6	$0.02071F_1^6$	F_9	$0.00085F_1^9$
F_4	$0.13739F_1^4$	F_7	$0.00741F_1^7$	F_{10}	$0.00028F_1^{10}$

does the series (26) join smoothly with the Baker expansion (25), this latter solution corresponding then to the slope at the origin of $a_2 = -1.58805$. This solution, which physically represents the neutral Thomas–Fermi atom, is shown schematically in Fig. 1, curve I. It is tabulated, for example, in the book by Gombás.[(5)]

As the slope at the origin, a_2, is made more negative than that corresponding to curve I, we generate solutions of type II which cut the x axis at a finite value, say x_0. These solutions correspond physically to positive ions, as can be seen from the following argument. If we examine the electron density $n(r)$, we see from equations (21)–(23) that this vanishes at $r_0 = bx_0$, since $\phi(x_0) = 0$. Applying Gauss' theorem to solutions of type II in Fig. 1, we readily obtain the further boundary condition

$$-x_0\phi'(x_0) = 1 - (N/Z) \tag{27}$$

This leads to the construction shown in Fig. 1 and we see how to read off the degree of ionization q, defined by $q = 1 - (N/Z)$, from a particular solution II. Thus, we have a universal, though approximate, solution for the potential field and electron density in atoms and in positive ions of a given degree of ionization q. There are no solutions of the dimensionless Thomas–Fermi equation (24) corresponding to stable negative ions; we shall say a little about the reason for that below. Nevertheless, in spite of this limitation, we shall see that the predictions of the Thomas–Fermi theory for the total energies of atoms and positive ions are already very valuable.

4.1. *Energy Relations*

We shall give below an explicit expression for the total energy $E(Z,N)$ of a positive ion of atomic number Z and with N ($\leq Z$) electrons. This turns out to depend only on the slope a_2 of $\phi(x)$ at the origin and on the semiclassical radius of the positive ion, namely, x_0. These quantities must be obtained, as

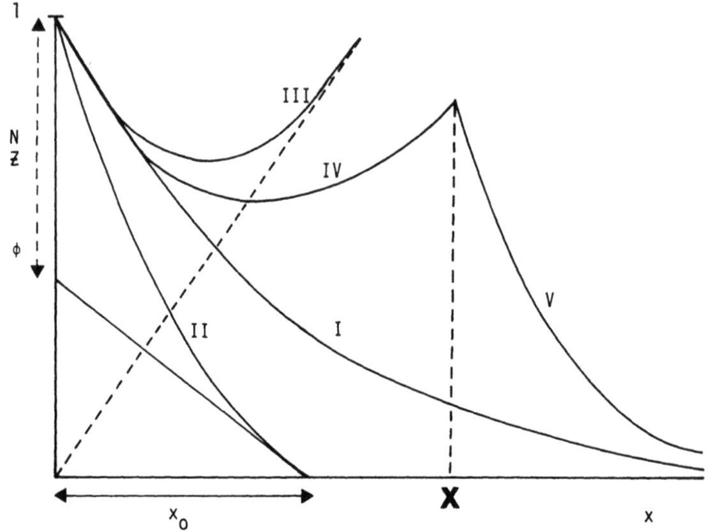

FIGURE 1. Types of solution of the dimensionless Thomas–Fermi equation (24). Essentially the function ϕ expresses the potential distribution in the system. Different solutions have the following physical interpretation: I Neutral atom; II, positive ions; III, cell model theory of material at high pressure; IV and V, model of tetrahedral and octahedral molecules.

is clear from Fig. 1, from the numerical solutions of type II of the Thomas–Fermi equation (24).

Before going on to the predictions for atomic ions of this simplest theory of the inhomogeneous electron gas, it is of interest to note some immediate consequences of the basic Euler equation (18) of the variation problem. Thus, let us multiply both sides of this equation by the electron density $n(\mathbf{r})$ and integrate over the whole of space. Since it has already been emphasized that the chemical potential μ is constant throughout the entire charge cloud, we find

$$N\mu = \tfrac{5}{3}T + U_{en} + 2U_{ee} \tag{28}$$

which follows directly from the result for the kinetic energy T in equation (11) and the definitions of the electron–nuclear and electron–electron potential energy terms, U_{en} and U_{ee}, respectively.

If we return to curve I of Fig. 1, it can be seen that since for the neutral atom $\phi(x)$ tends to zero for large x as x^{-3} and since $V(r)$ must tend to zero at infinity, then equation (22) predicts the chemical potential μ to be zero for the neutral Thomas–Fermi atom. Then we obtain from equation (28) that, for the neutral atom

$$-\tfrac{5}{3}T = U_{en} + 2U_{ee} \qquad \text{since } \mu = 0 \tag{29}$$

But the total potential energy $U = U_{en} + U_{ee}$ and since, as Fock[6] was the first to show, the virial theorem

$$2T + U = 0 \quad \text{or} \quad T = -E \tag{30}$$

is exactly satisfied by the Thomas–Fermi theory of atomic ions, it follows that

$$U_{en} = -\tfrac{7}{3}T = \tfrac{7}{3}E \tag{31}$$

and

$$U_{ee} = T/3 = -E/3 \tag{32}$$

These "scaling" relations between energy terms are found to be quite well obeyed when they are tested on numerical Hartree self-consistent field calculations for atoms and ions. We shall give explicit examples for molecules below which illustrate this, when we have effected the necessary generalizations to the multicenter case.

4.2. *Total Energy*

We can rewrite the electron–nuclear energy U_{en} as the nuclear charge Ze times the electrostatic potential at the nucleus due to the electron cloud, say $\psi_e(0)$. From the relation (32) between U_{en} and E, we can then write the neutral atom energy $E(Z,Z)$ as

$$E(Z,Z) = \tfrac{3}{7}Ze\psi_e(0) \tag{33}$$

and by relating the slope a_2 in expansion (25) to $\psi_e(0)$ we readily find

$$E(Z,Z) = -0.48a_2Z^{7/3}\, e^2/a_0 = -0.7687Z^{7/3}\, e^2/a_0 \tag{34}$$

a result due to Milne.[7]

For positive ions, we must clearly incorporate into the above argument the nonzero chemical potential. The electron density vanishes at the radius r_0 of the classically allowed region and hence from equation (22) we have

$$\mu = V(r_0) = \frac{-(Z-N)e^2}{r_0} = -\frac{Zq}{bx_0}\frac{e^2}{a_0} \tag{35}$$

where we have used the fact that just outside the charge cloud at r_0^+ the electrostatic potential of the spherical charge cloud is as though all the charge were lumped at the origin from Gauss' theorem. Both the fact that $q \to 0$ as N tends to Z and from Fig. 1, curve I, that x_0 tends to infinity as N tends to Z, lead back to the neutral atom result from equation (35) that $\mu = 0$ as in equation (29).

At this point, we can utilize the formula (19) for the chemical potential to obtain the total energy $E(Z,N)$ of a positive ion with nuclear charge Ze and N electrons by integration of μ with respect to N. The neutral atom energy can be regarded as the boundary condition in solving this first-order differential equation (19), using μ in equation (35). It then follows after a short calculation that $E(Z,N)$ can be written solely in terms of N, Z, a_2, and x_0 as

$$E(Z,N) = \frac{3}{7}\frac{(Ze)^2}{b}a_2 + \frac{3}{7}\frac{e^2}{bx_0}Z^2q^2 \tag{36}$$

This basic formula (36), combined with the knowledge from Fig. 1 that solutions of type II depend only on N/Z, leads immediately to the scaling property

$$E(Z,N) = Z^{7/3}f(N/Z) \tag{37}$$

where the form of the function $f(N/Z)$ is determined from the explicit solutions of equation (24) and is shown[8] in Fig. 2. For accurate tabular data for $f(N/Z)$, reference can be made to the work of Tal and Levy.[9] In Appendix A, the expansion of $E(Z,N)$ around the neutral atom value $E(Z,Z)$ is developed analytically.

4.3. *Relation to 1/Z Expansion*

As discussed by March and White,[8] the scaling property (37) of the Thomas–Fermi energy for positive ions gives valuable information on the asymptotic form for large N of the coefficients $\epsilon_n(N)$ of the $1/Z$ expansion.[10]

$$E(Z,N) = Z^2\left[\epsilon_0(N) + \frac{\epsilon_1(N)}{Z} + \frac{\epsilon_2(N)}{Z^2} + \cdots + \frac{\epsilon_n(N)}{Z^n} + \cdots\right] \tag{38}$$

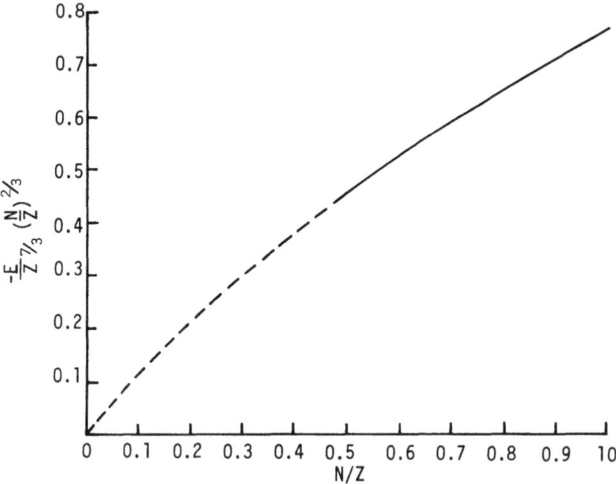

FIGURE 2. Form of function $f(N/Z)$ in equation (37). Quantity actually plotted is $-(E/Z^{7/3})(N/Z)^{2/3}$ against N/Z, with E in Hartree atomic units.

The above expansion has as its leading term the bare Coulomb energy and has been shown by Kato[11] to be such that convergence is assured for sufficiently large Z. Because of the statistical nature of the Thomas–Fermi theory, its predictions are correct in the limit of large Z and large N. Thus, the behavior of the coefficients $\epsilon_n(N)$ in equation (38), at large N, must be such that in this limiting situation the scaling property (37) is regained. This implies

$$\epsilon_n(N) \sim A_n N^{n+1/3}, \qquad N \to \infty \tag{39}$$

as pointed out by March and White.[8] These workers estimated numerically A_1 and A_2 and more accurate values were subsequently given by Dmitrieva and Plindov.[12] We have collected in Table 3 exact results for the coefficients $\epsilon_1(N)$ and $\epsilon_2(N)$ in the expansion (38). The asymptotic values have been estimated from equation (39), modified however to include exchange, a matter we shall return to below. The results from this asymptotic theory, of which equation (39) is the leading term (with $A_1 = 0.485$, $A_2 = -0.104$) are seen from Table 3 to join on quite well to the known exact results and to therefore constitute a valuable generalization of these to large N. We must stress here that Table 3, like the expansion (38), is based on the nonrelativistic Schrödinger equation. The relativistic generalization of the $1/Z$ expansion will be considered in Section 12.2 of this chapter.

TABLE 3. Coefficients in 1/Z Expansion of Equation (38)

N	$\epsilon_1(N)_{\text{exact}}$	$\epsilon_1(N)_{\text{asymptotic}}$	$-\epsilon_2(N)_{\text{exact}}$	$-\epsilon_2(N)_{\text{asymptotic}}$
2	0.625		0.1577	
6	3.2589		3.2880	
8	5.6619		8.1319	
10	8.7708		16.273	16.29
15	13.950			45.65
24	28.952			146.3
28	37.970	38.0		213.2
60	108.49	108.47		1342
110		247.44		5684

4.4. *Relation between Total Energy and the Sum of One-Electron Energies*

In theories based on Schrödinger's equation (1), we can associate with the potential energy $V(\mathbf{r})$ one-electron wave functions ψ_i and corresponding energies ϵ_i in addition to the electron density $n(\mathbf{r})$. It is clear that the sum E_s of the one-electron energies ϵ_i over occupied states is

$$E_s = \sum \epsilon_i = T + \sum_i \int \psi_i^* V \psi_i \, d\mathbf{r} = T + \int n(\mathbf{r}) V(\mathbf{r}) \, d\mathbf{r} \tag{40}$$

when we use equation (2). Introducing into this equation (40) the result of equation (18) that $V = V_N + V_e$ we find

$$E_s = \sum \epsilon_i = T + U_{en} + 2U_{ee} \tag{41}$$

where we have utilized the definitions (12) and (14) of U_{en} and U_{ee}. Combining equation (41) with the energy relations (31) and (32), it is readily shown that

$$E = \tfrac{3}{2} E_s \tag{42}$$

for neutral atoms, a result due to March and Plaskett.[13]

This result is of interest because elementary theories (cf. Section 5 for a discussion on molecules) assume the total energy to be given by the sum of the one-electron energies. It has, of course, been known for a long time in atomic theory that the sum of one-electron energies counts twice over the electron–electron Coulomb energy [cf. equation (41)], but the above arguments demonstrate that, in the Thomas–Fermi theory, the necessary correction for overcounting is directly related to E by equation (32). Thus the total energy and the sum of the orbital energies are related simply by the constant 3/2 in the neutral Thomas–Fermi atom.

The above argument is readily generalized to positive ions with the result[14] that

$$E = \tfrac{3}{2}(E_s - N\mu) \tag{43}$$

To show how the relation (42) for neutral atoms is quantitatively modified by the nonzero chemical potential for positive ions, some numerical results are collected in Table 4.

4.5. *Behavior of Positive Ions in Extremely Strong Magnetic Fields*

In concluding this discussion of atoms and ions, we shall briefly summarize here results for the total energy of positive ions in extremely strong magnetic fields B. Considerable interest has been shown in this problem because of astrophysical implications concerned with the emission of electrons and ions from the surface of pulsars, the abundance of the elements in cosmic radiation, and the properties of condensed matter forming the outer crust of magnetic neutron stars.

Extensive studies of the Thomas–Fermi theory of the atom, discussed fully above in zero magnetic field, have been made in the limit of very high magnetic fields.[15–19] Rather than presenting detailed numerical summaries of these investigations, for which the reader is referred to the original papers, we shall in the present context discuss the scaling of the total energy of positive ions at large B and the relation to the $1/Z$ expansion.[20]

The basic modifications of the Thomas–Fermi theory to take account of the presence of extremely strong magnetic fields are set out in Appendix B. The outcome is to replace the zero magnetic field equation (24) by a new dimensionless Thomas–Fermi equation which is shown from Appendix B to have the form

$$\phi'' = (x\phi)^{1/2} \tag{44}$$

with the scaling

$$r = b'x = a_0 2^{-3/5}\pi^{2/5}\alpha^{4/5}Z^{1/5}L^{-2/5}x \tag{45}$$

TABLE 4. Influence of Nonzero Chemical Potential μ of Positive Ions on Relation between Total Energy $E(Z,N)$ and Sum of Orbital Energies in Equation (43)

N/Z	0.478	0.777	0.879	1
$\tfrac{3}{2}\mu N/E(Z,N)$	0.207	0.059(6)	0.024(0)	0

in the strong-field case, where α is the fine structure constant, $L = eB/m^2 = B/B_c$, where $B_c = 4.4\times10^{13}$G. The energy is measured in the Rydberg unit $\frac{1}{2}m\alpha^2$ and using $\hbar = c = 1$ we have the total energy as

$$E_{\rm TF}(Z,N,B) = -2^{8/5}\pi^{-2/5}\alpha^{-4/5}Z^{9/5}L^{2/5}\epsilon\ ({\rm Ry}) \tag{46}$$

In this equation, ϵ is expressed in terms of the slope at the origin $\phi'(0)$ of the solution of the dimensionless Thomas–Fermi equation (44).

Clearly we must solve this for a positive ion subject to the appropriate boundary conditions. To ensure that there are N ($\leq Z$) electrons in the positive ion, the solution of equation (44) must satisfy

$$\phi(x_0) - x_0\phi'(x_0) = 1 - N/Z = q \tag{47}$$

and if the ion is in a free state then

$$\phi(x_0) = 0 \tag{48}$$

Then the quantity ϵ in equation (46) is explicitly

$$\epsilon = -\tfrac{5}{9}\,[\phi'(0) + q^2/x_0] \tag{49}$$

It is a straightforward matter to show that a solution satisfying the boundary conditions at x_0 can be expanded about x_0 as

$$\phi(x) = \frac{(x_0 - x)}{x_0}\,q + a_5\,(x_0 - x)^{5/2} + \text{higher-order terms} \tag{50}$$

where

$$a_5 = \tfrac{4}{5}q^{1/2} \tag{51}$$

In the limit of small N/Z, equation (50) is valid also into $x = 0$ and from the condition $\phi(0) = 1$ it can be demonstrated that, for small N/Z,

$$x_0 \sim \text{const}(N/Z)^{2/5} + \text{higher-order terms} \tag{52}$$

Furthermore, evaluating $\phi'(0)$ from equation (50) in the limit of small N/Z, it can be shown that

$$\epsilon \sim \text{const}(N/Z)^{3/5} + \cdots \tag{53}$$

Hence, the conclusion is that, for small N/Z,

$$E_{\mathrm{TF}}(Z,N,B) \sim Z^{9/5}(N/Z)^{3/5}L^{2/5} \tag{54}$$

in the high-magnetic-field regime. Developing ϵ to higher order shows that the next term in equation (53) is $O(N/Z)$ higher.

As for the zero-field case already treated in Section 4.3, it is again of interest to connect the above theory with the $1/Z$ expansion. We therefore first estimate the Coulomb energy $E_{\mathrm{Coulomb}}\,(Z,N,B)$ from the statistical theory by noting that the Thomas–Fermi density is given by (cf. Appendix B)

$$n(r) = \frac{m^{1/2}(eB)}{2^{1/2}\pi^2}\left(\frac{Ze^2}{r} - \mu\right)^{1/2} \tag{55}$$

Writing $\mu = Ze^2/r_0$, where $r_0 = b'x_0$ and using $\int n(\mathbf{r})d\mathbf{r} = N$ yields

$$\mu = \frac{Ze^2}{r_0} = \frac{Z^{6/5}L^{2/5}}{\alpha^{4/5}N^{2/5}}\ \mathrm{Ry} \tag{56}$$

Further, the electron–nuclear potential energy U_{en}, related to E by the virial theorem, is given by

$$\begin{aligned} E_{\mathrm{Coulomb}} &= \tfrac{5}{6}U_{en} = -\tfrac{5}{6}Ze^2\int\frac{n(r)\,d\mathbf{r}}{r} \\ &= -\tfrac{5}{3}\alpha^{-4/5}Z^{6/5}L^{2/5}N^{3/5}\ \mathrm{Ry} \end{aligned} \tag{57}$$

We now base the generalized $1/Z$ expansion on the (full) Coulomb field result by writing

$$E(Z,N,B) = E_{\mathrm{Coulomb}}\,(Z,N,B) + \frac{1}{Z}E_1(Z,N,B) + \cdots + \frac{1}{Z^n}E_n(Z,N,B) + \cdots \tag{58}$$

the idea again being to correct the Coulomb term by electron shielding effects in higher-order terms. The implication of the $1/Z$ expansion is that one can factor out the remaining Z dependence and in particular one finds

$$E_{\mathrm{TF}}(Z,N,B) = Z^{9/5}L^{2/5}\left[\frac{N}{Z}\right]^{3/5}\left[\epsilon_0 + \frac{N}{Z}\epsilon_1 + \cdots + \left[\frac{N}{Z}\right]^n\epsilon_n + \cdots\right] \tag{59}$$

It is now readily demonstrated from equations (46), (54), and (57) that ϵ_n is independent of N for large N and that explicitly we have

$$\epsilon_0 = -\tfrac{5}{3}\alpha^{-4/5} = -85.364 \tag{60}$$

March and Tomishima have estimated ϵ_1 and ϵ_2 by a least squares method as 36.9 and -2.78, respectively. The asymptotic behavior of E_n in equation (58) for large N is thereby established in the limit of extremely high magnetic fields B.

We turn now from single-center problems of atomic ions to multicenter systems.

5. Molecular Energies at Equilibrium

The use of such a statistical theory in the multicenter problem of a molecule is already greatly complicated in practice by the fact that there is no longer spherical symmetry, and therefore the basic differential equation (21) for the potential field is a nonlinear partial differential equation which can presently only be solved by numerical methods. This has been attempted approximately, in the pioneering work of Hund,[(21)] and subsequently by Townsend and Handler[(22)] and a number of later workers. For references to the early work, the reader may consult the earlier review by the writer.[(23)]

But even when one solves this equation (21) for a given nuclear configuration, the work of Teller[(24)] shows that molecular binding is not possible for molecules in the Thomas–Fermi theory. This seems, at first sight, to indicate that the theory will have little to say for molecules.

5.1. *Scaling of Potential Energy Terms by Kinetic Energy*

Fortunately, this is too pessimistic a view and we shall show below that regularities exist which the Thomas–Fermi theory can correctly predict. This can be done by making use of the Euler equation for the multicenter problem. Just as for atoms, this leads to relations between the energy terms [cf. equations (31) and (32)]. Though the Thomas–Fermi theory, as shown by Teller, does not give binding, we know that at equilibrium there is an exact relation again from the virial theorem. If we use this in conjunction with the Euler equation of the Thomas–Fermi theory, we obtain scaling relations for molecules[(25)] which generalize those given for atoms in the previous section. We must add, of course, the nuclear–nuclear energy U_{nn}, and hence in the notation used for atoms we can write

$$E = T + U_{en} + U_{ee} + U_{nn} \tag{61}$$

Multiplying again the Euler equation by $n(\mathbf{r})$ and integrating over the whole of space we have, as for atoms, equation (28). If, by analogy with the neutral Thomas–Fermi atom where the chemical potential is zero, the same result is assumed for neutral molecules, then we are led directly from equations (28), (30), and (61) to the following scaling relations:

$$\frac{U_{ee} - U_{nn}}{T} = \frac{1}{3} \tag{62}$$

$$\frac{U_{en} + 2U_{nn}}{T} = \frac{-7}{3} \tag{63}$$

and

$$\frac{U_{en} + 2U_{ee}}{T} = \frac{-5}{3} \tag{64}$$

only two of which are independent. The relation (63) was known earlier to Politzer,[26] though derived by a different line of argument from the above.

Mucci and March[25] have brought these predictions derived from the Euler equation of the Thomas–Fermi theory, plus use of the exact virial result (30) at equilibrium, into contact with self-consistent field results on a variety of light molecules. Their plots of equations (62) and (64) are reproduced in Figs. 3 and 4. It can be seen that the theoretical lines, with slopes 1/3 and 5/3, respectively, are passing through the points for the individual molecules, and the gross regularities predicted by the scaling relations are clearly in evidence. This is very encouraging for the use of the theory of the inhomogeneous electron gas in molecules, notwithstanding Teller's undoubtedly correct theorem.[24]

5.2. *Model of Tetrahedral and Octahedral Molecules*

Because of the success of the scaling relations (62)–(64), it is worthwhile to refer here to a central field model of tetrahedral and octahedral molecules. This might be appropriate to CH_4, GeH_4, etc., but it can hardly be expected to apply to molecules like CCl_4, UF_6, etc., with localized core electrons on the outer atoms. Its merit, as we shall show below, is that it affords a model of neutral molecules in which all the above assumptions are fulfilled and therefore the scaling relations are exactly obeyed.

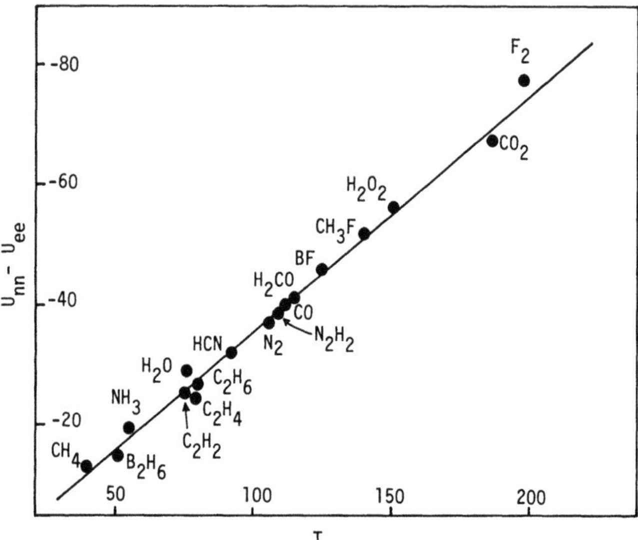

FIGURE 3. Difference $U_{nn} - U_{ee}$ between nuclear–nuclear and electron–electron potential energy versus total kinetic energy T for light molecules. Energies are in Hartree units.

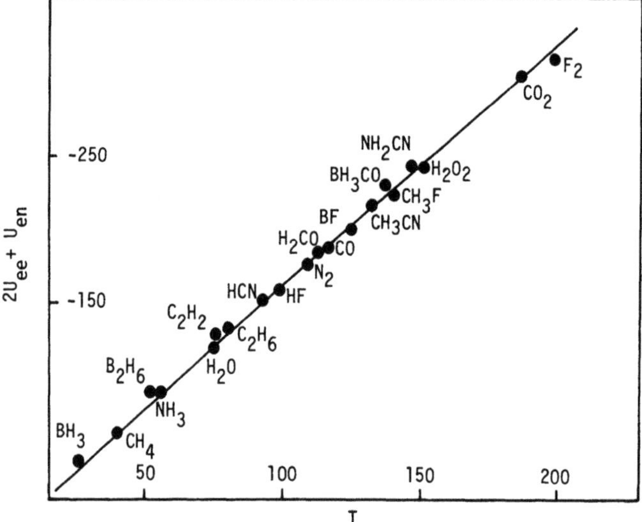

FIGURE 4. Quantity $2U_{ee} + U_{en}$, with U_{en} the electron–nuclear potential energy, versus total kinetic energy T for light molecules. Energies are in Hartree units.

The dimensionless Thomas–Fermi equation (24) must again be solved, but now the model is to smear the outer nuclei uniformly into a surface positive charge distribution on a sphere.[27] Figure 1 shows the types of solutions required, namely, curves IV and V, the discontinuity in slope at the sphere of surface charge reflecting precisely the discontinuity in electric field which must occur there.

Because one cannot, in this model, regain neutral isolated atoms as one lets the bond length tend to infinity, Teller's theorem is not applicable and it is easy to show that there is a minimum in the total energy. This leads to simple predictions for the scaling of the equilibrium bond length, but it is not to be expected that this model will be quantitatively applicable to real chemical systems.

5.3. *Nuclear–Nuclear Potential Energy for Tetrahedral and Octahedral Molecules*

Nevertheless, as stressed by Mucci and March,[25] it is of interest to construct from the above model the nuclear–nuclear potential energy at equilibrium. Using the correct point charge nuclear framework, this potential energy U_{nn} is readily written in terms of the bond length R as

$$U_{nn} = \frac{ne^2}{R}(z + cn) \tag{65}$$

where the central nucleus carries charge ze and the total outer charge is ne. The constant c has the values

$$\begin{aligned} c &= 3\sqrt{6}/32 \qquad \text{for tetrahedral cases} \\ &= (1 + 4\sqrt{2})/24 \qquad \text{for octahedral cases} \end{aligned} \tag{66}$$

Using the equilibrium bond length from the model, U_{nn} is plotted against the total number of electrons $N = n + z$ in Fig. 5. It will be seen that though there is a gross trend with N, tbe atomic number of the central atom is a significant additional variable in the plot. When we now take the bond length R from experiment and again use equation (65) to construct U_{nn}, we find an immediate correlation with the total number of electrons N as can be seen from Fig. 6. From this comparison between theory and experiment, it is clear that smearing the outer nuclei uniformly focuses too much attention on the attraction of the central atom for the molecular electrons. With point nuclei in the outer positions, the trend of U_{nn} with N is well represented by a constant $\times N^{4/3}$ (see Ref. 28).

It would obviously be of considerable interest if the density description

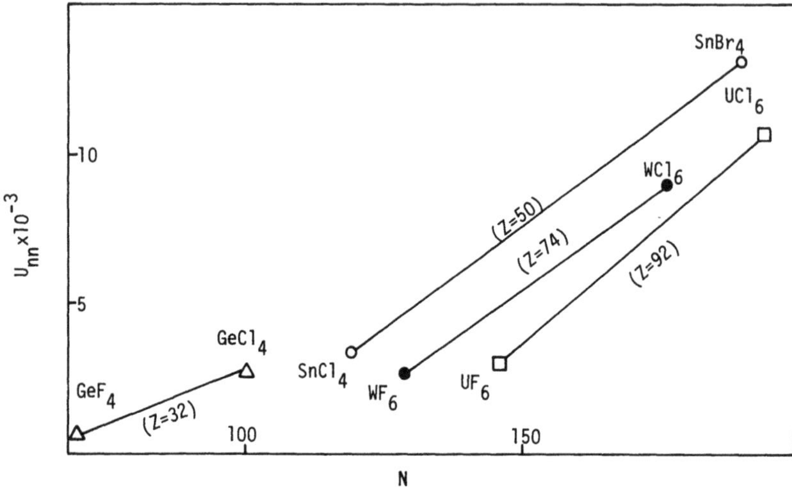

FIGURE 5. Nuclear potential energy U_{nn} for tetrahedral and octahedral molecules as given by equation (65).

could be used for point nuclei to predict the result $U_{nn} \propto N^{4/3}$ referred to above. But even if that, eventually, proves possible, we must stress that, for bond length predictions of chemical accuracy, a second essential step would be to discuss fluctuations about such a curve representing gross physical regularities. While the density description seems well suited to treating these gross features, it is not clear that all the chemical subtleties can be incorporated. What is known is that, at very least, a careful discussion of the chemical potential will be needed which transcends the statistical theory. Indeed Parr *et al.*[29] have argued convincingly for a direct relation between chemical potential and electronegativity. For further details the reader is referred to the review by the writer.[30]

5.4. *Total Energy and Eigenvalue Sum*

In concluding this discussion of molecules in the Thomas–Fermi approximation, it is relevant to note that in Hückel's theory of the π electrons in organic molecules, he assumed that the total energy E was equal to the sum of the one-electron energies E_s. For a spherical charge cloud such as in a heavy atom, we have seen from equation (42) that $E = \frac{3}{2}E_s$. Such a relation was also proposed for molecules at equilibrium by Ruedenberg,[31] who showed that it was quite well obeyed in self-consistent field calculations. The argument yielding equation (42) again from the statistical theory[32] for molecules at

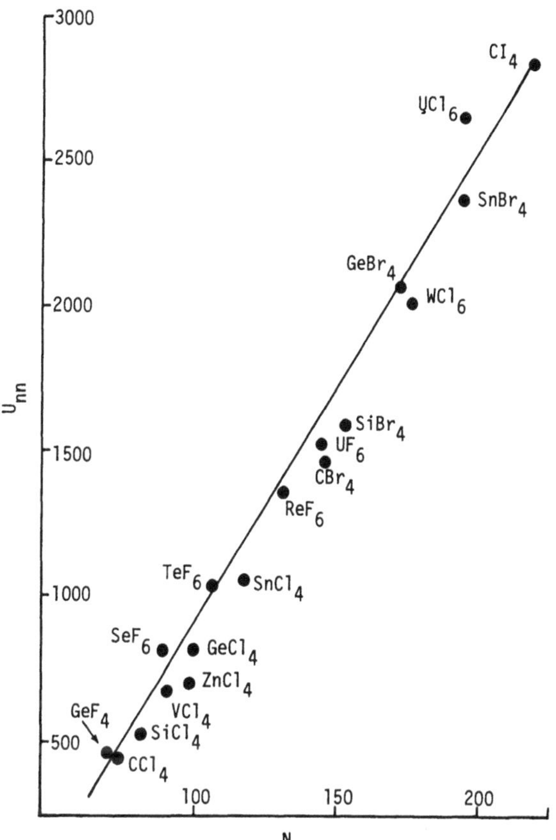

FIGURE 6. Nuclear–nuclear potential energy U_{nn} versus total number of electrons N for tetrahedral and octahedral molecules. In contrast to Fig. 5, empirical data for bond lengths are used to contract this figure.

equilibrium is briefly as follows. For molecules we have the total energy E in equation (61) and at equilibrium in an exact theory we have equation (30) expressing the virial theorem. Furthermore equation (41) is still valid as it stands for the sum of the orbital energies, E_s. Finally, equation (28) remains valid, and again we assume the chemical potential to be zero for neutral molecules. Putting these equations together yields again, for molecules at equilibrium, the result $E = \frac{3}{2}E_s$ in equation (42). Actually, the assumptions that are made in the above derivation are all exactly valid in the model of tetrahedral and octahedral molecules discussed in Section 5.2 and therefore $E = \frac{3}{2}E_s$ is exactly obeyed in this treatment.

Deviations from this relation have been shown to correlate with electronegativity by examining self-consistent field data on a variety of molecules, but the detailed explanation lies beyond the Thomas–Fermi theory and we must refer the reader to the original paper for this.[33]

Hückel theory was referred to above in connection with the assumption that $E = E_s$. Pucci and March[34] have made a start on the problem of basing Hückel theory on the electron density description. They point out that, for π electrons in benzene say, in which streamers of π-electron charge density exist in planes parallel to, and above and below, the plane of the carbon nuclei, the kinetic energy density should have features more like a two-dimensional than a three-dimensional electron gas. Assuming this, it is readily shown that the total energy is simply the sum of the one-electron energies.[34] Evidently this is only a starting point for a deeper discussion of π-electron theory in a manner which should eventually transcend Hückel theory without, however, losing the essential simplicity which makes that theory remain so attractive.

6. Introduction of Exchange

An important step in the electron density description was taken by Dirac[35] when he introduced exchange into the Thomas–Fermi theory. Though we shall adopt a variational argument below, the idea is the same as Dirac's, namely, that in the spirit of the use of free electron relations locally in Section 2.1, one should calculate the exchange energy of a uniform electron gas as a function of density and then take over the exchange energy density thus obtained into the theory of the inhomogeneous electron gas in an atom or molecule.

Before calculating precisely the exchange energy of a uniform electron gas, it is of interest to note, following the analogy with the kinetic energy density of the Thomas–Fermi theory, proportional to $n^{5/3}$, that if it is assumed that the exchange energy density is proportional likewise to a power s say of the density, then s can be determined by dimensional analysis.

Starting from the kinetic energy, involving the operator $-(\hbar^2/2m)\nabla^2$, it follows that the kinetic energy density t must be proportional to h^2/m. Then dimensional analysis confirms that $t \propto n^{5/3}$ if a power law dependence is assumed. For the exchange energy, which arises from the electron–electron interaction e^2/r_{ij} between electrons i and j at distance r_{ij}, we write

$$\text{exchange energy density} \propto e^2 n^s \tag{67}$$

Since an energy density has dimensions $ML^{-1}T^{-2}$, this must be equated to the dimensions of the right-hand side of equation (67). As for e^2, this has dimensions ML^3T^{-2} since e^2/L is an energy, while finally n is a number per unit volume with dimensions L^{-3}. From equation (67) it follows that

$$ML^{-1}T^{-2} = ML^3T^{-2}L^{-3s}$$

giving immediately

$$-1 = 3 - 3s$$

and hence in equation (67) the exponent $s = 4/3$. This basic result that the exchange energy density is proportional to $n^{4/3}$ has been taken up by Slater in an important paper on a simplification of the Hartree–Fock method which will be discussed later.

6.1. *Fermi Hole in Uniform Electron Gas*

Of course, dimensional analysis, though leading to the correct 4/3 power of the density in the exchange energy per unit volume, does not enable the magnitude of this energy density to be calculated. This we now calculate for the uniform electron gas by a physical argument.

As we shall do many times in this volume, let us imagine we are sitting on an electron, say with upward spin, and we examine the density of the other $N-1$ electrons in the uniform gas relative to this one. The resulting density can be written as the uniform density n_0 times a function $g(r)$, r being the distance measured from the electron on which we are sitting.

If we neglect Coulomb repulsions between electrons, then the only reason that $g(r)$ is not unity everywhere is because the Pauli exclusion principle forbids another electron with spin parallel to that of the electron at $r = 0$ to sit at the origin. Thus, there is zero probability of finding another upward spin electron at $r = 0$. But, since we are neglecting the Coulomb repulsions, there is no correlation between antiparallel spins. We are dealing with a Fermi sphere in which spins are paired and therefore it follows that $g(r)$ defined above goes from a value ½ at the origin to 1 as r tends to infinity. The latter value is due to the fact that when we are far from the electron at the origin the density must obviously revert to its mean value n_0.

The form of the "hole" surrounding the electron at the origin is discussed for all r in Appendix C. It can thus be shown that $g(r)$ is given by

$$g(r) = 1 - \frac{9}{2}\left[\frac{j_1(k_f r)}{k_f r}\right]^2, \qquad \frac{k_f^3}{3\pi^2} = n_0, \qquad j_1(x) = \frac{\sin x - x\cos x}{x^2} \tag{68}$$

a result that goes back to Wigner and Seitz.[36] When Coulomb repulsions are neglected, this hole, described quantitatively by equation (68), is referred to as the Fermi or exchange hole. When interparticle repulsions are incorporated, as they must be, then we refer to the exchange-correlation hole surrounding an electron in the gas. One vital property of both the Fermi and the exchange-correlation holes is that they contain precisely one electron; so to speak they screen out the Coulomb interaction of the electron at the origin with its fellows. This property has its direct analogue in the inhomogeneous electron gas. In only one example, a model of a metal surface, does it appear presently possible to get an analytical generalization of the result (68) for the uniform electron gas. This example is of sufficient interest to summarize therefore in Appendix D.

6.2. *Exchange Energy from Fermi Hole*

The quantity $g(r)$ is referred to as the pair correlation function between electrons. In the case of equation (68) representing the exchange (Fermi) hole, these correlations are solely due to the Fermi statistics whereas in the following section we shall consider the modifications due to the Coulomb repulsions which keep apart evidently also electrons with antiparallel spins.

However, we wish here to complete the calculation of the exchange energy for electrons in a uniform electron gas of density n_0. To do this, we merely note that there are no Coulombic potential energy terms of classical form in the uniform electron gas, because there is a neutralizing background of uniform positive charge (this is the so-called jellium model). The potential energy is solely exchange therefore in the Fermi hole approximation, and we can view this energy as the electrostatic interaction between the electron at the origin and the hole it digs around itself, of density $n_0[g(r) - 1]$. Thus the exchange energy per electron is evidently

$$\text{exchange energy per electron} = \frac{e^2}{2}\int \frac{d\mathbf{r}\; n_0[g(r)-1]}{r} \tag{69}$$

the factor ½ avoiding double counting of the interactions. Inserting the Fermi hole form, and making use of the definite integral

$$\int_0^\infty dx\, x\left[\frac{j_1(x)}{x}\right]^2 = \frac{\pi}{4}\frac{\Gamma(2)\Gamma(1)}{[2\Gamma(3/2)]^2} \tag{70}$$

one obtains the result

$$\text{exchange energy per electron} = \frac{-e^2 3k_f}{4\pi} \tag{71}$$

and hence the exchange energy density ϵ_x is immediately obtained. This calculation not only confirms the 4/3 power arrived at above, but it leads to the coefficient of proportionality in equation (67), and below we shall write

$$\epsilon_x = -c_e n^{4/3}, \qquad c_e = e^2 \frac{3}{4}\left(\frac{3}{\pi}\right)^{1/3} \tag{72}$$

6.3. *Thomas–Fermi–Dirac Density–Potential Relation*

To incorporate exchange into the Thomas–Fermi theory we merely return to the variational derivation in Section 3 and add the exchange energy A given by

$$A = -c_e \int n^{4/3}\, d\mathbf{r} \tag{73}$$

to obtain for the total energy E the result [cf. equation (15)]

$$E = c_k \int n^{5/3}\, d\mathbf{r} + \int n(\mathbf{r}) V_N(\mathbf{r})\, d\mathbf{r} + \frac{1}{2} e^2 \int \frac{n(\mathbf{r}) n(\mathbf{r}')}{|\mathbf{r} - \mathbf{r}'|}\, d\mathbf{r}\, d\mathbf{r}' - c_e \int n^{4/3}\, d\mathbf{r} \tag{74}$$

Minimizing E with respect to the electron density $n(\mathbf{r})$, subject only to the normalization constraint (16) leads to the new Euler equation including exchange as

$$\mu = \tfrac{5}{3} c_k \{n(\mathbf{r})\}^{2/3} + V_N(\mathbf{r}) + V_e(\mathbf{r}) - \tfrac{4}{3} c_e \{n(\mathbf{r})\}^{1/3} \tag{75}$$

Equation (75) is the Thomas–Fermi–Dirac density–potential relation. We shall solve it for atoms to obtain below the generalization of the dimensionless Thomas–Fermi equation (24) to include exchange. However, there are some difficulties with this equation and a more fruitful way of viewing equation (75), which stems from the work of Slater,[37] is to argue that the incorporation of exchange can be viewed equally properly by regarding equation (75) as a result of applying the Thomas–Fermi approximation to the one-electron Schrödinger equation (1), but now with the potential $V(\mathbf{r})$ as not simply the Hartree potential energy $V_N + V_e$ but modified by exchange to yield

$$V = V_N + V_e + V_X \tag{76}$$

where the exchange potential (energy) V_X is given from equations (72) and (75), as

$$V_X = -e^2 \left(\frac{3}{\pi}\right)^{1/3} [n(\mathbf{r})]^{1/3} \tag{77}$$

This so-called "Dirac–Slater exchange potential," as given in equation (77), has a coefficient differing by a factor from Slater's potential.[37] The Dirac coefficient in equation (77) is undoubtedly the correct one for pure exchange (see also the work of Kohn and Sham,[38] and the discussion in Chapters 2 and 3).

The idea that the nonclassical parts of the electron–electron interaction can be represented, for calculating the electron density, as a one-body potential is a very important one, which recurs throughout this volume. However, one must also take account of the fact that antiparallel spin electrons avoid one another, i.e., are also correlated in their motions and we shall discuss how this can be incorporated (approximately) in a later section. But, for the moment, let us turn to the calculation of the exchange energy in atoms. We shall see that this will lead us to an important correction term to the Milne total energy given in equation (34).

6.4. *Exchange Energy in Atoms*

The Thomas–Fermi–Dirac relation (75), as already remarked, has some difficulties associated with it, when applied to free atoms (see also comments in Chapter 5). Briefly, it does not lead to a self-consistent solution in which the electron density $n(\mathbf{r})$ either tends to zero at infinity for the neutral atom, nor indeed a solution of finite radius in which the boundary density is zero. Rather, if used for neutral atoms, there is inevitably a discontinuity in $n(\mathbf{r})$ at the boundary of the "atom" which is manifestly unsatisfactory.

Fortunately, as Scott[39] pointed out, there is a way round this difficulty which allows a satisfactory (if approximate) way of calculating the exchange energy in the Thomas–Fermi theory of neutral atoms. It is to argue that, since the energy E in equation (74) is stationary with respect to small variations in the electron density, we can estimate the total energy including exchange by calculating the total energy E in equation (74) with the neutral atom Thomas–Fermi density. Then, we merely must add to the Thomas–Fermi energy already calculated the exchange energy A in equation (73), but now calculated with the Thomas–Fermi density. This calculation was performed by Scott,[39] with the result

$$A = -0.221\ (e^2/a_0)\ Z^{5/3} \tag{78}$$

This shows that in the statistical limit of very large Z the Milne term proportional to $Z^{7/3}$ dominates the exchange term proportional to $Z^{5/3}$. Nevertheless, in the range of stable atoms, that is for $Z \lesssim 100$, the exchange energy represents a substantial correction to the Milne energy. However, this will be discussed numerically later, for it will turn out that there is a more important correction to be made to the Milne energy than that in equation (78), due to the rapid variation in the potential energy and electron density near the atomic nucleus. That is, the inhomogeneity of the electron cloud in an atom is really too large to apply free electron relations locally, and we must account for this correction in atomic energy calculations. Before turning to these corrections for inhomogeneity (or equivalently for density gradients), it will be convenient to briefly treat the incorporation of correlation in the Thomas–Fermi theory.

7. Correlation in Thomas–Fermi Framework

Before modern density functional theory, discussed in Chapter 2 of this volume, gave a rigorous foundation to the description of many-electron ground-state problems in terms of the electron density $n(\mathbf{r})$, Gombás and others[5] had recognized that electron correlation could be included within the Thomas–Fermi framework by applying relations derived for the uniform electron gas, but now incorporating Coulomb repulsion between antiparallel spin electrons, within a local density framework. Thus if we denote the exchange plus correlation energy density of the uniform electron gas by ϵ_{XC}, we can evidently rewrite the Euler equation (75), already incorporating exchange, in the more general form

$$\mu = \frac{5}{3} c_k\,[n(\mathbf{r})]^{2/3} + V_N(\mathbf{r}) + V_e(\mathbf{r}) + \frac{\delta\epsilon_{XC}}{\delta n} \tag{79}$$

The variation $\delta\epsilon_{XC}/\delta n$ in the exchange-correlation term in equation (79) is easily carried out in a local density Thomas–Fermi framework. A favorite representation of the correlation energy in the early work, due to Wigner,[40] leads to

$$\epsilon_{XC} = -\ c_e n^{4/3} + \epsilon_C^{\mathrm{Wigner}} \tag{80}$$

where the Wigner correlation energy may be written (in atomic units)

$$\epsilon_C^{\mathrm{Wigner}} = -\frac{0.056 n^{4/3}}{0.079 + n^{1/3}} \tag{81}$$

Other choices of correlation energy are now available and are discussed extensively in later chapters so we shall not go into further detail here except to summarize the arguments leading to the form (81) in the limit of very low density in Appendix C. We could think again of equation (79) as defining an exchange-correlation potential $\delta\epsilon_{XC}/\delta n$ within the local density framework, a very important point for the modern density functional theory of the inhomogeneous electron gas (cf. Chapter 2 for full details of this point).

We remark finally that for atoms, correlation energy[41] varies approximately proportional to Z, and hence is a lower-order term for large Z than the exchange energy in equation (78).

8. Density Gradient Corrections

In this section, we must now discuss the deficiencies in the Thomas–Fermi theory which arise due to the use of free electron relations locally. While such an assumption would clearly be appropriate in "almost homogeneous gases," the electron clouds to which we have so far applied the theory, namely those in atoms and molecules, show very strong spatial variations in potential energy $V(\mathbf{r})$ and in electron density $n(\mathbf{r})$. It is therefore essential to consider what are the errors that are thereby introduced into the Thomas–Fermi theory.

This topic, the effect of strong density gradients in the electron density description, is a matter that will come up in later chapters in various ways. But we can say at this point that the most serious limitation of the Thomas–Fermi theory is undoubtedly its description of the kinetic energy density by the local form (10) proportional to the five-thirds power of the electron density. In any fully quantitative theory, it is essential to relax this description, at any rate for atoms for which $Z \lesssim 100$, and for most molecules of current interest. It might turn out eventually that for really huge molecules of biological interest the Thomas–Fermi kinetic energy is a quantitative starting point, but even that should be approached with caution.

8.1. *Boundary Correction to Atomic Energies*

Before going on to describe how one might try, systematically, to correct the local Thomas–Fermi formula (10) for the kinetic energy for density gradients, let us return to the problem of atomic energies. To see how the Thomas–Fermi theory is deficient, let us consider briefly the problem of Z electrons moving independently in a bare Coulomb potential energy $-Ze^2/r$. Then the total energy $E = E_s$, the sum of the orbital energies, which are given by the usual hydrogenic levels

$$\epsilon_n = \frac{-Z^2}{2n^2}\frac{e^2}{a_0} \tag{82}$$

where n as usual is the principal quantum number. Since there are $2n^2$ electrons in a closed shell of principal quantum number n, we have immediately that the energy per closed shell is simply $-Z^2e^2/a_0$, independent of n. Thus for $\mathcal{N}$ closed shells the total energy $E = E_s$ is simply

$$E_{\text{Coulomb}} = -\frac{Z^2 e^2}{a_0}\mathcal{N} \tag{83}$$

But with Z electrons, we must have, for $\mathcal{N}$ closed shells,

$$Z = \sum_{1}^{\mathcal{N}} 2n^2 = \frac{\mathcal{N}(\mathcal{N}+1)(2\mathcal{N}+1)}{3} \tag{84}$$

and if we think literally of the statistical limit when not only Z is large but also $\mathcal{N}$ then we can invert equation (84) to find[42]

$$\mathcal{N} = (\tfrac{3}{2}Z)^{1/3} - \tfrac{1}{2} + \tfrac{1}{18}(\tfrac{3}{2})^{2/3}Z^{-1/3} + O(Z^{-5/3}) \tag{85}$$

Thus E_{Coulomb} is obtained from equations (83) and (85) as

$$E_{\text{Coulomb}} = \left[-(\tfrac{3}{2})^{1/3}Z^{7/3} + \tfrac{1}{2}Z^2 + \cdots\right]\frac{e^2}{a_0} \tag{86}$$

This is a formally exact result for closed shells. If we apply the Thomas–Fermi theory to this problem, that is we use the virial theorem $T = -E$ and calculate T from equations (10), (7), and $V(r) = -Ze^2/r$, we find first for the chemical potential μ from the normalization requirement $\int n(r)d\mathbf{r} = Z$ that

$$\mu_{\text{Coulomb}}^{\text{TF}} = -\frac{1}{18^{1/3}}Z^{4/3}\frac{e^2}{a_0} \tag{87}$$

and hence

$$E_{\text{Coulomb}}^{\text{TF}} = -(\tfrac{3}{2})^{1/3}Z^{7/3}\frac{e^2}{a_0} \tag{88}$$

The observation made by Scott was that the correction term $(\tfrac{1}{2})Z^2$ to equation (88) in the exact form (86) for the Coulomb field was coming mainly from the K shell, for which screening of the nucleus by the other electrons was least important. Thus he argued that the "boundary" correction due to the strong variation of potential and density near the atomic nucleus in the self-consistent Thomas–Fermi atom remained as $(\frac{1}{2})Z^2e^2/a_0$, and later work by Schwinger[43] supports Scott's arguments further. Thus, Scott's arguments lead to the binding energy of heavy atoms in a nonrelativistic framework as [cf. equations (34), (78), and (86)].

$$E(Z,Z) = [-0.7687Z^{7/3} + \tfrac{1}{2}Z^2 - 0.221Z^{5/3} + O(Z^{4/3})]\, e^2/a_0 \tag{89}$$

Work of March and Plaskett[13] modified the coefficient of $Z^{5/3}$ to -0.266, and if we adopt this value then Table 5 shows a comparison of the resulting atomic binding energies with "observed" values for a selection of values of Z. The "observed" values were given by Scott[39] and they were corrected in such a way that they are nonrelativistic as are all the formulas presented so far.

We conclude this discussion of atomic binding energies by mentioning that if one constrains the coefficient of $Z^{7/3}$ to be -0.7687, then least-squares fitting to Hartree–Fock data[28] yields

$$E = [-0.7687Z^{7/3} + 0.4904Z^2 - 0.2482Z^{5/3}]\, e^2/a_0 \tag{90}$$

which is in satisfactory agreement with the first principles results discussed here.

In view of the interest in density gradient corrections, we turn next to some numerical studies of explicit density gradient corrections to the Thomas–Fermi kinetic energy form (10).

8.2. *Test of Gradient Series in Atoms with Hartree–Fock Quality Densities*

Following the pioneering work of von Weizsäcker,[44,45] the lowest-order gradient correction to the Thomas–Fermi kinetic energy density was calculated quantitatively by Kirznits[46] who obtained (cf. Appendix E)

TABLE 5. Atomic Binding Energies (Nonrelativistic)[a]

Z	6	9	13	18	23
"Observed"	37.8	99.7	242.4	528	945
Theory, including Z^2 and $Z^{5/3}$ terms	37.6	99.4	240.1	524	941

[a]In Hartree units e^2/a_0; 1 Hartree or atomic unit equals approximately 27 eV.

$$t = c_k n^{5/3} + \frac{\hbar^2}{72m}\frac{(\nabla n)^2}{n} + \cdots \tag{91}$$

Hence for the total kinetic energy we can write

$$T = T_0 + T_2 + T_4 + \cdots \tag{92}$$

where T_0 is the Thomas–Fermi term, T_2 comes from the $(\nabla n)^2/n$ term in equation (91) while the fourth-order term, which we shall not display here, has been given by Hodges.[47]

Using wave-mechanical densities n of Hartree–Fock quality, Wang *et al.*[48] have calculated T_0, T_2, and T_4 numerically and a selection of their results are recorded in Table 6. The numerical convergence is seen to be quite good, but of course if one is interested in chemical problems the energies involved in the "correction" T_4 remain very large. Fortunately, modern density functional theory has now its own way of treating the single-particle kinetic energy essentially exactly, as discussed in Chapters 2, 4, and 5. Less fortunately, one is then led back to the Schrödinger equation (1) and the associated one-electron wave functions!

A direct application of equation (91) to the electron density profile at a planar metal surface is given in Appendix E. Various attempts have been made to sum subseries of gradient corrections, and some of this work is referred to in later chapters.

While the above discussion has focused on the kinetic energy, for which, as already mentioned, the Thomas–Fermi approximation is drastic, if we consider the exchange and correlation contribution to the total energy the local use of uniform gas relations turns out to have a much wider range of validity. As an example to show this, we have summarized in Appendix D the exact calculation of the exchange energy in an inhomogeneous electron gas model with strong density gradients present. In spite of this, it is demonstrated there that the exchange energy density is very well represented by the local form proportional to $n^{4/3}$. The same local use of uniform gas relations

TABLE 6. Total Energies of Closed-Shell Atoms Built up from Gradient Expansion of Equations (91) and (92)[a]

	T_0	T_2	T_4
He	2.56	0.32	0.08
Ne	117.8	10.1	1.9
Ar	490.6	34.3	6.2
Kr	2594	142	24

[a]Energies are in Hartree units.

seems to be valid for the correlation energy in many systems and some plausible reasons for this are presented in Chapters 2 and 4.

Our final comments on density gradients are as follows:

(i) Ma and Brueckner[49] have made basic many-body calculations of density gradient corrections in atoms; the reader is referred to the original paper or to the summary in Ref. 50 for details.

(ii) Various workers[51–53] have considered the question of the existence of the gradient expansion. There is agreement that it does not exist in Hartree–Fock theory, but that when exchange and correlation are treated together then the gradient expansion is recovered (see Chapter 2).

(iii) From Teller's theorem[24] that there is no binding in molecules in the Thomas–Fermi theory (or any purely local theory in fact), it follows that density gradients are crucial for molecular binding. It would seem of obvious interest to study T_2 defined in equations (91) and (92) for molecules, using the highest-quality wave-mechanical densities available. However, it should be stressed that while T_2 calculated in such a manner might be useful in characterizing molecular binding, solution of Euler equations for the density $n(\mathbf{r})$ in atoms and molecules using low-order gradient terms only in the kinetic energy as in equation (91) are almost certain to remain disappointing as they omit shell structure. Calculations of total energy corrections from density gradients, using high-quality densities, are what is being referred to here. Much more refined treatments of molecular bonding by the density method are summarized in Chapter 4 and in Ref. 30.

Another way of viewing density gradient corrections will appear when defects in metals are treated in Sections 9 and 10 below. To lead into this discussion of defects in metals, a closely related discussion to that in the present section will be given at this point. It concerns an alternative way of viewing the Thomas–Fermi theory; namely, as a perturbation theory, summed to all orders, but based on plane waves as the unperturbed problem.[54] This is very natural, when we recall that Thomas–Fermi theory is based on plane waves, but used locally. Obviously, as the final simplifying step in the plane wave perturbation theory, one must allow the "perturbation" $V(\mathbf{r})$ to vary slowly in the Thomas–Fermi sense. Then it will be demonstrated that this enables the perturbation series to be summed to all orders, yielding the nonlinear Thomas–Fermi density–potential relation (7).

9. Thomas–Fermi Theory as an Approximate Summation of One-Body Perturbation Theory Based on Plane Waves

Though density functional theory is discussed in Chapter 2 as one, formally exact, approach to solving the many-electron problem, it is of interest here to summarize the proof by March and Murray[54] that the Thomas–Fermi theory is an approximate summation to all orders of one-body perturbation theory based on plane waves.

In the proofs referred to above, the Dirac density matrix $\rho(\mathbf{r},\mathbf{r}')$, defined as an off-diagonal generalization of equation (2), i.e., $\rho(\mathbf{r},\mathbf{r}) \equiv n(\mathbf{r})$

$$\rho(\mathbf{r},\mathbf{r}') = \sum_{\text{occupied states}} \psi_i^*(\mathbf{r})\psi_i(\mathbf{r}') \tag{93}$$

was generated for a given one-body potential energy $V(\mathbf{r})$, where the wave functions ψ_i satisfy the single-particle Schrödinger equation (1). By using perturbation theory based on plane waves $\exp(i\mathbf{k}\cdot\mathbf{r})$, for which the unperturbed Dirac density matrix $\rho_0(\mathbf{r},\mathbf{r}')$ is readily obtained from equation (93) as (cf. Appendix C)

$$\rho_0(\mathbf{r},\mathbf{r}') = \frac{k_f^3}{2\pi^2}\frac{j_1(k_f|\mathbf{r}-\mathbf{r}'|)}{k_f|\mathbf{r}-\mathbf{r}'|}, \qquad j_1(x) = \frac{\sin x - x\cos x}{x^2} \tag{94}$$

k_f as usual denoting the radius of the Fermi sphere in $\mathbf{k}$ space, it was shown that the expansion had the form

$$\rho(\mathbf{r},\mathbf{r}') = \sum_{j=0}^{\infty} \rho_j(\mathbf{r},\mathbf{r}') \tag{95}$$

where

$$\rho_j(\mathbf{r},\mathbf{r}_0) = \frac{k_f^2}{2\pi^2}\int^j \prod_{l=1}^{j}\left[\frac{-d\mathbf{r}_l V(\mathbf{r}_l)}{2\pi}\right] j_1\left(k_f\sum_{l=1}^{j+1} s_l\right) \Bigg/ \prod_{l=1}^{j+1} s_l \tag{96}$$

and $s_l = |\mathbf{r}_l - \mathbf{r}_{l-1}|$, $\mathbf{r}_{j+1} = \mathbf{r}$. Explicitly, the first-order term is given by

$$\rho_1(\mathbf{r},\mathbf{r}_0) = \frac{-k_f^2}{2\pi^2}\int d\mathbf{r}_1 \frac{V(\mathbf{r}_1)}{2\pi}\frac{j_1(k_f|\mathbf{r}-\mathbf{r}_1|+k_f|\mathbf{r}_1-\mathbf{r}_0|)}{|\mathbf{r}-\mathbf{r}_1|\,|\mathbf{r}_1-\mathbf{r}_0|} \tag{97}$$

Summation for Slowly Varying Potentials

We now consider the summation of the diagonal element $\rho(\mathbf{r},\mathbf{r}) \equiv n(\mathbf{r})$ of the Dirac density matrix generated to all orders in perturbation theory in equations (95) and (96). To see how the argument goes, we take the diagonal element of the first-order term in equation (97), to find

$$\rho_1(\mathbf{r},\mathbf{r}) = \frac{-k_f^2}{2\pi^2}\int d\mathbf{r}_1 \frac{V(\mathbf{r}_1)}{2\pi}\frac{j_1(2k_f|\mathbf{r}-\mathbf{r}_1|)}{|\mathbf{r}-\mathbf{r}_1|^2} \tag{98}$$

Assuming $V(\mathbf{r})$ to vary by but a small fraction of itself over a de Broglie wavelength for an electron at the Fermi surface, then one can replace $V(\mathbf{r}_1)$ in equation (98) approximately by $V(\mathbf{r})$. The integration over $\mathbf{r}_1$ can then be carried out with the result

$$\rho_1(\mathbf{r},\mathbf{r}) = -k_f V(\mathbf{r})/2\pi^2 \tag{99}$$

In a similar manner, the higher-order terms of equation (95) may be calculated, when one obtains[54]

$$\rho(\mathbf{r},\mathbf{r}) = \frac{k_f^3}{6\pi^2}\left(1 - \frac{3V}{k_f^2} + \frac{3V^2}{2k_f^4} + \frac{V^3}{2k_f^6} + \cdots\right) = \frac{1}{6\pi^2}[k_f^2 - 2V(\mathbf{r})]^{3/2} \tag{100}$$

or

$$\rho(\mathbf{r},\mathbf{r}) = \frac{2^{3/2}}{6\pi^2}[\mu - V(\mathbf{r})]^{3/2}, \ \mu = \frac{k_f^2}{2} \tag{101}$$

This, apart from a factor of 2 due to the fact that the above argument has considered singly occupied energy levels, is the Thomas–Fermi density–potential relation (7).

Because the density matrix is also known through equations (95) and (96), the kinetic energy can easily be obtained,[55] and the results in the above perturbative framework are recorded in Appendix F.

10. Screening of Charges in Metals and Semiconductors

10.1. Screening of Test Charge in Metallic Medium

One of the areas in which the Thomas–Fermi theory, and its extensions, has played a valuable role is in screening theory in simple metals. This theory makes quantitative the well-known qualitative consequence of electrostatics that long-range electric fields cannot exist in a conducting medium.

Suppose we place a test charge ze at the origin in an originally uniform electron gas. Then the electrostatic potential $\psi(r)$ must tend to $ze/4\pi\epsilon_0 r$ as r tends to zero, r measuring evidently the position from the test charge, and must fall off more rapidly than $1/r$ as r tends to infinity. The potential energy felt by an electron is given by $-e\psi = V(\mathbf{r})$, and since ze is a test charge we can use the linearized form of the Thomas–Fermi theory in equation (99). Using units in which $4\pi\epsilon_0 = 1$ for convenience, we then find, from Poisson's equation, the result

$$\nabla^2 V = q^2 V, \qquad q^2 = \frac{4k_f}{\pi a_0} \tag{102}$$

a_0 being as usual the Bohr radius $\hbar^2/4\pi me^2$. Evidently the solution of equation (102) with spherical symmetry appropriate to the single test charge at the origin and satisfying the boundary conditions discussed above, namely,

$V(r) \to -ze^2/r$ as r tends to zero

$V(r)$ tends to zero faster than $1/r$ as $r \to \infty$

is given by

$$V(r) = (-ze^2/r)\exp(-qr) \tag{103}$$

Equation (102) and the solution (103) were first given by Mott.[56] It can be seen that according to the Thomas–Fermi theory, the Coulomb potential is screened out exponentially with distance r, with a screening length q^{-1} which in a good metal is found from equation (102) to be of the order of 1 Å.

While the Thomas–Fermi screening length q^{-1} derived above is of fundamental importance in treating charged perturbations in metals, there is one important feature which is missing from the above treatment. This omission is due to the semiclassical character of the Thomas–Fermi theory, and one really should be treating the scattering of electron waves from the test charge. Then the correct first-order treatment is given by combining equation (98) with Poisson's equation to obtain, following March and Murray[54]

$$\nabla^2 V = \frac{me^2}{\hbar^2}\frac{2k_f^2}{\pi^2}\int d\mathbf{r}' \frac{j_1(2k_f|\mathbf{r}-\mathbf{r}'|)}{|\mathbf{r}-\mathbf{r}'|^2}V(\mathbf{r}') \tag{104}$$

which is the correct wave-mechanical generalization of the Mott[56] equation (102).

Without solving equation (104) it is easy to see that the displaced charge can have a very different character at large r, depending on whether we use the wave theory result (104) or the semiclassical result (102). Thus, consider a simple case when $V(\mathbf{r})$ is very short range, i.e., write $V(\mathbf{r}) = \lambda\delta(\mathbf{r})$. Then the displaced charge according to equation (99) is of similar short range while equation (98) gives in contrast, if n_0 is as usual the unperturbed uniform density

$$n(\mathbf{r}) - n_0 \propto \frac{j_1(2k_f r)}{r^2}, \qquad \frac{k_f^3}{3\pi^2} = n_0 \tag{105}$$

From the definition (94) of the spherical Bessel function j_1 in equation (105), it follows that at large r the displaced charge decays as

$$n(r) - n_0 \propto \frac{\cos 2k_f r}{r^3} \tag{106}$$

This is the new feature of the wave theory based on equation (104) and cautions one to be careful to check that the conditions of validity of the Thomas–Fermi model are valid before using it in an application such as the present one. We shall see in Section 10.7 that the same theory, applied to ionized impurities in nondegenerate conditions in a semiconductor, is perfectly valid. That such "wiggles" as described in equation (106) exist in the displaced charge round a given fixed perturbation in a Fermi gas was first pointed out by Blandin *et al.*[57, 58]

We shall return to the above discussion below, when we treat the interaction energy between two charges in a Fermi gas. But we wish to study a little further the range of validity of the Thomas–Fermi solution in equation (103). In order to do so, it proves important to introduce the dielectric function $\epsilon(k)$, by working in Fourier transform.

10.2. Wave-Number-Dependent Dielectric Function

We therefore define the Fourier components of the screened potential energy $V(\mathbf{r})$ treated above by

$$\tilde{V}(\mathbf{k}) = \int d\mathbf{r}\, V(\mathbf{r}) \exp(i\mathbf{k}\cdot\mathbf{r}) \tag{107}$$

and if we use first the screened Coulomb potential (103) of the semiclassical Thomas–Fermi theory then we find

$$\tilde{V}_{\mathrm{TF}}(k) = \frac{-4\pi z e^2}{k^2 + q^2} \tag{108}$$

To see the limitations of this Thomas–Fermi description, let us return to the first-order wave theory equation (104). This can be solved analytically in **k** space, using the properties of the Fourier transform of a convolution, to yield

$$\tilde{V}_{\text{wave theory}} = \frac{-4\pi z e^2}{k^2 + (k_f g(k/2k_f)/\pi a_0)} \tag{109}$$

where the function g is defined by

$$g(x) = 2 + \frac{x^2 - 1}{x} \ln \left| \frac{1 - x}{1 + x} \right| \tag{110}$$

In the long wavelength limit k tends to zero, $g(x)$ tends to the value 4, and we find the same result

$$\tilde{V}(0) = -4\pi z e^2/q^2 \tag{111}$$

from both the Thomas–Fermi approximation and from the correct first-order wave theory.

Often it proves valuable to express these results in terms of dielectric function $\epsilon(k)$ of the Fermi gas already referred to. This is best defined in the present context by writing

$$\tilde{V}(k) = \frac{-4\pi z e^2}{k^2 \epsilon(k)} \tag{112}$$

It follows from equation (108) that in the Thomas–Fermi approximation we have

$$\epsilon_{\mathrm{TF}}(k) = \frac{k^2 + q^2}{k^2} \tag{113}$$

while from the wave theory result (109) one finds

$$\epsilon(k) = \frac{k^2 + (k_f g(k/2k_f)/\pi a_0)}{k^2} \tag{114}$$

This latter expression is due to Lindhard[59]; it becomes equivalent to the Thomas–Fermi result (113) only when we replace $g(x)$ by its value $g(0) = 4$, which is easily seen by using the expression (110). Thus, the Thomas–Fermi approximation to the dielectric function is restricted in its range of validity to the long-wavelength regime. The Friedel wiggles displayed in equation (106) arise from the kink in $\epsilon(k)$ in equation (114) at $k = 2k_f$, the diameter of the Fermi sphere. This kink has interesting implications for electron–phonon interaction and is the origin of the Kohn anomaly in the dispersion relation of lattice waves in metals (see for example, Ref. 50).

10.3. *Electrostatic Model of Interaction between Charges in Fermi Gas*

So far, we have been concerned with the screening out of the field of a single test charge. But in a number of important applications in condensed matter, we need to know the interaction energy between charges. We shall first use the semiclassical Thomas–Fermi theory below,[60] though the essential result obtained, namely the applicability of the so-called "electrostatic model," is valid also in the wave theory discussed above.[61]

Let us suppose that ions 1 and 2, embedded in the bath of conduction electrons, are screened such that the potential due to ion 1 alone is V_1 and that due to ion 2 is V_2. Then in the linear approximation of equation (102), it is immediately clear that the total potential V of the two-center problem is just the superposition potential given by

$$V = V_1 + V_2 \tag{115}$$

Hence each ion is surrounded by its own displaced charge and this is not affected by bringing up further ions. Incidentally, this is then the condition that the total potential energy of the ions can be written as a sum of pair potentials in the multicenter case.[61, 62]

The interaction energy between the ions, separated by a distance R say, may now be obtained directly by calculating the difference between the total energy of the metal when the ions are separated by an infinite distance and when they are brought up to separation R. Clearly this is all to be done in the Fermi sea of constant density n_0, say, and the result for the interaction energy will depend on the density n_0, or equivalently from equation (4) on the Fermi energy $E_f = p_f^2/2m \equiv \mu$.

Let us consider the changes in the kinetic and potential energy separately. The act of introducing an ion carrying a charge ze into the Fermi gas changes the kinetic energy in the manner discussed in Appendix F. If T_0 denotes the total kinetic energy of the unperturbed Fermi sea, then in the Thomas–Fermi theory the result in the perturbed inhomogeneous gas, measured relative to the unperturbed value, is from equation (F.4) given by

$$T - T_0 = E_f \int \Delta n \, d\mathbf{r} + \frac{E_f}{3n_0} \int d\mathbf{r} \, (\Delta n)^2 + O(\Delta n^3) \tag{116}$$

where Δn denotes the displaced charge $n - n_0$. We now observe that the first term involves the normalization condition for the displaced charge, or in other words the condition that the ionic charge ze is screened completely by the electron distribution. Clearly, this term will make no contribution to the energy difference between infinitely separated ions and the ion pair at separation R.

We are now in a position to calculate the changes in both kinetic and potential energy contributions when we bring the ions together to separation R from infinity. We may write down the following contributions:

i. The interaction energy between the charge ze of one impurity and the perturbing potential, say V_2, due to the other.
ii. The interaction energy between the displaced charge $(q^2/4\pi e^2)\, V_1$ round the first ion [cf. equation (99)] and the potential V_2 due to the other.
iii. The change in kinetic energy.

These three terms are evidently given by

$$\text{(i)} \quad z^2 e^2 \exp(-qR)/R \tag{117}$$

$$\text{(ii)} \quad \frac{-q^2}{4\pi e^2} \int d\mathbf{r} \, V_1 V_2 \tag{118}$$

$$\text{(iii)} \quad \left(\frac{-q^2}{4\pi e^2}\right)^2 \frac{E_f}{3n_0} \left[\int d\mathbf{r} \, [(V_1 + V_2)^2 - V_1^2 - V_2^2] \right]$$

$$= \frac{q^2}{4\pi e^2} \int d\mathbf{r} \, V_1 V_2 \tag{119}$$

where (iii) follows from equations (116), (99), and (115). Thus it can be seen that contribution (ii) is precisely canceled by the change in kinetic energy (iii) and we are left with the final result for the interaction energy ΔE as

$$\Delta E = z^2 e^2 \exp(-qR)/R \tag{120}$$

But this is simply the electrostatic energy of an ion of charge ze sitting in the electrostatic potential $(ze/R)\exp(-qR)$ of the second ion. This result, namely the electrostatic model, has been shown to hold also in the wave theory,[61] though now, of course, the correct screened potential with the Friedel wiggles must be used.

The above argument is very valuable for indicating the type of interaction energy between charged defects in a metal, as well as the nature of the ion–ion interaction in condensed metallic media. Of course we are then usually dealing with ions of finite size, but if we can represent these by a bare pseudopotential $v_b(k)$ say in Fourier transform, then it can be shown[62] that the interaction energy takes the form

$$\Delta E(R) = \int \frac{k^2 v_b^2(k)}{\epsilon(k)} \exp(i\mathbf{k}\cdot\mathbf{R})\, d\mathbf{k} \tag{121}$$

This reduces to the r space form (120) for point ions, with $v_b(k)$ replaced by $4\pi ze^2/k^2$, if for $\epsilon(k)$ we use the Thomas–Fermi dielectric constant. To obtain the Friedel oscillations in the long-range form of the interaction energy $\Delta E(R)$, we must use instead for $\epsilon(k)$ the Lindhard dielectric function $\epsilon(k)$, or its refinements to include electron exchange and correlation (see Chapters 2 and 4; see also the summary in Ref. 63).

These applications on interaction between charges caution us that while the Thomas–Fermi theory remains valuable for obtaining some important general results (for example the screening length q^{-1} and the above derivation of the electrostatic model for the interaction between charges in a Fermi gas) its assumption that the screened potential varies slowly is too drastic in metals. The wave theory is therefore essential for quantitative work in this area. Nevertheless, there are two further interesting applications we wish to discuss: the problem of the vacancy formation energy in a simple close-packed metal, and its relation to lattice properties, and secondly the important problem as to the way the Fermi level E_f varies with concentration in a binary metallic alloy in which there is a valence difference between the components (e.g., a Cu–Zn or an Al–Mg alloy), in which the Thomas–Fermi screening theory remains valuable.

10.4. *Relation of Vacancy Formation Energy to Debye Temperature in Simple Metals*

The energy required to create a vacancy in a metal can be related to phonon properties, and in particular to the Debye temperature, through the linear screening theory discussed above. The argument goes as follows. Let us consider first a potential energy $V(\mathbf{r})$ created by the vacancy screened by the charge it displaces. If we represent the vacancy in a simple metal of valency Z by a negative charge $-Ze$ at the vacant site, $V(\mathbf{r})$ is given by the (now repulsive) screened Coulomb form (103) in the Thomas–Fermi approximation. Fortunately, in a linear theory, the change in the sum of the one-electron energies ϵ from their free electron values can be estimated without choosing such a detailed form of $V(\mathbf{r})$, but by merely asserting that it is sufficiently weak to allow the application of first-order perturbation theory. In the unperturbed free electron metal the wave functions are the plane waves $\mathcal{V}^{-1/2}\exp(i\mathbf{k}\cdot\mathbf{r})$, $\mathcal{V}$ being the total volume of the metal, and we can therefore write the change in the one-electron energy $\epsilon_{\mathbf{k}} = k^2/2$ as

$$\Delta\epsilon_{\mathbf{k}} = \mathcal{V}^{-1}\int e^{-i\mathbf{k}\cdot\mathbf{r}}\, V(\mathbf{r})\, e^{i\mathbf{k}\cdot\mathbf{r}} d\mathbf{r} = \mathcal{V}^{-1}\int V(\mathbf{r})\, d\mathbf{r} \tag{122}$$

which is evidently independent of $\mathbf{k}$. Summing this over the N electrons in the metal gives for the change ΔE_s in the sum of the one-electron energies the result

$$\Delta E_s = n_0 \int V(\mathbf{r})\, d\mathbf{r} \tag{123}$$

where n_0 as usual is the unperturbed electron density. But now, in linear theory, it is easy to show that this integral is determined merely by the requirement that the excess charge is perfectly screened, the result for the change in the one-electron energy sum being [from equations (99) or (98)]

$$\Delta E_s = \tfrac{2}{3} Z E_f \tag{124}$$

where E_f is the Fermi energy. If we neglect ionic relaxation there is, in the process of taking an atom from the bulk and placing it on the surface, a reduction in kinetic energy of the Fermi electrons, due to the increase of one atomic volume, which is readily calculated from equation (10) to be [64]

$$\Delta T = \tfrac{2}{5} Z E_f \tag{125}$$

This admittedly elementary theory of the vacancy formation energy E_v yields

$$E_v \doteqdot \Delta E_s - \Delta T = \tfrac{4}{15} Z E_f \tag{126}$$

Though such a theory appears to be significant in the sense of extrapolating measured vacancy formation energies for $Z \geq 1$ to the limit Z tends to zero, it turns out that ZE_f is not the appropriate unit in which to measure E_v. That this is so was implicit in the work of Mukherjee,[65] who pointed out an

TABLE 7. Relation between Debye Temperature Θ and Vacancy Formation Energy E_v for Some Close-Packed Metals

Metal	Debye temperature (K)	Vacancy formation energy (eV)	$\Theta/(E_v/M\Omega^{2/3})^{1/2}$
Cu	245	1.17	32
Ag	225	1.10	32
Au	165	0.95	34
Mg	406	0.89	34
Al	428	0.75	33
Pb	94.5	0.5	33
Ni	441	1.5	33
Pt	229	1.4	37

empirical relation between E_v and the Debye temperature Θ, his results for some close-packed metals being summarized in Table 7.

To see how such a relation arises, let us use the same linear theory of screening which led to equation (126) to discuss the velocity of sound in a simple metal. This can be done by first noting that if the ions in a metal were embedded in a completely uniform electron gas, then they would vibrate with the ionic plasma frequency

$$\omega_p^{\text{ion}} = \left[\frac{4\pi n_i(Ze)^2}{M}\right]^{1/2} \tag{127}$$

determined by the ionic number density n_i, the ionic charge Ze, and ion mass M. Rewriting n_i through the electron density $n_0 = Zn_i$ yields immediately

$$\omega_p^{\text{ion}} = \left(\frac{4\pi n_0 Ze^2}{M}\right)^{1/2} \tag{128}$$

But this represents an optic mode and to obtain the desired acoustic mode of the form

$$\omega = v_s k \tag{129}$$

at small wave numbers k, where v_s is the velocity of sound, one must allow the electrons to pile up round the positive ions and screen them. This can be allowed for using the dielectric function of Section 10.2 and the consequence is that the bare Coulomb potential, written in k space as $4\pi Ze^2/k^2$, must be screened such that

$$\frac{4\pi Ze^2}{k^2} \rightarrow \frac{4\pi Ze^2}{k^2 + q^2} \qquad \text{or as } k \rightarrow 0, \qquad Z \rightarrow Zk^2/q^2 \tag{130}$$

Making the appropriate substitution (130) in equation (128) leads to the so-called Bohm–Staver formula[66] for the velocity of sound

$$v_s = (Zm/3M)^{1/2}\, v_f \tag{131}$$

where v_f is the Fermi velocity of the electrons. Between equations (126) and (131) we can eliminate the quantity ZE_f, and with it at least some of the dependence of the argument on linear theory, to find[67]

$$E_v = \text{const } Mv_s^2 \tag{132}$$

Using in equation (132) the usual (average) relation between v_s and Debye temperature,[68] namely,

$$\Theta = \frac{v_s}{\Omega^{1/3}} \left(\frac{3}{4\pi}\right)^{1/3} \frac{h}{k_B} \tag{133}$$

where Ω is the atomic volume and k_B is Boltzmann's constant, leads to the Mukherjee relation

$$\Theta = CE_v^{1/2}/\Omega^{1/3}M^{1/2} \tag{134}$$

The constant C given in Table 7 would be correctly obtained by the above argument if, for example, the number 4/15 in equation (126) were replaced by 1/6, so that the theory is already semiquantitative for the constant C.

In fact, the theory of vacancy energy in the polyvalent metals Al and Pb is a nonlinear problem and there is gross cancellation of energy terms, as a fully nonlinear calculation using the theory of the inhomogeneous electron gas shows.[69] The relation to the phonons in equation (132) appears then to be a deeper way of tackling the problem of the vacancy formation energy. What the admittedly oversimplified linear theory given above shows is that, at this level, the problem of the response of the electron gas to the removal of an atom is related to that of introducing a phonon into the ionic lattice, via the dielectric function $\epsilon(k)$. It should be cautioned that the argument presented above should not be used for open body-centered-cubic metals and in particular the alkali metals, because relaxation of the ions around the vacant site now makes the dominant contribution. Nevertheless, it is remarkable that Mukherjee's relation (134) is again found to hold.[70]

10.5. *Variation of Fermi Level with Concentration in Dilute Binary Metallic Alloy*

The question comes up in an alloy such as Cu–Zn, with say a finite concentration of Zn atoms in a Cu matrix, as to the way in which the Fermi level E_f varies with the impurity concentration c. The theory for binary metallic alloys was given by Friedel[71] in the case of a valence difference Z between the constituent atoms. The elementary physical argument, which we make quantitative below, goes as follows. Whereas, in the so-called rigid band model, we would merely fill in the additional Zn electrons into the Cu density of states curve, at least at dilute concentration, and therefore the Fermi level would show a linear increase with concentration, in the screening theory presented in Section 10.1, each one-electron level would be lowered in energy by the attractive Zn impurities. It turns out that filling electrons in to the resulting density of states merely preserves the Fermi level intact to first order in the concentration.

To see how this result comes quantitatively from the Thomas–Fermi theory, and in the process to generalize it away from the dilute concentration regime, we follow Friedel[71] in assigning to each impurity its own volume, which for simplicity one takes to be a sphere of radius R, this being related to the concentration of impurities c by $(qR)^{-3} = (qr_s)^{-3}\,c$.†

We next use the Thomas–Fermi equation, but being careful to allow the Fermi level E_f in the unperturbed metal to shift by an amount ΔE_f. Then assuming the necessary condition $|\Delta E_f - V| << E_f$ for linearization we can write for the screened potential energy $V(\mathbf{r})$ round an impurity at the origin of its "own" sphere of radius R the Poisson equation

$$\nabla^2 V = q^2\,(V - \Delta E_f) \tag{135}$$

This has now to be solved subject to the boundary conditions[71]

$$\left(\frac{dV}{dr}\right)_R = 0, \qquad V(R) = 0, \qquad V \to \frac{-Ze^2}{r}, \qquad r \to 0 \tag{136}$$

the first condition expressing electrical neutrality of the cell through Gauss' theorem. The independent solutions of equation (135) being of the form $(Ze^2/r)\exp(\pm qr)$ it is easy to obtain the solution satisfying the boundary condition that the electric field vanish at R as

$$V(r,R) - \Delta E_f = -\frac{Ze^2}{r}\,\frac{qR\cosh[q(R-r)] - \sinh[q(R-r)]}{qR\cosh qR - \sinh qR} \tag{137}$$

†For typical metals, $(qr_s)^{-3} \sim 0.01$ to 0.05, r_s being the radius of the atomic sphere.

The second boundary condition (136) is then used to determine the shift in the Fermi level as

$$\Delta E_f = Ze^2 q/(qR \cosh qR - \sinh qR) \tag{138}$$

and a plot of this is shown in Fig. 7. It will be seen that the slope at the origin $c = 0$ is zero, and therefore that there is no term linear in the concentration. Experimental support for such a movement in the Fermi level is afforded by the optical experiments of Biondi and Rayne.[72]

Of course, such an argument as presented above is gross, and could only apply in simple metallic systems in at best a semiquantitative way. Modern alloy theory, discussed in Chapter 4, has of course now progressed far beyond such an approach in yielding results of quantitative accuracy. Nevertheless, the result that there is no shift in the Fermi level to $O(c)$ is an important qualitative point in alloy theory. As one example which can be cited, Bhatia and March[73] have used it in a discussion of the surface tension of liquid binary alloys, which generalizes the relation (E.14) of Appendix E for a pure metal to include the concentration fluctuations in a liquid binary metal alloy.

10.6. Dielectric Function of a Semiconductor: Thomas–Fermi Model

As emphasized by Resta,[74] the wave-number-dependent dielectric function $\epsilon(k)$ of a semiconductor can be approximated usefully by the linearized Thomas–Fermi model discussed above. Whereas, in the metal, the dielectric

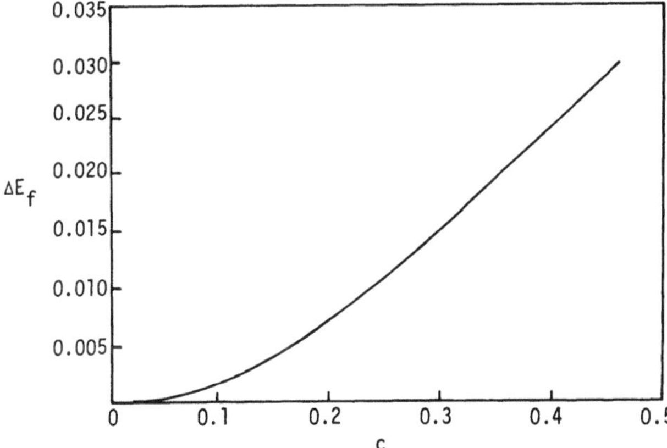

FIGURE 7. Shift ΔE_f in Fermi level in binary metallic alloy as a function of concentration, according to equation (138). Atomic units are used.

function $\epsilon(k)$ diverged as k^{-2} as k tended to zero, in the semiconductor the value must tend to the static dielectric constant.

Essentially Resta solves equation (135), namely,

$$\nabla^2 V = q^2(V - A) \tag{139}$$

The model semiconductor is thereby being treated as an electron gas, and as usual we denote the unperturbed electron density by n_0.

Resta now argues for the existence of a finite screening radius R around a test charge Ze, such that the density $n(r)$ becomes equal to n_0 at distance R from the test charge, that is,

$$n(R) = n_0 \tag{140}$$

which gives for the constant A in equation (139) the value $V(R)$. Beyond this screening distance R, the screened potential energy $V(r)$ of the point charge Ze is given by

$$V(r) = -Ze^2/\epsilon(0)r, \qquad r > R \tag{141}$$

$\epsilon(0)$ being the static dielectric constant of the semiconductor. Following the solution discussed in the previous section, the general solution for the potential energy from equation (139) may be written

$$V(r) = -(Ze^2/r)\,[\alpha \exp(qr) + \beta \exp(-qr)] + A, \qquad r \leq R \tag{142}$$

By imposing continuity at $r = R$, using $V \to -(Ze^2/r)$ as $r \to 0$, and putting $A = V(R)$ as discussed above, equation (142) takes the form

$$V(r) = -(Ze^2/r)\frac{\sinh q(R-r)}{\sinh qR} - \frac{Ze^2}{\epsilon(0)R}, \qquad r \leq R \tag{143}$$

The screening distance R is found, by requiring continuity of the electric field at $r = R$, to be related to $\epsilon(0)$ through

$$\sinh qR/qR = \epsilon(0) \tag{144}$$

Equation (144) yields a finite solution for R for any $\epsilon(0) > 1$. The metallic limit, discussed fully in earlier sections, formally corresponds to $\epsilon(0) \to \infty$, when we find $R \to \infty$, as expected.

The wave-number-dependent dielectric function $\epsilon(k)$ can be calculated readily from the above treatment and takes the form[74]

$$\epsilon(k) = \frac{k^2 + q^2}{[q^2/\epsilon(0)] \sin kR/kR + k^2} \tag{145}$$

A merit of equation (145), in common with other usable models[74] of $\epsilon(k)$, is that the only input data required to evaluate $\epsilon(k)$ from equation (145) are the static dielectric constant and the Fermi momentum k_f of the valence electrons. Comparison of equation (145) with other approximate theories is made by Resta,[74] to whose paper the interested reader is referred.

The final point to be made, which adds to one's confidence in Resta's treatment, is that the numerical values of R found from equation (144) are very close to the nearest-neighbor distances in the crystals in the cases of the valence semiconductors diamond, silicon, and germanium.

10.7. Ionized Impurities in Semiconductors

The theory of impurity scattering in semiconductors has a good deal of interest from a practical point of view, and can be tackled along rather similar lines to the corresponding problem in metals discussed in Section 10.1. However, it is evident that a number of modifications of that treatment must be made. For example, it is no longer permissible to suppose that in semiconductors we are dealing with a completely degenerate gas of free electrons. Indeed, in many cases the assumption that the electrons obey Maxwell–Boltzmann statistics is more appropriate. This latter case has been discussed by Debye and Conwell,[75] who give a formula for the mobility, derived by Herring. The generalization to arbitrary degeneracy has been considered independently by Dingle[76] and Mansfield.[77]

In order to calculate the electric field around an ionized impurity centre, use is made of the generalized Thomas–Fermi theory, first formulated for low temperatures by Marshak and Bethe[78] and for arbitrarily high temperatures by Sakai[79] and independently Feynman *et al.*[80] in the course of their work on equations of state of elements at high pressures and temperatures (see Section 11.4 below). In this generalized theory, the restriction made so far that the electron gas is completely degenerate is removed and the Fermi–Dirac distribution function must be introduced into the theory.

10.7.1. Density–Potential Relation in Generalized Thomas–Fermi Theory. The number of electrons with momenta of magnitude between p and $p + dp$ in a volume element $d\tau$ must now be written as

$$\frac{4\pi p^2\, dp\, d\tau\, (2/h^3)}{\exp[(p^2/2m^* + V)/k_BT - \eta^*] + 1}$$

where V is as usual the potential energy. m^* is the effective electronic mass and η^* is a constant, the reduced chemical potential of the electrons, which is to be determined as usual by normalization requirements. Then it follows that the electron density $n(\mathbf{r})$ is given by

$$n(\mathbf{r}) = \int_0^\infty (8\pi p^2\, dp/h^3) \Big/ \exp[(p^2/2m^* + V)/k_BT - \eta^*] + 1 \tag{146}$$

This is the modified relation between density and potential in the generalized Thomas–Fermi approximation.

10.7.2. Self-Consistent Field in Terms of Fermi–Dirac Functions. It is convenient at this point to introduce the Fermi–Dirac functions $I_k(\eta^*)$ defined by

$$I_k(\eta^*) = \int_0^\infty \frac{y^k\, dy}{\exp(y - \eta^*) + 1} \tag{147}$$

Then equation (146) becomes

$$n(\mathbf{r}) = \frac{4\pi}{h^3} (2m^* k_BT)^{3/2} I_{1/2}\left(\eta^* - \frac{V}{k_BT}\right) \tag{148}$$

Next we wish to write down the self-consistent field equation determining the screened potential $V(r)$ around an ionized impurity center. We note in doing so that if n_0 is the carrier density in the unperturbed lattice, then since n_0 is also the density of positive charge, Poisson's equation gives us

$$\nabla^2 V = \frac{-4\pi e^2}{\epsilon}(n - n_0) \tag{149}$$

where ϵ is the static dielectric constant. Assuming that η^* is unchanged by the introduction of the impurity we have therefore

$$-\nabla^2 V = \frac{16\pi^2 e^2}{\epsilon h^3} (2m^* k_BT)^{3/2} \left[I_{1/2}\left(\eta^* - \frac{V}{k_BT}\right) - I_{1/2}(\eta^*) \right] \tag{150}$$

since n_0 is evidently given by

$$n_0 = \frac{4\pi}{h^3}(2m^*k_BT)^{3/2}I_{1/2}(\eta^*) \tag{151}$$

The boundary conditions appropriate to a single impurity center carrying charge e, and taken as the origin of coordinates are evidently, in the model of the dielectric medium adopted here

$$\begin{aligned} &V \to -e^2/\epsilon r \qquad \text{as } r \to 0 \\ &V \to 0 \text{ faster than } r^{-1} \qquad \text{as } r \to \infty \end{aligned} \tag{152}$$

In order to obtain an approximate solution, Dingle and Mansfield assume that

$$|\eta^* k_B T| >> |V|$$

and write equation (150) in the approximate form

$$\nabla^2 V = q^2 V \tag{153}$$

with solution

$$V = \frac{-e^2}{\epsilon r}\exp(-qr) \tag{154}$$

where the screening length q^{-1} is now determined by

$$q^2 = \frac{16\pi^2 e^2}{\epsilon h^3}(2m^*)^{3/2}(k_BT)^{1/2}\, I'_{1/2}(\eta^*) \tag{155}$$

The expression for q^2 reduces in the case of complete degeneracy to the result in equation (102), if we replace ϵ by unity and put $m^* = m$. In the nondegenerate case,

$$q^2_{\text{nondegenerate}} = \frac{4\pi e^2 n_0}{\epsilon k_B T} \tag{156}$$

and as an example to which equation (156) is applicable, taking one impurity phosphorus atom for every 10^5 Si atoms, $1/q$, the screening radius, is found to be about 60 Å at room temperature.

Immediately below, we shall use the generalized Thomas–Fermi theory developed above to treat material under high pressure.

11. Material under High Pressure

We must emphasize that the Thomas–Fermi theory is limited in its usefulness when we are dealing with perfect crystals under normal conditions since it cannot account for the Periodic Table effects. However, when pressures sufficient to obliterate the detailed influence of the outer electronic structure are applied, the results of the method should be valid for all elements. The information thus obtained on the behavior of material at high pressures has proved of very considerable value in astrophysics and, to a lesser extent, in geophysics.

We shall confine ourselves here to three aspects:

i. The equations of state of elements as given by the Thomas–Fermi method;
ii. The generalization to the high temperatures often encountered in astrophysical problems; and
iii. The relation between the predictions of the theory and existing experimental information obtained from laboratory experiments.

11.1. *Equations of State from Thomas–Fermi Theory*

In principle, the problem of describing the electronic density and the electric field in a crystal in the Thomas–Fermi theory can be reduced immediately to that of solving the Thomas–Fermi equation within the Wigner–Seitz cellular polyhedron surrounding a particular atom in the crystal, subject to the requirement of periodicity that the normal derivative of the potential shall vanish over the surface of the atomic polyhedron. However, for many crystal structures, the Wigner–Seitz cell has high symmetry and it is a very useful approximation to replace the atomic polyhedron by a sphere of equal volume, as suggested by Slater and Krutter[81] in their pioneering paper on the Thomas–Fermi method for metals. The problem then reduces to solving the dimensionless Thomas–Fermi equation (24), already discussed fully for atoms and ions, subject to the different boundary conditions

$$\phi(0) = 1, \qquad \frac{\phi(x_0)}{x_0} = (\phi')_{x_0} \tag{157}$$

where x_0 is the radius of the atomic sphere, say R, in dimensionless units, i.e., $R = bx_0$. Such a solution is shown in curve III of Fig. 1. The second boundary condition accounts for periodicity in the sphere model, and ensures charge neutrality. Thus, for any particular element, one can obtain a set of solutions of the Thomas–Fermi equation corresponding to different values of the atomic volume (lattice parameter).

Though we have argued that the approximation is valid when the influence of the outer structure is obliterated, it is worth mentioning in passing the relevance to the cohesion of solids. To discuss this, one wishes to obtain the derivative of the energy with respect to the radius of the atomic sphere, and the easiest way to calculate this is to compute the pressure p, which, at absolute zero, is given by

$$p = \frac{-dE}{dv} = \frac{-1}{4\pi R^2}\frac{dE}{dR} \tag{158}$$

where v is the atomic volume, and E is the energy. The pressure can be got from the intuitive argument (and confirmed by direct calculation) that, since the Thomas–Fermi theory is derived by applying free electron relations locally, the pressure due to the bombardment of the electron gas on the boundary of the atomic sphere would then appear to be that due to a free electron gas equal in density to that of the actual inhomogeneous electron cloud on the cell boundary. If $n(R)$ is the boundary density obtained by solution of the Thomas–Fermi equation, the pressure is easily shown to be

$$p = \tfrac{2}{3}c_k[n(R)]^{5/3} \tag{159}$$

and this can be verified from the theory without difficulty. It can be seen that the pressure is never zero, except when $n(R) = 0$, which is only true for the isolated atom solution corresponding to an infinite value of R. Therefore from equation (159) there is no minimum in the energy curve for the Thomas–Fermi theory.

Introducing exchange, arguing intuitively as before, but treating the free electron gas by the Hartree–Fock approximation (cf. Appendix C), the pressure is now given by

$$p = \tfrac{2}{3}c_k[n(R)]^{5/3} - \tfrac{1}{3}c_e\,[n(R)]^{4/3} \tag{160}$$

where $n(R)$ is the boundary density calculated from the Thomas–Fermi–Dirac equation. In a fully wave-mechanical treatment, there is no doubt about the importance of the boundary density in treatments of cohesion (see Chapter 4 and, for example, Ref. 82).

In the completely degenerate case, the equation of state given by the

Thomas–Fermi theory in equation (159) can conveniently be written in terms of the boundary value of the dimensionless description $\phi(x)$ [cf. equation (24)] of the self-consistent potential, at x_0, namely,

$$pv = \frac{2}{15}\frac{Z^2e^2}{b} x_0^{1/2} [\phi(x_0)]^{5/2} \tag{161}$$

where v is the atomic volume.

For small values of v, an expansion can be generated in which the first term is the free electron gas pressure, leading to the equation of state[83]

$$pv = \frac{h^2}{5m}\left[\frac{3}{8\pi}\right]^{2/3} \frac{Z^{5/3}}{v^{2/3}}\left[1 - \frac{2\pi me^2}{h^2}(4Zv)^{1/3} + \cdots\right] \tag{162}$$

A convenient representation of the Thomas–Fermi equation of state can be written, following Gilvarry[84] as

$$pv^{2/5}\left[\sum_{n=2}^{6} A_n \left[\frac{3v}{4\pi b^3}\right]^{(n+2)/6}\right] = \left[\frac{Z^2e^2}{10\pi b^4}\right]^{2/5} \tag{163}$$

where the coefficients A_n are recorded in Table 8. This fit should be sufficiently accurate for most purposes.

11.2. *Introduction of Exchange in Equation of State*

Equation (160) in principle defines the equation of state in the Thomas–Fermi–Dirac theory, and in terms of the boundary value $\phi(x_0)$ found by solving the Thomas–Fermi–Dirac equation

$$\frac{d^2\phi}{dx^2} = x\left[\alpha + \left(\frac{\phi}{x}\right)^{1/2}\right]^3, \qquad \alpha = \frac{6^{1/3}}{4}(\pi Z)^{2/3} \tag{164}$$

TABLE 8. Coefficients A_n in Representation (163) of Thomas–Fermi Equation of State

A_2	4.8075×10^{-1}
A_3	0
A_4	6.934×10^{-2}
A_5	9.700×10^{-3}
A_6	3.3704×10^{-3}

we find

$$pv = \frac{2}{15}\frac{Z^2e^2}{b}x_0^3\left\{\left[\frac{\phi(x_0)}{x_0}\right]^{1/2} + \alpha\right\}^5\left\{1 - \frac{5\alpha/4}{[\phi(x_0)/x_0]^{1/2} + \alpha}\right\} \quad (165)$$

Equation (164) is the generalization of equation (24) obtained by using equation (75), ϕ and x again being defined by equations (22) and (23). Unfortunately the Thomas–Fermi–Dirac equation (164) must be solved separately for each atomic number Z. The analog of the extreme high-pressure result (162) can be shown to be[83]

$$pv = \frac{h^2}{5m}\left[\frac{3}{8\pi}\right]^{2/3}\frac{Z^{5/3}}{v^{2/3}}\left[1 - \frac{2\pi me^2}{h^2}(4Zv)^{1/3} - \frac{10\pi me^2}{3^{2/3}h^2}(4Zv)^{1/3}\alpha + \cdots\right] \quad (166)$$

For many purposes, the graphical presentation of the equations of state by Feynman *et al.*[80] is adequate, but we shall not reproduce their plots here.

11.3. *Equations of State for Case of Incomplete Degeneracy*

In this discussion of material under high pressure, it has so far been assumed that the electrons form a completely degenerate gas. However, in astrophysics one wants to relax this assumption and for the low-temperature case, first discussed by Marshak and Bethe,[78] results for the equations of state can again be expressed in a convenient analytical form applicable to all elements.[85–87] For this reason we shall consider these results briefly here, although the discussion in Section 11.4 is applicable to arbitrary temperatures and therefore in principle covers the case dealt with here.

If the pressure in the case of incomplete degeneracy is denoted by P, then this can be expressed in terms of the pressure p in the completely degenerate case, by the equation

$$P = p[1 + \tfrac{5}{2}(\sigma + 2\tau)\,\zeta\,(k_BT)^2] \quad (167)$$

Here

$$\zeta = \pi^2b^2/8Z^2e^4 \quad (168)$$

while τ is defined in terms of the Thomas–Fermi boundary value $\phi(x_0)$ by

$$\tau[\phi(x_0)]^2 = x_0^2 \tag{169}$$

The quantity σ can be represented approximately by

$$\sigma\phi(x_0) = \sum C_n x_0^n \tag{170}$$

where in the summation, $n = 3, n'$, and 5. The values of n' and of the coefficients C_n are recorded in Table 9, taken from the work of Gilvarry and Peebles.[86]

These results can be used whenever the temperature is low in comparison with the maximum kinetic energy of electrons near the boundary of the atomic sphere, or when the inequality

$$k_B T << \frac{Ze^2}{b}\left[\frac{\phi(x_0)}{x_0}\right]$$

is satisfied. High temperatures can in fact fall within the range of this treatment in the limit as x_0 tends to zero.

11.4. Generalized Thomas–Fermi Theory for Arbitrary Degeneracy

We have already outlined the argument showing how the relation (148) between density and potential in the generalized Thomas–Fermi theory is derived. Combining this with Poisson's equation we find

$$-\nabla^2 V = \frac{16\pi^2 e^2}{h^3}(2mk_BT)^{3/2} I_{1/2}\left(\eta^* - \frac{V}{k_BT}\right) \tag{171}$$

to be solved for the potential energy V. In the sphere approximation we have again the boundary conditions

TABLE 9. Parametrization of Equations of State for Case of Incomplete Degeneracy by Equation (167). Coefficients C_n in Equation (170)

n	C_n
3	-3.205×10^{-1}
$n' = 4.215$	-2.331×10^{-2}
5	-2.519×10^{-3}

$$V \to -Ze^2/r \qquad \text{as } r \to 0, \qquad \left(\frac{dV}{dr}\right)_R = 0 \tag{172}$$

The spherically symmetrical solutions of equation (171) satisfying the boundary conditions (172) were first examined by Feynman *et al.*[80] and later by Latter.[88] The pressure p can be calculated either by considering the rate of transfer of momentum between the electrons and the surface of the atomic sphere or from the thermodynamic relation

$$p = -\left(\frac{\partial F}{\partial v}\right)_T \tag{173}$$

where F is the Helmholtz free energy. The result may be written

$$p = \frac{8\pi}{3h^3}(2mk_BT)^{3/2}(k_BT)\, I_{3/2}\left[\eta^* - \frac{V(R)}{k_BT}\right] \tag{174}$$

and once the self-consistent potential energy V is known the pressure p may then be obtained from equation (174) as a function of temperature T and volume v. Results are given, for example, by Latter[88] for the pressure–volume relation for a series of temperatures.

In concluding this discussion of material under high pressure, we merely show one, rather arbitrarily selected, comparison of the pressure–density relationships of the Thomas–Fermi–Dirac theory with experimental results of Bridgman in Fig. 8. The theoretical curves, inevitably, show a steady increase of density with increasing atomic number for a given pressure, whereas in Bridgman's results at 10^5 atmospheres (10^6dyne/cm^2 = 0.9869 atmospheres) such a correlation of density with atomic number is fair, but far from complete. This emphasizes the point we made at the outset, that the Thomas–Fermi predictions can only become fully quantitative at pressures sufficient to obliterate the outer electronic structure effects.

12. Relativistic Thomas–Fermi Theory

We shall conclude this chapter by summarizing briefly the relativistic generalization of the Thomas–Fermi theory, due to Vallarta and Rosen[89] and some of its consequences for atomic binding energies. In particular, we shall exhibit the scaling properties of the total energy[90] in terms of atomic number Z, number of electrons N, and the fine structure constant $\alpha = e^2\hbar c$ = 1/137.

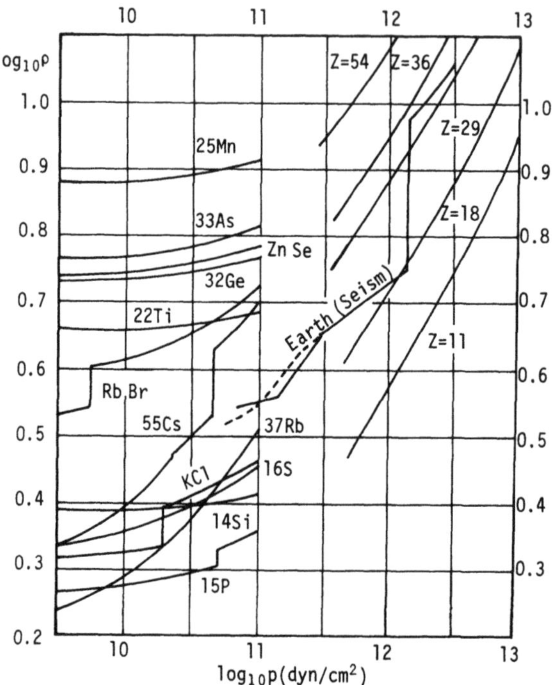

FIGURE 8. Pressure–density relationships. Curves on left-hand side: experimental results of Bridgman. Curves on right-hand side: results of the Thomas–Fermi–Dirac theory. (Redrawn from Ref. 23.)

Following the nonrelativistic approach, but subtracting, from the outset, the rest mass energy mc^2, we can write down the semiclassical energy equation for the fastest electron. This maximum energy is equivalent to the chemical potential μ, which, as emphasized in the nonrelativistic theory, must be a constant throughout the entire electronic cloud of the positive ion. Otherwise, electrons could redistribute in space to lower the total energy. If $p_f(\mathbf{r})$ is the momentum of this fastest electron at position $\mathbf{r}$, while $V(\mathbf{r})$ is, as usual, the self-consistent potential energy in which the electrons move, then using the customary relativistic expression for the kinetic energy we have

$$\mu = [c^2 p_f^2(\mathbf{r}) + m^2c^4]^{1/2} - mc^2 + V(\mathbf{r}) \tag{175}$$

The maximum momentum $p_f(\mathbf{r})$ is again related to the electron density $n(\mathbf{r})$ through equation (5). Thirdly, we must add to equations (175) and (5) the requirement of self-consistency embodied in the Poisson equation

$$\nabla^2 V = -\nabla^2 (\mu - V) = -4\pi n(\mathbf{r})e^2 \tag{176}$$

where we have obviously utilized the constancy of the chemical potential in space.

Following the scaling used in the nonrelativistic theory of Section 4, we work with dimensionless quantities $\phi(x)$ and x defined precisely as before through equations (22) and (23). Equation (176) can then be rewritten in spherical symmetry as

$$\frac{1}{r}\frac{d^2}{dr^2}[r(\mu - V)] = \frac{Ze^2}{b^3 x}\frac{d^2\phi}{dx^2} = 4\pi n e^2 \tag{177}$$

At this stage we return to equation (175). Rearranging this, and then squaring it, yields

$$(\mu - V)^2 + 2mc^2(\mu - V) = c^2 p_f^2 \tag{178}$$

and hence in terms of ϕ, x, and n we find

$$\begin{aligned}\frac{p_f^2(r)}{2m} &= \frac{1}{2m}\left(\frac{3h^3}{8\pi}\right)^{2/3}[n(r)]^{2/3} = \frac{(\mu - V)^2}{2mc^2} + (\mu - V) \\ &= \frac{1}{2mc^2}\left(\frac{Ze^2}{b}\right)^2 \frac{\phi^2}{x^2} + \frac{Ze^2}{b}\frac{\phi}{x}\end{aligned} \tag{179}$$

Utilizing equation (177) for the electron density $n(r)$ and writing $d^2\phi/dx^2$ as ϕ'', we obtain after a little rearrangement the result

$$\frac{\text{const } Z^{4/3}}{mc^2}\left(\frac{\phi}{x}\right)^2 + \frac{\phi}{x} = \left(\frac{\phi''}{x}\right)^{2/3} \tag{180}$$

where the constant is readily written down but is not, in fact, needed for our purpose below. Equation (180) evidently reduces to the correct dimensionless equation (24) in the nonrelativistic limit c tends to infinity, using the fact that the constant referred to above is independent of c.

12.1. *Scaling Property of Energy of Positive Ions*

What is important for the present discussion of the scaling properties of the relativistic Thomas–Fermi theory of heavy positive ions is that, since the fine structure constant α is proportional to c^{-1} and $\epsilon = \alpha^2 Z^2$ is proportional to $c^{-2}Z^2$, it is therefore quite clear from equation (180) that the fine structure constant always appears in relativistic Thomas–Fermi theory in the combination $\alpha^2 Z^{4/3} = \epsilon / Z^{2/3}$.

Having established this scaling property, we next note that, just as for the nonrelativistic theory discussed in Section 4, the boundary conditions are given by $\phi(0) = 1$ and by equation (27), where the positive ion again has a finite radius $r_0 = bx_0$. The equation for the chemical potential in terms of x_0 is identical with equation (35). Therefore it follows that the scaling of the chemical potential is determined by the scaling of x_0. Whereas in nonrelativistic theory we have $x_0 = x_0(N/Z)$, in the relativistic theory it follows from the above line of argument that

$$x_0 = x_0(N/Z;\ \epsilon/Z^{2/3}) \tag{181}$$

and hence the chemical potential has the scaling property

$$\mu = Z^{4/3} F(N/Z;\ \epsilon/Z^{2/3}) \tag{182}$$

This is the basic result needed to discuss the scaling properties of the total binding energies of heavy positive ions in the relativistic Thomas–Fermi theory.

It should be stressed that the result (182) has been obtained by focusing on the outer region of the positive ion, i.e., on the positive ion radius. This avoids a severe difficulty, known to Vallarta and Rosen, namely, that with a point nucleus the Thomas–Fermi relativistic electron density cannot be normalized. It is known that the introduction of the finite size of the nucleus will allow one to obtain a normalizable electron density. Since the size of the positive ion is orders of magnitude greater than the nuclear radius, it is evident that the radius x_0 must be extremely insensitive to the nuclear radius. Thus, it would be quite appropriate to evaluate the positive ion radius in the point nucleus limit, even though we know that just at this limiting point the electron density is not normalizable. This dependence of the normalization on the nuclear radius has been discussed explicitly by Ferreirinho *et al.*[(91)] for neutral atoms, which, however, have infinite radius and zero chemical potential.

Having established the important scaling property (182) of the relativistic Thomas–Fermi theory, it will be fruitful at this point to make contact with the relativistic generalization of the 1/Z expansion (38) due to Layzer and Bahcall.[(92)]

12.2. *Generalization of 1/Z Expansion to Include Fine Structure Constant*

Their proposed generalization of equation (38) takes the form

$$E(Z,N) = Z^2 \sum_{n=0}^{\infty} \sum_{m=0}^{\infty} E_{nm}(N)\, \epsilon^m Z^{-n} \tag{183}$$

It is necessary to stress here that the expansion (183) has some difficult points associated with it which are not shared by the nonrelativistic expansion (38). Relevant work discussing some of these difficulties is that of Ermolaev and Jones.[93] Nevertheless, it can be argued that equation (183), which reduces to equation (38) if we neglect all terms except $m = 0$, is the natural relativistic expansion, in terms of the parameter $\epsilon = \alpha^2 Z^2$. We shall see below that one can obtain from it meaningful results in the limit of large N. As to the difficult points referred to above, one might mention here that if one uses the Dirac equation as a basis, then with a point nucleus there are singularities occurring when $\alpha Z = 1$, with appropriate modifications in this value for finite values of the nuclear radius as discussed, for example, in Ref. 94. Secondly, and less basic, one ought strictly to allow for the fact that the coefficients $E_{nm}(N)$ can also depend on ϵ, as discussed for example by Doyle.[95]

Our final task below will be to determine the form of $E_{nm}(N)$ for large N. To do so, we use again the relation (19) between chemical potential and total energy $E(Z,N)$. Then it follows from equation (183) that we can write the chemical potential in the form

$$\mu = Z^2 \sum_{n=0}^{\infty} \sum_{m=0}^{\infty} \frac{dE_{nm}(N)}{dN} \epsilon^m Z^{-n} \tag{184}$$

and to compare with the above discussion of the relativistic Thomas–Fermi theory let us introduce the variables N/Z and $\epsilon/Z^{2/3}$ into equation (184) to obtain

$$\mu = Z^{4/3} \sum_{n=0}^{\infty} \sum_{m=0}^{\infty} \frac{dE_{mn}(N)}{dN} \frac{1}{N^{n-2m/3-2/3}} \left(\frac{N}{Z}\right)^{n-2m/3-2/3} \left(\frac{\epsilon}{Z^{2/3}}\right)^{m} \tag{185}$$

It follows that to obtain the scaling property (182) at large N and Z we must choose

$$\frac{dE_{nm}(N)}{dN} \propto N^{n-2m/3-2/3} \tag{186}$$

or, by integration

$$E_{nm}(N) = c_{nm}^{(0)} N^{n-2m/3+1/3} + \text{higher-order terms} \tag{187}$$

This result reduces, as it must, to the nonrelativistic result (39) when $m = 0$. The higher order terms in equation (187) cannot be discussed solely on the basis of relativistic Thomas–Fermi theory. However, they have been examined in Ref. 90, to which the reader is referred. The result displayed in equation (187) can be reinserted in equation (183) to show that the Thomas–Fermi energy scales as

$$E_{TF}(Z,N) = Z^{7/3}F_1(N/Z;\ \epsilon/N^{2/3}) \tag{188}$$

which reduces to the form (37) in the nonrelativistic limit. References to relativistic generalizations of density functional theory are given in Chapter 2.

Appendix A: Expansion of Binding Energy of Positive Ions about Neutral Atom Limit

A basic property of the Thomas–Fermi theory of positive ions is that $E(Z,N)/Z^{7/3}$ is simply a function of N/Z, as expressed in equation (37) and displayed in Fig. 2.

In this appendix, an expansion of $E(Z,N)/Z^{7/3}$ will be developed about the neutral atom energy,[96] the expansion parameter being the degree of ionization q defined by

$$q = 1 - N/Z \tag{A.1}$$

To achieve such an expansion, we first note that the exact solution $144/x^3$ of the dimensionless Thomas–Fermi equation (24), which does not satisfy the boundary condition $\phi(0) = 1$ appropriate to atomic ions, can be extended inwards to smaller x by the asymptotic expansion (26). This large x expansion is such that the coefficients F_n can be expressed in terms of F_1 as in Table 2. It is known that to match this solution with the neutral atom solution satisfying $\phi(0) = 1$ requires F_1 to take the value 13.27.

The development of the energy about the neutral atom point $q = 0$ can now be carried out using a method given by Fermi[97] in his pioneering work, by writing the solution for a positive ion with q small, or equivalently with large semiclassical radius x_0, as

$$\phi(x) = \phi_0(x) + \eta\phi_1(x) \tag{A.2}$$

where $\phi_0(x)$ now denotes the neutral atom solution with $q = 0$.

Then to first order in η, the perturbation $\phi_1(x)$ can be expressed in terms of quadrature on $\phi_0(x)$, the result being

$$\phi_1(x) = (\phi_0 + \tfrac{1}{3}x\phi_0') \int_0^x \frac{dx}{(\phi_0 + \tfrac{1}{3}x\phi_0')^2} \tag{A.3}$$

The parameter η is then determined by the boundary condition that $\phi(x)$ must vanish at the semiclassical radius x_0. One can now insert the series (26) to complete the integration in equation (A.3) and hence to determine x_0. The result in lowest order is given by

$$x_0 = 10.36/q^{1/3} \tag{A.4}$$

which, since the chemical potential is determined solely by x_0 through equation (35), yields[98]

$$\mu = -0.109\frac{e^2}{a_0}(Zq)^{4/3} \tag{A.5}$$

Since equation (19) allows $E(Z,N)$ to be obtained by integration of μ with respect to N, we can use the neutral atom energy $E(Z,Z)$ as a boundary condition to write

$$E(Z,N) = E(Z,Z)\,(1 - 0.0608q^{7/3} + \cdots) \tag{A.6}$$

Of course, such an expansion is only proved in nonrelativistic theory in the statistical limit in which both N and Z become large in such a way that N/Z remains finite and less than or equal to unity. Calculation of the next term in the series (A.6) shows it to be proportional to $q^{7+c/3}$, with $c = 0.772$. The fact that the expansion (A.6) has no term proportional to q is due to the fact that the chemical potential of the neutral Thomas–Fermi atom is identically zero [cf. equation (29)]. In turn, this is related to no stable negative ions in the statistical theory.

Appendix B: Thomas–Fermi Equation in Extremely Strong Magnetic Field

We shall follow below the treatment of Banerjee *et al.*[17] in deriving the form of the Thomas–Fermi theory appropriate to an atom in an extremely strong magnetic field. In expressing the kinetic energy of the atom in terms of the electron density $n(\mathbf{r})$, explicit use is made by them of the adiabatic hypothesis.

To relate the electron density to the Fermi momentum [cf. equation (5) in zero magnetic field] one counts the number of states inside a thin cylindrical shell of radius ρ, thickness $\Delta\rho$, and height Δz along the magnetic field direction. The transverse states correspond (in the adiabatic approximation) to cyclotron orbits having radii given by

$$\rho_n = [(2n + 1)/eB]^{1/2} \tag{B.1}$$

B being the magnetic field strength. Hence the number of transverse states inside $\Delta\rho$ is $\Delta N_\perp = \Delta n = eB\rho\Delta\rho$. The longitudinal motion is treated statistically. As the electron density in its variation along z is a smooth function, the motion can be approximated, inside the small interval Δz, by a superposition of plane waves

$$f_p(z) = (\Delta z)^{-1/2}\exp(ipz) \tag{B.2}$$

with $|p| \leqslant p_f(\mathbf{r})$, where $p_f(\mathbf{r})$ is the Fermi momentum. The number of longitudinal states inside Δz is then $\Delta N_\| = (p_f/\pi)\,\Delta z$. Thus, according to the Pauli exclusion principle, the cell of volume $2\pi\rho\,\Delta\rho\,\Delta z$ can accommodate a number of electrons $\Delta N = \Delta N_\|\Delta N_\perp = (eBp_f/2\pi^2)\;2\pi\rho\Delta\rho\Delta z$. Thus one obtains the density of electrons as

$$n(\mathbf{r}) = \frac{eB}{2\pi^2}\,p_f(\mathbf{r}) \tag{B.3}$$

in the extreme high-field limit.

Turning to the kinetic energy expression, one keeps only the contribution of the longitudinal motion. The energy of the atom is the energy of all the electrons bound by the Coulomb attraction of the nucleus, minus their energy when they are moving freely in the magnetic field. This latter quantity cancels, in the adiabatic approximation employed, the kinetic energy of their transverse motion inside the atom. Thus the kinetic energy of the electrons in the cell under consideration is

$$\begin{aligned}\Delta T &= \Delta N_\perp \frac{\Delta Z}{2\pi}\int_{|p|\leq p_f} dp\,\frac{p^2}{2m}\\ &= \frac{eB}{(2\pi)^2}\,\frac{p_f^3}{3m}\,2\pi\rho\Delta\rho\Delta z\end{aligned} \tag{B.4}$$

Therefore the kinetic energy density t is evidently

$$t = \frac{(2\pi)^4}{3m(eB)^2}\,n^3(\mathbf{r}) \tag{B.5}$$

As in the zero magnetic field case, the total energy E can be written as a sum of the kinetic energy $T = \int t\, d\mathbf{r}$ calculated above and the electrostatic potential energy terms. Minimizing the sum with respect to $n(\mathbf{r})$, subject to the normalization condition, yields the Thomas–Fermi relation in a very strong magnetic field as

$$\mu = \frac{2\pi^4}{m(eB)^2}\, n^2(\mathbf{r}) + V(\mathbf{r}) \tag{B.6}$$

where μ is again the Lagrange multiplier taking care of normalization, while $V(\mathbf{r})$, as in the case $B = 0$, is given by

$$V(\mathbf{r}) = \frac{-Ze^2}{r} + e^2 \int \frac{n(\mathbf{r}')}{|\mathbf{r} - \mathbf{r}'|}\, d\mathbf{r}' \tag{B.7}$$

Using Poisson's equation and writing

$$(\mu - V) = \frac{Ze^2}{r}\, \phi(x) \tag{B.8}$$

one obtains the dimensionless Thomas–Fermi equation (44) with the dimensionless scaled variable x defined in equation (45).

Appendix C: Exchange and Correlation in Jellium Model

In this appendix we summarize some of the more elementary aspects of the model of jellium, that is, that model of a metal in which the positive ions are smeared out into a uniform background of positive charge which is just sufficient to neutralize the electronic negative charge.

In the high-density limit of this model, the kinetic energy dominates over the potential energy. This is evident if we introduce the mean interelectronic spacing r_s, related to the mean electron density n_0 by

$$n_0 = {}^3\!/\!_4 \pi r_s^3 \tag{C.1}$$

Since the kinetic energy depends on the operator ∇^2, it will scale like r_s^2 per particle, while the potential energy per particle will be proportional to r_s^{-1}. Thus, in the high-density limit of very large n_0, or from equation (C.1) $r_s \to 0$, the kinetic energy will dominate.

Therefore, the picture of the noninteracting electron gas used in Section

2 is the appropriate starting point. All states with momentum p less than the Fermi momentum are occupied, and these states correspond to the plane waves $\exp(i\mathbf{p}\cdot\mathbf{r})$. The total wave function is then an antisymmetrized product of these plane wave states, or what is equivalent, a single Slater determinant of these free-particle wave functions.

Now a property of such a single Slater determinant is that if the orbitals are denoted by ψ_i in general, the single-particle density matrix $\rho(\mathbf{r}, \mathbf{r}')$, defined precisely in equation (93), is sufficient to determine the pair correlation function $\Gamma(\mathbf{r}_1, \mathbf{r}_2)$. In terms of the many-electron wave function Ψ, this pair correlation function is defined, apart from a normalization factor, by

$$\Gamma(\mathbf{r}_1\mathbf{r}_2) = \int \Psi^*(\mathbf{r}_1\mathbf{r}_2 \cdots \mathbf{r}_N)\Psi(\mathbf{r}_1 \cdots \mathbf{r}_N)d\mathbf{r}_3\, d\mathbf{r}_4 \cdots d\mathbf{r}_N \tag{C.2}$$

With Ψ approximated by a single Slater determinant, it can be shown from the definition (C.2) that

$$\Gamma(\mathbf{r}_1, \mathbf{r}_2) = n(\mathbf{r}_1)\, n(\mathbf{r}_2) - \tfrac{1}{2}\{\rho(\mathbf{r}_1, \mathbf{r}_2)\}^2 \tag{C.3}$$

where the general definition of the first-order density matrix ρ from the many-electron wave function is (apart from normalization)

$$\rho(\mathbf{r}_1\mathbf{r}_1') = \int \Psi^*(\mathbf{r}_1\mathbf{r}_2 \cdots \mathbf{r}_N)\Psi(\mathbf{r}_1'\mathbf{r}_2 \cdots \mathbf{r}_N)d\mathbf{r}_2 \cdots d\mathbf{r}_N \tag{C.4}$$

The form (93) is again the special form following from the definition (C.4) when the many-electron wave function is a single Slater determinant built from orbitals ψ_i.

Calculating equation (93) when the orbitals ψ_i become the plane waves $\exp(i\mathbf{p}\cdot\mathbf{r})$ is a straightforward matter, the summation over occupied states i being replaced by an integration over momenta through the occupied Fermi sphere. This is how the result (94) for free particles is calculated, as is readily verified. Turning to the pair function in the noninteracting high-density limit of jellium, one uses this result (94) in equation (C.3), the density n being simply the constant value n_0. In this way, the pair function in equation (68) follows; the Fermi hole we discussed at some length in the main text. This leads to the exchange energy per particle through equation (69). Writing the results for the kinetic and exchange energy in terms of the interelectronic separation r_s yields

$$\frac{E}{N} = \frac{2.21}{r_s^2} - \frac{0.916}{r_s} \tag{C.5}$$

which is the Hartree–Fock energy for the jellium model. The correlation energy is defined as the lowering of the energy beyond this value, for a chosen r_s. In equation (C.5) the energy is in Rydbergs if r_s is in units of the Bohr radius a_0.

Many-body perturbation theory can be applied to calculate the correlation energy in the high-density limit[99] discussed above, when equation (C.5) becomes generalized to

$$\frac{E}{N} = \frac{2.21}{r_s^2} - \frac{0.916}{r_s} + A \ln r_s + C + \cdots \tag{C.6}$$

where $A = (2/\pi^2)(1 - \ln 2)$, $C = -0.096$.

The above exact result for the correlation energy in the high-density limit was not available to Wigner[40] when he constructed formula (81). The main physical reasoning on which this was based was to pass to the limit of strong coupling, that is to the low-density limit r_s tends to infinity.

Wigner Electron Crystal: Low-Density Limit. Wigner[40] noted that though the Hartree–Fock theory based on plane waves is inappropriate in the strong-coupling limit, nevertheless the inference from the different dependence of kinetic and potential energy on r_s is that as r_s tends to infinity the potential energy must dominate [cf. equation (C.5)]

Therefore, one must choose a description of the electronic configuration which minimizes the potential energy, and Wigner argued that this would be achieved if the electrons went onto a lattice, thereby avoiding one another optimally. The calculation of the potential energy is then equivalent to the calculation of the Madelung energy of a lattice of point electrons in a uniform positive background of neutralizing charge. Of the lattices examined the body-centered-cubic lattice turns out to have the lowest Madelung energy, the energy per electron E/N in this extreme low-density limit being

$$\frac{E}{N}(r_s \to \infty) = \frac{-1.79}{r_s} \tag{C.7}$$

This leads to a correlation energy per particle given by subtracting the exchange energy in equation (C.5) from equation (C.7), and this difference is embodied in the Wigner formula (81) in the limit $n \to 0$. As r_s is reduced, the electrons vibrate about their (body-centered-cubic) lattice sites, and one can calculate the next term in equation (C.7) by essentially phonon theory applied to the Wigner electron crystal. The next term is found to be proportional to $r_s^{-3/2}$ but we shall refer the reader to a review on Wigner crystallization for the details.[100]

We conclude by remarking that the Wigner electron crystal is an insulator at absolute zero, since the electrons are localized on lattice sites. Obviously the high-density limit, in which the wave function is the Slater determinant of plane waves, is a delocalized state and is metallic. For a long time it was a difficult matter to decide at what density the transition from the metal to the insulator occurred in the ground state. That matter has been settled by a Monte Carlo computer calculation by Ceperley and Alder.[101] The critical value of r_s is near $100a_0$, which can be contrasted with the lowest-density metal, Cs, with an r_s of $5.5a_0$.

Appendix D: Pair Correlation Function and Exchange Energy Density in Infinite Barrier Model of Metal Surface

The purpose of this appendix is to present the exact result[102] which describes the Fermi hole, or pair correlation function, around an electron moving in a metal with a surface represented by the infinite barrier model introduced by Bardeen.[103]

As Bardeen showed in his original paper, the electron density profile $n(z)$ as a function of distance z from the infinite barrier is given by

$$\begin{aligned} n(z) &= n_0\left[1 - \frac{3}{2}\frac{j_1(2k_f z)}{k_f z}\right] \quad \text{for } z > 0 \\ &= 0 \qquad \text{otherwise} \end{aligned} \tag{D.1}$$

n_0 being the electron density in the bulk metal. The positive background density in the model, $n_b(z)$ say, is defined by

$$n_b(z) = -\, n_0\, \Theta(z - \xi) \tag{D.2}$$

Θ being the usual Heaviside step function and ξ being fixed by the condition of electrical neutrality, namely,

$$\int dz\, [n(z) + n_b(z)] = 0 \tag{D.3}$$

which yields

$$\xi = 3\pi/8k_f \tag{D.4}$$

Though in Chapter 5 fully self-consistent density profiles are described, which avoid the drastic infinite barrier assumption of the Bardeen model, the merit of this model is that the off-diagonal density matrix defined in equation (93) can also be calculated exactly in addition to the diagonal density (D.1). Through the relation (C.3) given in the previous appendix, this allows the Fermi hole or pair correlation function to be obtained. The explicit form of this is given by

$$\Gamma(\mathbf{r}, \mathbf{r}') = n(z)n(z') - \tfrac{1}{2}[\rho(\mathbf{r}, \mathbf{r}')]^2$$
$$= n(z)n(z') - \frac{9}{2} n_0^2 \left\{ \frac{j_1(k_f|\mathbf{r} - \mathbf{r}'|)}{k_f|\mathbf{r} - \mathbf{r}'|} - \frac{j_1[k_f(|\mathbf{r} - \mathbf{r}'|^2 + 4zz')^{1/2}]}{k_f(|\mathbf{r} - \mathbf{r}'|^2 + 4zz')^{1/2}} \right\}^2 \tag{D.5}$$

which evidently also defines the first-order density matrix $\rho(\mathbf{r}, \mathbf{r}')$ for this model. The expression (D.5) for the pair function $\Gamma(\mathbf{r}, \mathbf{r}')$ is such that in the curly brackets the first term alone would give back the uniform gas result (68). The distance $|\mathbf{r} - \mathbf{r}'|$ appearing there is, in the second term in the curly brackets, replaced by the distance between $\mathbf{r}'$ and the image of the point $\mathbf{r}$ in the planar metal surface.[104]

If the total electron density, equal to $n(z) + n_b(z)$, is denoted by $n_t(z)$, then the electrostatic energy density is evidently given by $\tfrac{1}{2}e^2 \int d\mathbf{r}'\, n_t(\mathbf{r})\, n_t(\mathbf{r}')/|\mathbf{r}-\mathbf{r}'|$ and this has been calculated analytically in Ref. 104. Our interest here is the form of the exchange energy density $\epsilon_X(z)$, defined in terms of the first-order density matrix, via the Fermi hole (D.5) as [cf. equations (69) and (72) for the uniform electron gas].

$$\epsilon_X(z) = \frac{-3e^2 n_0 k_f}{4\pi} J(2k_f z) \tag{D.6}$$

The function J is given by the rather complicated expression[104]

$$\begin{aligned} J(2k_f z) = {} & \left(\frac{2}{3x^2} + \frac{1}{6x^4} + \frac{4\cos x}{x^4}\right)\cos(2x) + \left(\frac{1}{3x^3} + \frac{4\sin x}{x^4}\right)\sin(2x) \\ & + \left[-\frac{2}{15} + \frac{64}{15x^2} - \frac{36}{5x^4} + \frac{4}{x^4}f(x) + \frac{4\,\mathrm{si}(2x)}{x^3}\right]\cos x \\ & + \left[-\frac{2}{15x} - \frac{56}{5x^3} + \frac{4}{x^3}f(x) - \frac{4\,\mathrm{si}(2x)}{x^4}\right]\sin x \\ & + 1 + \frac{1}{3x^2} + \frac{91}{30x^4} - \frac{2x}{15}\,\mathrm{si}(x) + \frac{4}{3x}\,\mathrm{si}(2x) \end{aligned} \tag{D.7}$$

with $x = 2k_f z$. Here $f(x)$ is defined by

$$f(x) = C + \ln(2x) - \text{ci}(2x) \tag{D.8}$$

C being Euler's constant and ci$(2x)$ the cosine integral. The quantity si(x) is explicitly given by

$$\text{si}(x) = -\int_x^\infty \frac{\sin t}{t}\, dt \tag{D.9}$$

The point to be emphasized here is that we have an example of a truly inhomogeneous gas with a strong density gradient in which we can test the approximate electron gas form (72) of the exchange energy density against the exact result given by equations (D.6) and (D.7), while $n(z)$ is available in equation (D.1).

In Fig. A.1 we show plots of the exchange energy density in units of

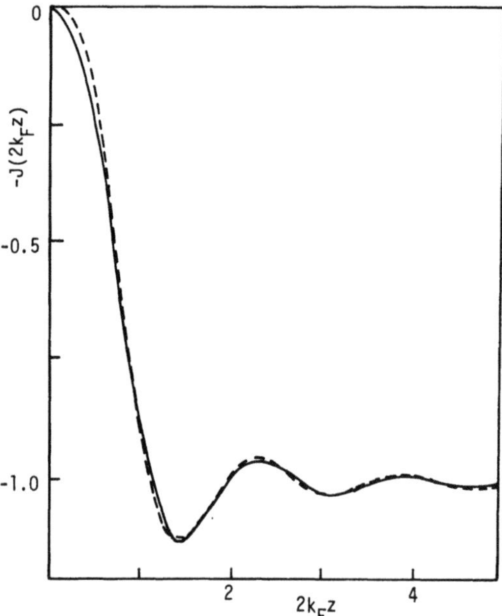

FIGURE A.1. Exchange energy density (in units of $3e^2 n_0 k_f / 4\pi$) versus $2k_f z$ for infinite barrier model of metal surface. Solid line: exact result. Dashed line: local density approximation.

$3e^2n_0k_f/4\pi$. The full line is the exact result of equation (D.6), while the broken line shows the local density approximation of equation (72) calculated using the exact density profile (D.1). It can be seen that there is remarkable agreement between the exact result and the local density approximation for all z. Nevertheless, in spite of this, when the curves are integrated to obtain the total exchange energy the exact and local theories differ by some 30%, the local density approximation underestimating the the magnitude of the total exchange energy. But there can be no doubt that this example is very encouraging for the local density theory of the exchange energy, even in the presence of a strongly varying density. In contrast, as Moore and March[102] demonstrate, the kinetic energy density given by the local formula (10) in this same model is an approximation of much poorer quality than the local exchange term discussed fully above.

Appendix E: Density Gradient Treatments of "Almost Homogeneous" Electron Gas and of Model of Metal Surface

In this appendix, we discuss two soluble problems to lowest order in the density gradient.

(i) "Almost Homogeneous" Electron Gas. We follow Jones and Young[105] in discussing the response of an initially uniform electron gas to the presence of a weak perturbing potential ΔV. Then, in the usual linear response theory [cf. equation (98)], we have for the density change

$$\Delta n(\mathbf{r}) = \int F(\mathbf{r}-\mathbf{r}')V(\mathbf{r}')d\mathbf{r} \tag{E.1}$$

We now write the kinetic energy density t in the form of the gradient expansion, effecting a slight generalization of equation (91) as

$$t = c_k n^{5/3} + \frac{\lambda}{8}\frac{(\nabla n)^2}{n}\frac{\hbar^2}{m} \tag{E.2}$$

The response function follows from this equation, for a general λ, as[105]

$$F_\lambda(k) = -\frac{k_f}{\pi^2}\left[\frac{1}{1+3\lambda\eta^2}\right], \qquad \eta = \frac{k}{2k_f} \tag{E.3}$$

However, it is readily shown from equation (98) in Fourier transform that the exact response function in the noninteracting case is [cf. equation (110)].

$$F_{\text{exact}}(k) = -\frac{k_f}{2\pi^2}\left(1 + \frac{1-\eta^2}{2\eta} \ln \left|\frac{1+\eta}{1-\eta}\right|\right) \tag{E.4}$$

If we compare these two results at small k, then they agree when λ is chosen to be 1/9. This can be regarded as a derivation of the Kirznits result in equation (91), since this is evidently the limit of slowly varying potential and is therefore the proper value in the spirit of gradient expansions. On the other hand, for large k, the two response functions agree if λ is chosen as 1. This is the original von Weizsäcker value[(44)]; this is not a gradient expansion result but valid for rapidly varying spatial perturbations (e.g., in the exponentially decaying tail in an atom).

(ii) Electron Density near a Planar Metal Surface. As a further application of the low-order density gradient expansion (91), we follow Brown and March[(106)] in calculating the electron density profile near a metal surface. Their work was done with a liquid metal in mind, where the positive ions and the electrons both spill out from the metal surface and the classical electrostatic terms almost cancel. One is then left with the problem of electrons moving in their own exchange potential which we take below to be of the Dirac–Slater form (77).

The aim therefore is to solve for the density profile using the Kirznits form (91) of the kinetic energy density. Then we can write for the total energy in the presence of the surface

$$E = \int \left[\epsilon(n) + \frac{\lambda\hbar^2}{8m}\frac{(\nabla n)^2}{n}\right] d\mathbf{r}, \qquad \lambda = \frac{1}{9} \tag{E.5}$$

Here, $\epsilon(n)$ is simply the bulk metal energy density, this functional form being corrected by the Kirznits term for the kinetic energy due to the inhomogeneity at the liquid metal surface. Writing $\psi = n^{1/2}$, using the fact that n varies only along the z axis perpendicular to the planar metal surface, and employing the variation principle (17), the Euler equation corresponding to equation (E.5) is

$$-\frac{\lambda\hbar^2}{2m}\frac{d^2\psi}{dz^2} + \left(\frac{\partial\epsilon}{\partial n} - \mu\right)\psi = 0 \tag{E.6}$$

In the above model, the surface energy σ takes the form[(106)]

$$\sigma = \frac{\lambda\hbar^2}{m}\int_{-\infty}^{\infty}\left(\frac{d\psi}{dz}\right)^2 dz \tag{E.7}$$

Denoting the bulk density by n_0, and assuming that the electrons move only in their own exchange potential as mentioned above, one can write ϵ in the form

$$\epsilon(n) = An\left[\left(\frac{n}{n_0}\right)^{2/3} - 2\left(\frac{n}{n_0}\right)^{1/3}\right] \tag{E.8}$$

Making the substitution

$$f(z) = (n_0/n)^{1/3} \tag{E.9}$$

the differential equation (E.6) can be solved to yield

$$f(z) = 1 + B\exp(z/l) \tag{E.10}$$

where B is a constant of order unity, chosen to ensure charge neutrality. The quantity l in equation (E.10) evidently measures the surface thickness and is given by

$$l = (9\lambda\hbar^2/8mA)^{1/2} \tag{E.11}$$

Using equations (E.7), (E.9), and (E.10) we obtain

$$\sigma = \frac{n_0}{4}(A\lambda\hbar^2/2m)^{1/2} \tag{E.12}$$

Bearing the earlier remarks in mind that $\epsilon(n)$ is the bulk metal energy density, we can construct the compressibility K of the homogeneous metal as

$$K = 9/2An_0 \tag{E.13}$$

and combining equations (E.13), (E.12), and (E.11) gives

$$K\sigma = 3l/4 \tag{E.14}$$

Since l is expected to be ~1 Å in simple metals, and not to vary widely, this relation indicates that $K\sigma$ should be rather constant. That such a constancy was an empirical fact was known to Frenkel[107] for a whole class of liquids, both nonmetals and metals, and such a relationship $K\sigma \doteqdot$ const was rediscovered by Egelstaff and Widom.[108]

In Chapter 5, much more refined treatments than the above of the inhomogeneous electron gas at metal surfaces will be discussed, both for the electron density profiles and for surface energies. However, crude as the use of the low-order gradient expansion is, it does exhibit clearly how surface energy is correlated with a bulk property, compressibility, via the surface thickness l. The argument of Brown and March[(106)] given above has been applied by Singwi and Tosi[(109)] to discuss the surface tension of the electron–hole droplet. We shall not go into detail here, but we note that other aspects of the electron–hole droplet are treated in Chapter 2.

Appendix F. Kinetic Energy in Perturbative Series in One-Body Potential

In Section 9, the density–potential relation of the Thomas–Fermi theory was shown to follow from an approximate summation of the perturbation series (95) for the (diagonal element of) the Dirac density matrix. Since it was shown in Section 3 that an alternative way to derive this density–potential relation was from a variation principle, in which the kinetic energy density was taken as that given by free electron gas theory, in a local spatially varying theory, it is of interest to summarize here the results of the kinetic energy density to all orders in perturbation theory, and to see how the Thomas–Fermi theory can be regained.

The kinetic energy density is given in terms of the Dirac density matrix by

$$t = -\frac{\hbar^2}{2m}\sum \psi_i^* \nabla_i^2 \psi_i = -\frac{\hbar^2}{2m}\left[\nabla_{\mathbf{r}}^2 \rho(\mathbf{r},\mathbf{r}')\right]_{\mathbf{r}'=\mathbf{r}} \tag{F.1}$$

Using the full perturbation expansion of the Dirac matrix in equation (95) it is straightforward to show that the kinetic energy density Δt measured relative to the unperturbed Fermi gas is given by[(55)]

$$\Delta t = \sum_{j=1}^{\infty} \frac{j}{j+1} V(\mathbf{r})\rho_j(\mathbf{r}) \tag{F.2}$$

Since it was shown in the approximate summation (100) that $\rho_j(\mathbf{r}) = \text{const}(V)^j$ for slowly varying $\mathbf{r}$, it is easy to complete the summation in equation (F.2) in terms of the one-body potential, to obtain

$$t = \text{const}[E_f - V(r)]^{5/2}, \qquad E_f = k_f^2/2 \tag{F.3}$$

Using the density–potential relation (2.7) one regains the Thomas–Fermi form (10) for the kinetic energy density, proportional to $\{n(\mathbf{r})\}^{5/3}$.

As discussed in Refs. 60 and 61, the kinetic energy change from a free electron gas can be expressed explicitly in terms of the displaced charge $(n - n_o)$ in low order, using the perturbation series (95) in conjunction with equation (F.1). This result has been utilized in Section 10.3 to discuss the interaction between charged defects in an electron gas. The Thomas–Fermi limit of the kinetic energy change ΔT obtained from the above procedure is

$$\Delta T = E_f \int (n - n_0)\, d\mathbf{r} + \frac{E_f}{3n_0} \int (n - n_0)^2\, d\mathbf{r} + \cdots \tag{F.4}$$

which is readily verified to be the low-order form of $c_k[\int (n^{5/3} - n_0{}^{5/3}) d\mathbf{r}]$ in terms of an expansion in the displaced charge.

References

1. L.H. Thomas, *Proc. Camb. Phil. Soc.* **23**, 542 (1926).
2. E. Fermi, *Z. Phys.* **48**, 73 (1928).
3. E.B. Baker, *Phys. Rev.* **36**, 630 (1930).
4. C.A. Coulson and N.H. March, *Proc. Phys. Soc.* **A63**, 367 (1950).
5. P. Gombás, *Die Statistiche Theorie des Atoms und Ihre Andwendungen* (Springer-Verlag, Vienna, 1949).
6. V. Fock, *Phys. Z. Sowjetunion* **1**, 747 (1932).
7. E.A. Milne, *Proc. Camb. Phil. Soc.* **23**, 794 (1927).
8. N.H. March and R.J. White, *J. Phys. B* **5**, 466 (1972).
9. Y. Tal and M. Levy, *Phys. Rev. A* **23**, 408 (1981).
10. D. Layzer, *Ann. Phys. (N.Y.)* **8**, 271 (1959).
11. T. Kato, *Commun. Pure. Appl. Math.* **10**, 151 (1957); *J. Fac. Sci. Tokyo Univ.* **16**, 145 (1951).
12. I.K. Dmitrieva and G.I. Plindov, *Phys. Lett. A* **55**, 3 (1975).
13. N.H. March and J.S. Plaskett, *Proc. R. Soc. London, Ser. A* **235**, 419 (1956).
14. N.H. March, *J. Chem. Phys.* **72**, 1994 (1980).
15. B.B. Kadomtsev, *Sov. Phys. JETP* **31**, 945 (1970).
16. R.O. Mueller, A.R.P. Rau, and L. Spruch, *Phys. Rev. Lett.* **26**, 1136 (1971).
17. B. Banerjee, D.H. Constantinescu, and P. Rehak, *Phys. Rev. D* **10**, 2384 (1974).
18. A.R.P. Rau, R.O. Mueller, and L. Spruch, *Phys. Rev. A* **11**, 1865 (1975).
19. Y. Tomishima and K. Yonei, *Prog. Theor. Phys.* **59**, 683 (1978).
20. N.H. March and Y. Tomishima, *Phys. Rev. D* **19**, 449 (1979).
21. F. Hund, *Z. Phys.* **77**, 12 (1932).
22. J.R. Townsend and G.S. Handler, *J. Chem. Phys.* **36**, 3325 (1962).

23. N.H. March, *Adv. Phys.* **6**, 1 (1957).
24. E. Teller, *Rev. Mod. Phys.* **34**, 627 (1962).
25. J.F. Mucci and N.H. March, *J. Chem. Phys.* **71**, 5270 (1979).
26. P. Politzer, *J. Chem. Phys.* **64**, 4239 (1976).
27. N.H. March, *Proc. Camb. Phil. Soc.* **48**, 665 (1952).
28. N.H. March and R.G. Parr, *Proc. Natl. Acad. Sci. USA*, **77**, 6285 (1980).
29. R.G. Parr, R.A. Donnelly, M. Levy, and W.E. Palke, *J. Chem. Phys.* 68, 3801 (1978).
30. N.H. March, "Electron Density Description of Atoms and Molecules," in *Specialist Periodical Reports, Royal Society of Chemistry* (Burlington House, London); *Theoretical Chemistry*, Volume 4, p. 92 (1981).
31. K. Ruedenberg, *J. Chem. Phys.* **66**, 375 (1977).
32. N.H. March, *J. Chem. Phys.* **67**, 4618 (1977).
33. J.F. Mucci and N.H. March, *J. Chem. Phys.* **71**, 1495 (1979).
34. R. Pucci and N.H. March, *J. Chem. Phys.* **74**, 2936 (1981).
35. P.A.M. Dirac, *Proc. Camb. Phil. Soc.* **26**, 376 (1930).
36. E.P. Wigner and F. Seitz, *Phys. Rev.* **43**, 804 (1933).
37. J.C. Slater, *Phys. Rev.* **81**, 385 (1951).
38. W. Kohn and L.J. Sham, *Phys. Rev. A* **140**, 1133 (1965).
39. J.M.C. Scott, *Phil. Mag.* **43**, 859 (1952).
40. E.P. Wigner, *Phys. Rev.* **46**, 1002 (1934); *Trans. Faraday Soc.* **34**, 678 (1938).
41. E. Clementi, *J. Chem. Phys.* **39**, 175 (1963).
42. R.A. Ballinger and N.H. March, *Phil. Mag.* **46**, 246 (1955).
43. J. Schwinger, *Phys. Rev. A* **22**, 1827 (1980).
44. C.F. Von Weizsäcker, *Z. Phys.* **96**, 431 (1935).
45. See N.H. March and W.H. Young, *Proc. Phys. Soc.* **72**, 182 (1958), for a discussion of Von Weizsäcker's work as a variational method based on the density matrix.
46. D.A. Kirznits, *Sov. Phys. JETP* **5**, 64 (1957).
47. C.H. Hodges, *Can. J. Phys.* **51**, 1428 (1973).
48. W.P. Wang, R.G. Parr, D.R. Murphy, and G.A. Henderson, *Chem. Phys. Lett.* **43**, 409 (1976).
49. S. Ma and K.A. Brueckner, *Phys. Rev.* **165**, 18 (1968).
50. W. Jones and N.H. March, *Theoretical Solid State Physics* (Interscience-Wiley, London, 1973), Vols. I and II.
51. A.M. Beattie, J.C. Stoddart, and N.H. March, *Proc. R. Soc. London Ser. A* **326**, 97 (1971).
52. A. Sjölander, G. Niklasson, and K.S. Singwi, *Phys. Rev. B* **11**, 113 (1975).
53. D.J.W. Geldart and M. Rasolt, *Phys. Rev. B* **13**, 1477 (1976).
54. N.H. March and A.M. Murray, *Proc. R. Soc. London Ser. A* **261**, 119 (1961); *Phys. Rev.* **120**, 830 (1960).
55. J.C. Stoddart and N.H. March, *Proc. R. Soc. London Ser. A* **299**, 279 (1967).
56. N.F. Mott, *Proc. Camb. Phil. Soc.* **32**, 281 (1936).
57. A. Blandin, E. Daniel, and J. Friedel, *Phil. Mag.* **4**, 180 (1959).
58. See also N.H. March and A.M. Murray, *Proc. R. Soc. London Ser. A* **256**, 400 (1960).
59. J. Lindhard, *Kgl. Danske Mat. Fys. Medd.* **28**, 8 (1954).
60. L.C.R. Alfred and N.H. March, *Phil. Mag.* **2**, 985 (1957).
61. G.K. Corless and N.H. March, *Phil. Mag.* **6**, 1285 (1961).
62. J.M. Ziman, *Adv. Phys.* **13**, 89 (1964).
63. N.H. March and M.P. Tosi, *Atomic Dynamics in Liquids* (Macmillan, London, 1975).
64. F.G. Fumi, *Phil. Mag.* **46**, 1007 (1955).

65. K. Mukherjee, *Phil. Mag.* **12**, 915 (1965).
66. D. Bohm and T. Staver, *Phys. Rev.* **84**, 836 (1952).
67. N.H. March, *Phys. Lett.* **20**, 231 (1966).
68. N.F. Mott and H. Jones, *Theory of the Properties of Metals and Alloys* (Clarendon Press, Oxford, 1936).
69. M.J. Stott, S. Baranovsky, and N.H. March, *Proc. R. Soc. London Ser. A* **316**, 201 (1970).
70. F. Flores and N.H. March, *J. Phys. Chem. Solids* **42**, 439 (1981).
71. J. Friedel, *Adv. Phys.* **3**, 446 (1954).
72. M.A. Biondi and J.A. Rayne, *Phys. Rev.* **115**, 1522 (1959).
73. A.B. Bhatia and N.H. March, *J. Chem. Phys.* **68**, 4651 (1978).
74. R. Resta, *Phys. Rev. B* **16**, 2717 (1977).
75. P.P. Debye and E.M. Conwell, *Phys. Rev.* **93**, 693 (1954).
76. R.B. Dingle, *Phil. Mag.* **46**, 831 (1955).
77. R. Mansfield, *Proc. Phys. Soc. B* **69**, 76, 862 (1956).
78. R.E. Marshak and H.A. Bethe, *Astrophys. J.* **91**, 239 (1940).
79. T. Sakai, *Proc. Phys. Math. Soc. Jpn* **24**, 254 (1942).
80. R.P. Feynman, N. Metropolis, and E. Teller, *Phys. Rev.* **75**, 1561 (1949).
81. J.C. Slater and H.M. Krutter, *Phys. Rev.* **47**, 559 (1935).
82. J.A. Alonso and N.H. March, *Phys. Lett.* **A83**, 455 (1981).
83. N.H. March, *Proc. Phys. Soc.* **68**, 726 (1955).
84. J.J. Gilvarry, *Phys. Rev.* **95**, 71 (1954).
85. J.J. Gilvarry, *Phys. Rev.* **96**, 934 (1954).
86. J.J. Gilvarry and G.H. Peebles, *Phys. Rev.* **99**, 550 (1955).
87. N.H. March, *Proc. Phys. Soc.* **A68**, 1145 (1955).
88. R. Latter, *Phys. Rev.* **99**, 1854 (1955).
89. M.S. Vallarta and N. Rosen, *Phys. Rev.* **41**, 708 (1932).
90. U. Marini Bettolo Marconi and N.H. March, *Int. J. Quantum Chem.* **20**, 693 (1981).
91. J. Ferreirinho, R. Ruffini, and L. Stella, *Phys. Lett.* **B91**, 314 (1980).
92. D. Layzer and J. Bahcall, *Ann. Phys. (N.Y.)* **17**, 177 (1962).
93. A.M. Ermolaev and M. Jones, *J. Phys. B* **6**, 1 (1973).
94. L.P. Fulcher, J. Rafelski, and A. Klein, *Sci. Amer.* **Dec.**, 120 (1979).
95. H.T. Doyle, *Advances in Atomic and Molecular Physics*, D.R. Bates and I. Eshermann, editors (Academic Press, New York, 1969).
96. N.H. March, *J. Chem. Phys.* **76**, 1430 (1982).
97. E. Fermi, *Mem. Acc. Italia* **1**, 1 (1930).
98. N.H. March, *Phys. Lett.* **82A**, 73 (1981).
99. M. Gell-Mann and K.A. Brueckner, *Phys. Rev.* **106**, 364 (1957).
100. C.M. Care and N.H. March, *Adv. Phys.* **24**, 101 (1975).
101. D.M. Ceperley and B.J. Alder, *Phys. Rev. Lett.* **45**, 566 (1980).
102. I.D. Moore and N.H. March, *Ann. Phys.* **97**, 136 (1976).
103. J. Bardeen, *Phys. Rev.* **49**, 653 (1936).
104. L. Miglio, M.P. Tosi, and N.H. March, *Surf. Sci.* **111**, 119 (1981).
105. W. Jones and W.H. Young, *J. Phys. C* **4**, 1322 (1971).
106. R.C. Brown and N.H. March, *J. Phys. C* **6**, L363 (1973).
107. J. Frenkel, *Kinetic Theory of Liquids* (Oxford University Press, Oxford, 1942).
108. P.A. Egelstaff and B. Widom, *J. Chem. Phys.* **53**, 2667 (1970).
109. K.S. Singwi and M.P. Tosi, *Phys. Rev. B* **23**, 1640 (1981); *Solid State Commun.* **34**, 209 (1980).

CHAPTER 2

GENERAL DENSITY FUNCTIONAL THEORY

W. KOHN AND P. VASHISHTA

1. Introduction

Over the course of the last 15 years density functional theory (DFT)* has evolved as a conceptually and practically useful method for studying the electronic properties of many-electron systems. In this chapter we present a critical review of the general theory together with some illustrative examples.† More complete discussions of several important areas of application may be found in other chapters of this book.

The well-known Thomas–Fermi method,[2,3] which can be considered as the simplest version of this theory, has been discussed in detail in Chapter 1. For many important questions, such as trends of the energies and density distributions of atoms, it provides a quick and often surprisingly good first orientation. Various corrections, to allow in an approximate manner for the effects of exchange, correlation, and density gradients, have often—but not consistently—led to improved results.[4] However, since Thomas–Fermi theory and its refinements had been derived heuristically and not as approximations to an exact formulation of electronic theory in terms of the electron

*An alternative designation is theory of the inhomogeneous electron gas.
†This chapter is limited to nonrelativistic physics. Relativistic density functional theory is included in a recent review by A. K. Rajagopal.[1]

W. KOHN • Institute for Theoretical Physics, University of California, Santa Barbara, California 93108. P. VASHISHTA • Solid State Science Division, Argonne National Laboratory, Argonne, Illinois 60439.

density distribution $n(\mathbf{r})$, it was difficult to assess their validity and to devise further improvements.

A firm and exact theoretical foundation for dealing with interacting electronic systems in their ground state in terms of the density $n(\mathbf{r})$ was provided in 1964 in a paper by Hohenberg and Kohn (HK).[5] These authors demonstrated that all aspects of the electronic structure of a system in a nondegenerate ground state are completely determined by its electron density $n(\mathbf{r})$. They also derived a formal stationary expression for the energy E of such a system, as a functional of the density $n(\mathbf{r})$. This functional contains a term, $F[n(\mathbf{r})]$, which represents the kinetic and interaction energies, and for which—in view of the impossibility of exactly solving a many-electron problem—no exact expressions are known. However, several useful approximations for $F[n]$ have been proposed and used, the simplest of which leads back to the Thomas–Fermi equation and its refinements.

Starting from the energy variational principle of HK, Kohn and Sham (KS)[6] derived a system of self-consistent one-particle equations for the description of electronic ground states. The effective one-particle potential at the point $\mathbf{r}$, $v_{\text{eff}}(\mathbf{r})$, depends on the entire density distribution, $n(\mathbf{r}')$, and formally takes into account all many-body effects. It may be written in the form

$$v_{\text{eff}}(\mathbf{r}) = \phi(\mathbf{r}) + v_{xc}(\mathbf{r}) \tag{1}$$

where $\phi(\mathbf{r})$ is the total electrostatic potential and $v_{xc}(\mathbf{r})$ is, by definition, the exchange-correlation contribution to $v_{\text{eff}}(\mathbf{r})$. The expression $v_{xc}(\mathbf{r})$, like $v_{\text{eff}}(\mathbf{r})$, depends in a complicated way on the entire density distribution, $n(\mathbf{r}')$. Again, in practice, approximations must be made. The most useful has been the so-called local density approximation in which $v_{xc}(\mathbf{r})$ depends in a simple manner only on the local density $n(\mathbf{r})$ at the point $\mathbf{r}$.

The merit of the KS theory in the local density approximation is that, although it is nearly as simple to carry through as the Hartree theory, it gives in most cases a good semiquantitative account of exchange and correlation effects. In practice it is found that typically KS results are better than those of Hartree–Fock calculations, even though the latter are computationally very much more demanding and, for solids, almost prohibitive. The reason is that the KS equations include, fairly accurately, *both* exchange and correlation while the Hartree–Fock equations include exchange exactly but neglect correlation completely.

The reliability of DFT in the local approximation is very rarely better than 1%, and currently available refinements generally do not alter this situation substantially. If a higher degree of accuracy is required, one is forced to attempt a sufficiently precise calculation of the many-electron wave function by the Rayleigh–Ritz variational method or other schemes. However, such calculations generally become prohibitive for systems containing more than a few electrons, while DFT can be applied without any great difficulty also to systems containing large numbers of electrons.

The simplest systems to which DFT has been applied are systems of nonrelativistic electrons in a nondegenerate and spin-compensated ground state: DFT has made useful contributions to the theory of atoms and molecules in singlet ground states; cohesive properties of metals; defects in metals; and to metal and semiconductor surface physics. In addition, several generalizations of the simplest form of DFT have been developed. These include the following: atoms and molecules with finite net spin; relativistic effects; paramagnetism and ferromagnetism of metals; multicomponent systems such as electron–hole droplets in semiconductors; systems at finite temperature; excited states; and transport coefficients in weakly inhomogeneous systems.*

The plan of the present chapter is the following. In Section 2 we present the basic DFT for one-component, nonmagnetic systems in their ground state; we also make comparisons with other many-electron theories. Extension to magnetic systems is described in Section 3, and extension to multicomponent systems in Section 4. Section 5 deals with the generalization to finite temperature ensembles, and Section 6 discusses the application of DFT concepts to excited states and to transport coefficients.

2. Basic Theory

In this section we present the basic DFT for one-component spin-compensated electronic systems in their ground state. We develop and discuss the fundamental variational formulation in Section 2.1 and the KS self-consistent equations in Section 2.2. Comparison with other methods is made in Section 2.3 and illustrations are presented in Section 2.4.

2.1. Variational Formulation

2.1.1. Formal Theory. (a) The Density as Basic Variable. We present the formal theory following the original paper of HK.[5]

We consider a system of electrons enclosed in a large box and moving

*While this chapter deals only with electronic systems, it may be remarked that DFT has also been applied to inhomogeneous nuclear matter and the surface of liquid He^4.

under the influence of some external potential $v(\mathbf{r})$ and of their mutual Coulomb repulsion. In atomic units the Hamilton operator has the form

$$\mathsf{H} = \mathsf{T} + \mathsf{V} + \mathsf{U} \tag{2}$$

where

$$\mathsf{T} = \tfrac{1}{2} \int \nabla\psi^*(\mathbf{r})\cdot\nabla\psi(\mathbf{r})\, d\mathbf{r} \tag{3}$$

$$\mathsf{V} = \int v(\mathbf{r})\psi^*(\mathbf{r})\psi(\mathbf{r})\, d\mathbf{r} \tag{4}$$

$$\mathsf{U} = \frac{1}{2} \int \frac{1}{|\mathbf{r}-\mathbf{r}'|}\, \psi^*(\mathbf{r})\psi^*(\mathbf{r}')\psi(\mathbf{r}')\psi(\mathbf{r})\, d\mathbf{r}\, d\mathbf{r}' \tag{5}$$

The density operator is given by

$$\mathsf{n}(\mathbf{r}) = \psi^*(\mathbf{r})\, \psi\, (\mathbf{r}) \tag{6}$$

Summation over spin indices is implied. We consider systems of N electrons, moving in an external potential $v(\mathbf{r})$, which have a unique, nondegenerate ground state Ψ. Clearly then, Ψ is a unique functional of $v(\mathbf{r})$ and therefore so is the expectation value of the electron density,

$$n(\mathbf{r}) \equiv (\Psi, \mathsf{n}(\mathbf{r})\Psi) \tag{7}$$

We shall denote by C the class of all densities $n(\mathbf{r})$ associated with a nondegenerate ground state corresponding to some potential $v(\mathbf{r})$. All our subsequent considerations are logically restricted to systems whose density distribution $n(\mathbf{r})$ are in this class. For brevity of writing we shall, however, generally not reiterate this fact. Experience indicates that this restriction is not of practical importance since it appears that all, or almost all, physically and mathematically reasonable densities $n(\mathbf{r})$ are members of class C.*

We shall now derive the important conclusion that $v(\mathbf{r})$ and Ψ are uniquely determined by the knowledge of the density distribution $n(\mathbf{r})$.†

*Obvious restrictions are: $n(\mathbf{r}) \geq 0$; $\int n(\mathbf{r})\, d\mathbf{r}$ = integer; $n(\mathbf{r})$ continuous.
†More precisely, $v(\mathbf{r})$, is determined only to within a trivial additive constant.

The proof proceeds by *reductio ad absurdum.* Assume that a different potential $v'(\mathbf{r})$ with a ground state Ψ' gives rise to the same density $n(\mathbf{r})$. Clearly [unless $v'(\mathbf{r}) - v(\mathbf{r}) = \text{const}$], $\Psi' \neq \Psi$, since these states are eigenstates of different Hamiltonians. Let us denote the Hamiltonians and ground-state energies associated with Ψ and Ψ' by H, H′ and E, E', respectively. Then, because of the minimal property of the ground-state energy E and since Ψ' is not an eigenstate of H, we have

$$E = (\Psi, \mathsf{H}\Psi) < (\Psi', \mathsf{H}\Psi') = (\Psi', (\mathsf{H}' + \mathsf{V} - \mathsf{V}')\Psi') \tag{8}$$

Therefore, using the assumed identity of the two densities, equation (8) leads to

$$E < E' + \int [v(\mathbf{r}) - v'(\mathbf{r})]\, n(\mathbf{r})\, d\mathbf{r} \tag{9}$$

Similarly, interchanging primed and unprimed quantities, we find

$$E' = (\Psi', \mathsf{H}'\Psi') < (\Psi, \mathsf{H}'\Psi) = (\Psi, (\mathsf{H} + \mathsf{V}' - \mathsf{V})\Psi) \tag{10}$$

leading to

$$E' < E + \int [v'(\mathbf{r}) - v(\mathbf{r})]\, n(\mathbf{r})\, d\mathbf{r} \tag{11}$$

Addition of equations (9) and (11) yields the inconsistent result

$$E + E' < E + E' \tag{12}$$

Thus we conclude that the density distribution $n(\mathbf{r})$ associated with a ground state corresponding to an external potential $v(\mathbf{r})$ cannot be reproduced by the ground-state density corresponding to a different potential $v'(\mathbf{r})$ [except for $v'(\mathbf{r}) - v(\mathbf{r}) = \text{const}$]. *Further it follows clearly that, since $n(\mathbf{r})$ determines the potential $v(\mathbf{r})$, it also determines the ground state (assumed nondegenerate) as well as all other electronic properties of the system.*

(b) The Energy Variational Principle. Since, as just shown, Ψ is a functional of $n(\mathbf{r})$, so is the quantity

$$F[n(\mathbf{r})] \equiv (\Psi, (\mathsf{T} + \mathsf{U})\Psi) \tag{13}$$

which represents the sum of the kinetic and Coulomb interaction energies. This functional, whose existence for densities in class C has been demonstrated above, plays a key role in DFT, as we now show.

Let us consider a system of N interacting electrons in a given external potential $v(\mathbf{r})$. For such a system we define the following energy functional of density $n'(\mathbf{r})$:

$$E_v[n'(\mathbf{r})] \equiv \int v(\mathbf{r})n'(\mathbf{r})\, d\mathbf{r} + F[n'(\mathbf{r})] \tag{14}$$

where $v(\mathbf{r})$ is considered as prescribed and not as a functional of $n'(\mathbf{r})$.

We shall now show that $E_v[n']$ assumes a lower value for the correct $n(\mathbf{r})$ than for any other density distribution, $n'(\mathbf{r})$, with the same total number of particles N. Because of the assumed nondegeneracy of Ψ, it is well known that the conventional Rayleigh–Ritz functional of Ψ',

$$\mathscr{E}_v[\Psi'] \equiv (\Psi', \mathsf{V}\Psi') + (\Psi', (\mathsf{T}+\mathsf{U})\Psi') \tag{15}$$

has a lower value for the correct Ψ than for any other Ψ' with the same number of particles. Let Ψ' be the ground state associated with some other density $n'(\mathbf{r})$ in class C. Then

$$\begin{aligned} \mathscr{E}_v[\Psi'] &= \int v(\mathbf{r})n'(\mathbf{r})\, d\mathbf{r} + F[n'] \\ &> \mathscr{E}_v[\Psi] = \int v(\mathbf{r})n(\mathbf{r})\, d\mathbf{r} + F[n] \end{aligned} \tag{16}$$

This establishes the asserted minimum principle for $E_v[n']$,

$$E_v[n] < E_v[n'] \tag{17}$$

It is also clear from equations (14) and (13) that this minimum value, $E_v[n]$, is the correct ground-state energy associated with $v(\mathbf{r})$ and N.

It is customary to extract from $F[n]$ the classical Coulomb energy and write

$$G[n] \equiv F[n] - \frac{1}{2}\int \frac{n(\mathbf{r})\, n(\mathbf{r}')}{|\mathbf{r}-\mathbf{r}'|}\, d\mathbf{r}\, d\mathbf{r}' \tag{18}$$

so that the fundamental energy functional becomes

$$E_v[n'] = \int v(\mathbf{r})n'(\mathbf{r})\, d\mathbf{r} + \frac{1}{2}\int \frac{n'(\mathbf{r})n'(\mathbf{r}')}{|\mathbf{r}-\mathbf{r}'|}\, d\mathbf{r}\, d\mathbf{r}' + G[n'] \tag{19}$$

The stationary functional $E_v[n']$ allows *in principle* a much simpler determination of the ground-state energy E and density $n(\mathbf{r})$, than the conventional Rayleigh–Ritz minimal principle for $\mathscr{E}[\Psi']$. For, regardless of the number, N, of electrons involved, the required minimization of $E_v[n']$ is always with respect to the density $n'(\mathbf{r})$ which is a function of only three coordinate variables, while $\mathscr{E}[\Psi']$ must be minimized with respect to a function of $3N$ variables.

In practice the usefulness of the new minimal principle depends on whether approximations for $G[n]$ exist which are both simple to obtain and to apply, and are sufficiently accurate for the purpose at hand. For moderate accuracy (typically 1%–10%) such approximations do exist. If higher accuracy is needed the difficulty of calculating and using sufficiently accurate forms for $G[n]$ increases, however, extremely rapidly. Therefore DFT is generally not used when accuracies better than, say, 0.1% or 1% are required.

The next subsection deals with approximate forms of the functional $G[n]$.

2.1.2. Approximations for the Functional $G[n]$. (a) The Electron Gas of Almost Constant Density. We consider a system whose density is given by

$$n(\mathbf{r}) = n_0 + \tilde{n}(\mathbf{r}) \tag{20}$$

where n_0 is a constant and $\tilde{n}(\mathbf{r})/n_0 << 1$, with

$$\int \tilde{n}(\mathbf{r})\, d\mathbf{r} = 0 \tag{21}$$

A series expansion of $G[n]$ in powers of $\tilde{n}/n_0$ can, in view of equation (21), contain no term linear in $\tilde{n}$ and, because of the homogeneity and rotational invariance of the unperturbed system, has the form

$$G[n] = G[n_0] + \int K\,(|\mathbf{r}-\mathbf{r}'|)\, \tilde{n}(\mathbf{r})\tilde{n}(\mathbf{r}')\, d\mathbf{r}\, d\mathbf{r}' + O(\tilde{n}^3) \tag{22}$$

Here $G[n_0]$ is the sum of kinetic, exchange and correlation energies of a uniform gas of density n_0, which can of course be written as

$$G[n_0] = \int g_0(n_0)\, d\mathbf{r} \tag{23}$$

In atomic units the kinetic and exchange parts of $g_0(n)$ are given by

$$g_{0k}(n) = \frac{1.105n}{r_s^2} \left(= \frac{3}{10}\, k_F^2 n \right) \tag{24}$$

$$g_{0x}(n) = -\frac{0.458}{r_s}\, n \tag{25}$$

where r_s is the Wigner–Seitz radius given by

$$\frac{4\pi}{3} r_s^3 = n^{-1} \tag{26}$$

and k_F is the Fermi momentum

$$k_F = (3\pi^2 n)^{1/3} \tag{27}$$

The correlation contribution to $g_0(n)$ has been calculated by many authors with agreement and a presumed accuracy of a few percent. Among frequently used approximations are those due to Wigner[7]

$$g_{0c}(n) = -\frac{0.44}{r_s + 7.8} n \tag{28}$$

the high-density expression of Gell-Mann and Brueckner[8]

$$g_{0c}(n) = 0.031 \ln r_s - 0.047 + O(r_s \ln r_s) \tag{29}$$

the Nozieres–Pines interpolation result[9]

$$g_{0c}(n) = (-0.0575 + 0.0155 \ln r_s)n \tag{30}$$

and tabulated values of Gaskell[10] and of Vashishta and Singwi.[11]

The kernel K of the second-order term in equation (22) can be expressed in terms of the static q-dependent dielectric constant, $\epsilon(q)$, of a gas of density n_0[5]:

$$K(\mathbf{r} - \mathbf{r}') = \frac{1}{\Omega} \sum_q K(q)\, e^{i\mathbf{q}\cdot(\mathbf{r}-\mathbf{r}')} \tag{31}$$

where

$$K(\mathbf{q}) = \frac{2\pi}{q^2} \frac{1}{\epsilon(q) - 1} \tag{32}$$

At the present time the calculations of $\epsilon(q)$ by Singwi and collaborators[11–13] are generally considered as the best available approximations.

Calculations for nearly uniform systems carried out with the form (22) for $G[n]$ are equivalent to treating the effect of spatial variations of $v(\mathbf{r})$ by self-consistent linear response theory. Examples of systems for which such a procedure is reasonable are the alkali metals in which the ions are described by weak pseudopotentials in a uniform electron gas.

(b) The Electron Gas of Slowly Varying Density: Gradient Expansions. Let us now consider systems in which the density is slowly varying on the scale of the local value of $[k_F(\mathbf{r})]^{-1}$, i.e.,

$$\left|\frac{\partial}{\partial x_i} n(\mathbf{r})\right| \Big/ \left|n(\mathbf{r})\right| << k_F(\mathbf{r}), \qquad \left|\frac{\partial^2}{\partial x_i \partial x_j} n(\mathbf{r})\right| \Big/ \left|\frac{\partial n(\mathbf{r})}{\partial x_k}\right| << k_F(\mathbf{r}), \text{ etc.} \tag{33}$$

where $k_F(\mathbf{r})$ is the "local" Fermi momentum, i.e.,

$$k_F(\mathbf{r}) = [3\pi^2 n(\mathbf{r})]^{1/3} \tag{34}$$

We emphasize that the condition (33) does *not* exclude *large* variations of n, provided they occur over sufficiently large distances.

For such systems (unless pathological circumstances prevail*), the functional $G[n]$ must have a gradient expansion of the form[5]

$$\begin{aligned} G[n] = \int d\mathbf{r}\, [g_0(n) &+ g_2^{(2)}(n)|\nabla n|^2 + g_4^{(2)}(n)(\nabla^2 n)^2 \\ &+ g_4^{(3)}(n)\, \nabla^2 n |\nabla n|^2 + g_4^{(4)}(n)|\nabla n|^4 + \cdots O(\nabla_i^6)] \end{aligned} \tag{35}$$

The terms kept are the only independent terms, up to order ∇_i^4, satisfying the requirement of rotational invariance, which follows from the fact that $G[n]$ is a universal functional of n. The coefficients $g_2^{(2)}(n)$ and $g_4^{(2)}(n)$ can be obtained by considering a system in which $n(\mathbf{r})$ is slowly varying and *also* $n(\mathbf{r})$ is nearly constant in magnitude. Such a system is covered by both the approximations (22) and (35). This allows one to identify $g_2^{(2)}$ and $g_4^{(2)}$ with small-q expansion coefficients of the reciprocal of the static dielectric constant $\epsilon(q)$. If we write

$$\epsilon^{-1}(q;n) = \gamma_2(n)q^2 + \gamma_4(n)q^4 + \gamma_6(n)q^6 + \cdots \tag{36}$$

*There is in fact no pathology for the full $G[n]$ nor for its kinetic energy and exchange-correlation energy parts [see equation (39)]. However, L. Kleinman[14] has shown that neither the exchange energy nor the correlation energy *separately* has a gradient expansion.

we find

$$g_2^{(2)} = 2\pi(\gamma_4 + \gamma_2^2) \tag{37}$$

$$g_4^{(2)} = 2\pi(\gamma_6 + 2\gamma_2\gamma_4 + \gamma_2^3) \tag{38}$$

etc. The other two terms of order ∇_i^4 cannot be obtained from a knowledge of $\epsilon(q;n)$ but would require a knowledge of nonlinear response functions. Plots of $g_0(n)$ and $g_2^{(2)}(n)$ are shown in Figs. 1 and 2.

Nothing certain is known about the general convergence properties of the gradient expansion (35). In typical real systems, consisting of single or several atoms, the condition of slow density variation, (33), is usually *not* met and it is therefore a priori unclear how useful gradient expansions are. There are numerous calculations available, especially for the jellium model of a surface, treated fully in Chapter 5, which have a bearing on this question. We shall come back to these in Section 2.2.2.

It has been found convenient (see especially Section 2.2) to divide the functional $G[n]$ into two parts,

$$G[n] = T_s[n] + E_{xc}[n] \tag{39}$$

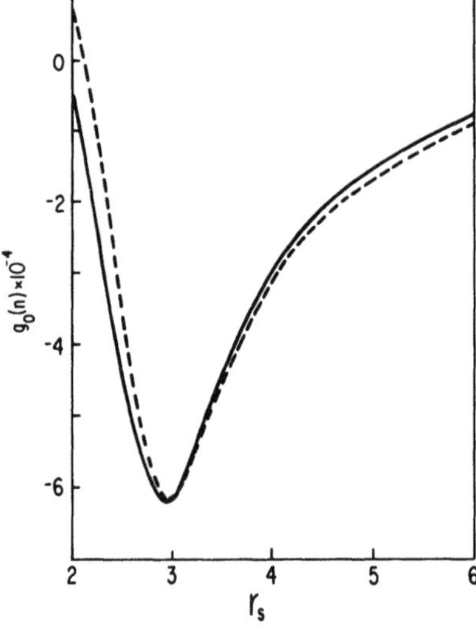

FIGURE 1. Energy per unit volume of a homogeneous electron gas in atomic units as a function of the Wigner–Seitz radius r_s, where $n = (4\pi r_s^3/3)^{-1}$. The correlation energies used in obtaining the dotted and the continuous curves are taken from Ref. 7 and 11, respectively.

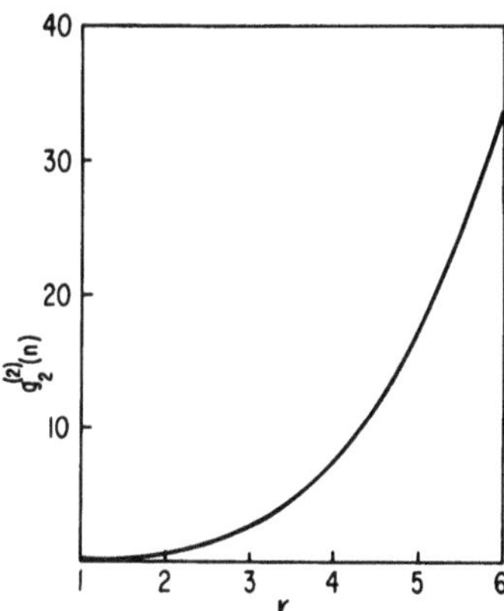

FIGURE 2. Coefficient of the leading gradient correction to the energy in a.u. per unit volume of an electron gas of local density n [$= (4\pi r_s^3/3)^{-1}$] as defined in equation (35). (After Ref. 22.)

where $T_s[n]$ is the kinetic energy of a *noninteracting* electron gas of density $n(\mathbf{r})$, in its ground state and the remainder, $E_{xc}[n]$, is called the exchange and correlation energy.

One can now address separately the validity of gradient expansions for T_s and E_{xc}. T_s refers to noninteracting electrons, and therefore its gradient expansion can be compared to exact calculations. The gradient expansion up to second order for noninteracting electrons is given by

$$T_s[n(\mathbf{r})] = \int d\mathbf{r}\, n(\mathbf{r}) t_s[n(\mathbf{r})] \tag{40a}$$

where

$$t_s[n(\mathbf{r})] = \frac{3}{10}(3\pi^2 n)^{2/3} + \frac{1}{72}\frac{|\nabla n|^2}{n^2} \tag{40b}$$

The first term is the kinetic energy per particle of a uniform system. The term proportional to $|\nabla n|^2$ is the von Weizsäcker correction.[15,16] Equation (40) has been checked against the soluble case of a nearly uniform electron gas with

$$n(\mathbf{r}) = n_0 + n_1 \cos qz \qquad (n_1 << n_0) \tag{41}$$

and it is found that in this case the expansion (40) is valid provided $q \lesssim k_F$. This suggests that for a general inhomogeneous system, not limited to the special form (41), the approximation (40) will be valid provided the local scale of variation [cf. equation (33)] is not faster than the local Fermi momentum.* However, since the von Weizsäcker correction in equation (40) has been derived for a noninteracting Fermi gas in which $\epsilon_F - v > 0$ (ϵ_F = Fermi energy), there is no reason for assuming that it has validity in "tail" regions, for which $\epsilon_F - v < 0$, such as occur at the edges of atoms, molecules, and solid surfaces.

In Section 2.2.1 we shall see how $T_s[n]$ can be treated *exactly* by the solution of appropriate self-consistent equations.

Approximations for $E_{xc}[n]$ are discussed in Section 2.2.2.

2.2. *Self-Consistent Equations*

2.2.1. Formal Theory. In equation (39) we introduced the quantities $T_s[n(\mathbf{r})]$, the kinetic energy of a noninteracting electron gas in its ground state with density distribution $n(\mathbf{r})$, and $E_{xc}[n(\mathbf{r})]$, the exchange-correlation energy, defined by

$$\begin{aligned} E_{xc}[n(\mathbf{r})] &= G[n(\mathbf{r})] - T_s[n(\mathbf{r})] \\ &= \langle \mathrm{T} \rangle + \langle \mathrm{U} \rangle - T_s[n(\mathbf{r})] - \frac{1}{2}\int \frac{n(\mathbf{r})n(\mathbf{r}')}{|\mathbf{r}-\mathbf{r}'|}\,d\mathbf{r}\,d\mathbf{r}' \end{aligned} \tag{42}$$

With these quantities we can rewrite E_v as

$$\begin{aligned} E_v[n'(\mathbf{r})] = {} & T_s[n'(\mathbf{r})] + \int v(\mathbf{r})n'(\mathbf{r})\,d\mathbf{r} \\ & + \frac{1}{2}\int \frac{n'(\mathbf{r})n'(\mathbf{r}')}{|\mathbf{r}-\mathbf{r}'|}\,d\mathbf{r}\,d\mathbf{r}' + E_{xc}[n'(\mathbf{r})] \end{aligned} \tag{43}$$

If the last term were omitted, the problem of minimizing E_v with respect to n' would evidently be identical to the minimization of the Hartree energy. Hence the designation of E_{xc} as exchange-correlation energy.

From the minimal property of $E_v[n]$, subject to the condition

$$\int n(\mathbf{r})\,d\mathbf{r} = N \tag{44}$$

we obtain the Euler condition

*For inhomogeneous systems of the special form (41) and *large* q, $T_s[n]$ is given correctly by the form (40), with the factor 1/72 replaced by the factor 1/8. This does of course not imply that for a *general* $n(\mathbf{r})$ which is rapidly varying a gradient correction term $\frac{1}{8}|\nabla n|^2/n$ will yield the correct $T_s[n]$.

$$\frac{\delta T_s[n]}{\delta n(\mathbf{r})} + \phi(\mathbf{r}) + v_{xc}(\mathbf{r}) - \mu = 0 \tag{45}$$

where $\phi(\mathbf{r})$ is the total classical potential energy

$$\phi(\mathbf{r}) \equiv v(\mathbf{r}) + \int \frac{n(\mathbf{r}')}{|\mathbf{r}-\mathbf{r}'|} d\mathbf{r}' \tag{46a}$$

$v_{xc}(\mathbf{r})$ is the functional derivative

$$v_{xc}(\mathbf{r}) \equiv \frac{\delta E_{xc}[n]}{\delta n(\mathbf{r})} \tag{46b}$$

and μ is the Lagrange parameter associated with the subsidiary condition (44).* Equations (45) and (46) must be solved self-consistently. If the term v_{xc} were absent, this would be accomplished by solving the self-consistent Hartree equations. Hence, in the presence of v_{xc}, it is plausible that we must simply solve self-consistently the equations

$$[-\tfrac{1}{2}\nabla^2 + v_{\text{eff}}(\mathbf{r})]\psi_i(\mathbf{r}) = \epsilon_i\psi_i(\mathbf{r}) \tag{47}$$

where

$$v_{\text{eff}}(\mathbf{r}) = \phi(\mathbf{r}) + v_{xc}(\mathbf{r}) \tag{48}$$

[see equation (46)] and

$$n(\mathbf{r}) = \sum_{i=1}^{N} |\psi_i(\mathbf{r})|^2 \tag{49}$$

where the sum is to be carried out over the N lowest occupied eigenstates. The equations (47)–(49) constitute the so-called Kohn–Sham (KS) "self-consistent equations."[6] They are derived in detail in Appendix A.

*In view of the restriction (44), the quantities $\delta T_s/\delta n(\mathbf{r})$ and $\delta E_{xc}/\delta n(\mathbf{r})$ can—for fixed N—be defined only to within $\mathbf{r}$-independent constants, since $\int C\delta n(\mathbf{r})d\mathbf{r} = 0$. These constants may be considered absorbed in the Lagrange parameter μ. For large systems, in which N may be regarded as continuously variable, the constants can be determined by considering systems with slightly different values for N. For finite temperature grand ensembles (Section 5) in which the mean value of N is strictly a continuous variable, this technical nuisance does not occur.

To calculate the total energy, we note that

$$\sum_i \epsilon_i = \sum_i (\psi_i, [-\tfrac{1}{2}\nabla^2 + v_{\text{eff}}(\mathbf{r})]\,\psi_i) = T_s[n] + \int v_{\text{eff}}(\mathbf{r})n(\mathbf{r})\,d\mathbf{r} \tag{50}$$

Hence, by equation (43),

$$E_{\text{tot}} = \sum_i \epsilon_i - \frac{1}{2}\int \frac{n(\mathbf{r})n(\mathbf{r}')}{|\mathbf{r}-\mathbf{r}'|}\,d\mathbf{r}\,d\mathbf{r}' + E_{xc}[n] - \int v_{xc}(n)n(\mathbf{r})\,d\mathbf{r} \tag{51}$$

Remarks about KS Eigenvalues. Solution of the KS equation leads to occupied energy levels, ϵ_i ($i = 1, \ldots, N$), and unoccupied levels, ϵ_i ($i = N + 1, \ldots$). Do the occupied energy levels represent physical energies required to excite electrons to the continuum? The answer in general is no. For while the ϵ_i are the real eigenvalues of the self-consistent KS Hamiltonian, equation (47), the excitation energies $\tilde{\epsilon}_i$ are solutions of the Dyson equation,

$$-\tfrac{1}{2}\nabla^2\tilde{\psi}_i(\mathbf{r}) + \int \Sigma(\mathbf{r}, \mathbf{r}'; \tilde{\epsilon}_i)\tilde{\psi}_i(\mathbf{r}')\,d\mathbf{r}' = \tilde{\epsilon}_i\tilde{\psi}_i(\mathbf{r}) \tag{52}$$

where $\Sigma(\mathbf{r},\mathbf{r}';\epsilon)$ is the nonlocal mass operator.[17] Certainly, the nonlocal operator Σ is not identical to the local $v_{\text{eff}}(\mathbf{r})$ of the KS equation. In contrast to the real eigenvalues ϵ_i, the $\tilde{\epsilon}_i$ of equation (52) are in general complex, reflecting the finite lifetime of the ionized state. Examples show that, in general, the real parts of ϵ_i and $\tilde{\epsilon}_i$ are also different.

There is, however, an important special case. For an infinite system, whose highest occupied KS orbitals are extended, the highest occupied eigenvalue, ϵ_N, equals the physical chemical potential $\mu = \tilde{\epsilon}_N$. This fact, the density functional analog of Koopmans' theorem,[18] is proved in Appendix B.

A second important question arises in connection with infinite periodic solids. Here the KS eigenfunctions carry a crystal momentum $\mathbf{k}$ and a KS Fermi surface in $\mathbf{k}$-space may be defined by the equation

$$\epsilon_{\mathbf{k}} \equiv \mu \tag{53}$$

where μ is the real chemical potential with the property that the total number of states, with KS eigenvalues $\leq \mu$, is equal to the number of electrons N. Is the KS Fermi surface the physical Fermi surface? This has not been proved and is probably in general not the case. For the physical Fermi surface is defined by those values of **k** for which the Dyson eigenvalues, equation (52), satisfy

$$\tilde{\epsilon}_{\mathbf{k}} = \mu \tag{54}$$

Since the operators v_{eff} and Σ are different we do not expect the two surfaces defined by equations (53) and (54) to be identical. Of course, by construction, they enclose the same volume in **k** space. Hence, in the special case of a uniform system, the KS and physical Fermi surfaces, which are both spherical, must be identical. In nonuniform cases we expect the differences between the two Fermi surfaces to be small since the major source of anisotropy of the Fermi surface is the electrostatic potential $\phi(\mathbf{r})$ which is common to both v_{eff} and Σ.

2.2.2. *Approximations for the Exchange-Correlation Energy* $E_{xc}[n]$. *(a) The Local Density Approximation.* For a system with slowly varying density, we can make the local density approximation (LDA)[6]

$$E_{xc}[n] = \int \epsilon_{xc}(n(\mathbf{r}))n(\mathbf{r})d\mathbf{r} \tag{55}$$

where $\epsilon_{xc}(n)$ is the exchange and correlation energy per particle of a uniform electron gas of density n. In LDA

$$v_{xc}(\mathbf{r}) = \frac{d}{dn}\{\epsilon_{xc}(n(\mathbf{r}))n(\mathbf{r})\} \equiv \mu_{xc}(n(\mathbf{r})) \tag{56}$$

where $\mu_{xc}(n)$ is the exchange and correlation contribution to the chemical potential of a uniform system. Then equation (51) becomes, in the LDA,

$$\begin{aligned} E_{\text{tot}} \approx \sum_i \epsilon_i - \frac{1}{2}\int \frac{n(\mathbf{r})n(\mathbf{r}')}{|\mathbf{r}-\mathbf{r}'|}\, d\mathbf{r}\, d\mathbf{r}' \\ + \int n(\mathbf{r})\,\{\epsilon_{xc}(n(\mathbf{r})) - \mu_{xc}(n(\mathbf{r}))\}\, d\mathbf{r} \end{aligned} \tag{57}$$

Plots of $\epsilon_{xc}(n)$ and $v_{xc}(n)$ are shown in Figs. 3 and 4. They are based on the correlation energy of von Barth and Hedin.[19]

Several analytic fits to $\epsilon_{xc}(n)$ have been devised which are of comparable accuracy (about 1%–2%). Gunnarsson and Lundqvist give, in atomic units,[20]

$$\epsilon_{xc}(n) = -\frac{0.458}{r_s} - 0.0666\, G\frac{r_s}{11.4} \tag{58}$$

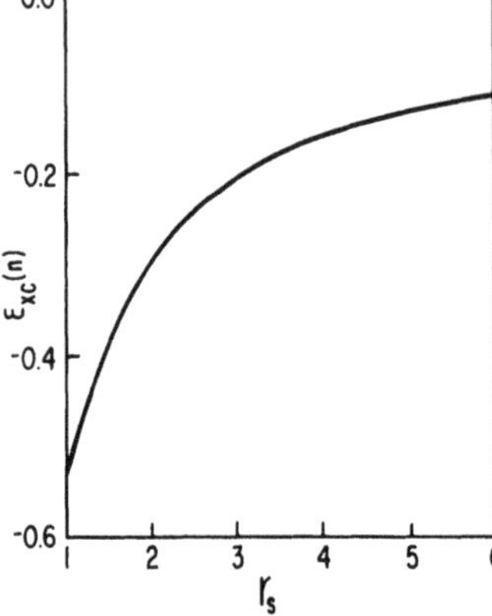

FIGURE 3. The exchange-correlation energy per particle of a uniform electron gas in a.u. (After Ref. 19.)

where r_s is given in terms of n by equation (26) and the function $G(x)$ is defined as

$$G(x) \equiv \tfrac{1}{2}\left[(1 + x^3)\log(1 + x^{-1}) - x^2 + x/2 - \tfrac{1}{3}\right] \tag{59}$$

The great majority of all practical calculations with the KS equations have been carried out with the use of the LDA, which has yielded surprisingly good results even in cases where the density is *not* slowly varying. Later in this section we shall present a partial rationalization of the success of the LDA by the observation that it satisfies an important sum rule.

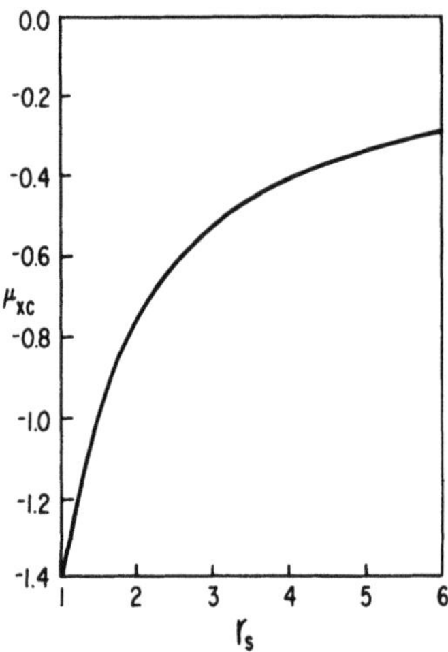

FIGURE 4. The exchange-correlation potential in the local density approximation in a.u. (After Ref. 19.)

(b) Gradient Corrections to $E_{xc}[n]$. Including the lowest-order gradient correction, $E_{xc}[n]$ has the form

$$E_{xc}[n] = \int \epsilon_{xc}[n(\mathbf{r})]\, n(\mathbf{r})\, d\mathbf{r} + \int B_{xc}(n(\mathbf{r}))\, |\nabla n(\mathbf{r})|^2\, d\mathbf{r} \tag{60}$$

The function $B_{xc}(n)$ has been calculated in the random phase approximation.[21,22] It is conveniently represented in the form

$$B_{xc}(n) = \frac{C(r_s)}{n^{4/3}} \text{ a.u.} \tag{61}$$

where $C(r_s)$ is a slowly varying function of r_s shown in Fig. 5. For high densities, $C(0) = 2.57 \times 10^{-3}$ and for $r_s = 6$ it has dropped by about 30%. It is believed that the random phase approximation has an accuracy of about 10% for $r_s \leq 6$.

Opinion about the usefulness of including the lowest gradient correction in $E_{xc}[n]$ in real condensed matter systems (in which usually density gradients are *not* small) is divided. Ma and Brueckner[21] report that for heavy atoms the correlation part of the gradient correction is too high by a factor of 5. Other checks have been made for metal surfaces. Gupta and Singwi[23] estimate that, with the use of the first gradient correction, the remaining error in the surface energy is only a few percent. However, Lau and Kohn[24] and especially Perdew *et al.*[25] conclude that addition of the first gradient correction to $E_{xc}[n]$ substantially overestimates inhomogeneity effects. The present authors incline towards the pessimists. In our view the usefulness of the first

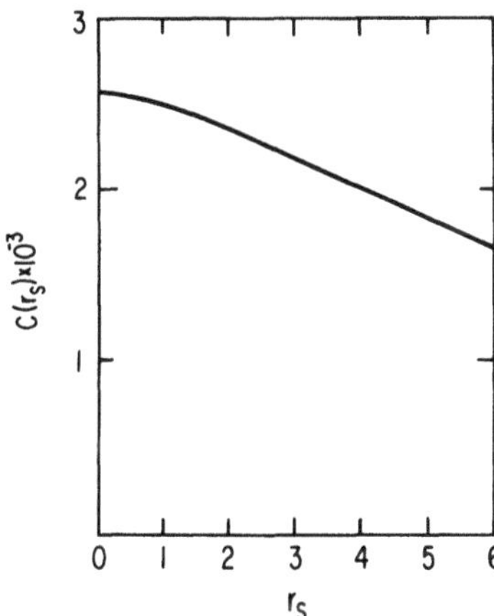

FIGURE 5. The dimensionless function $C(r_s)$ which determines the coefficient B_{xc} of the leading gradient correction of the exchange-correlation energy, equations (60) and (61). (After Ref. 22.)

gradient correction to E_{xc} is sufficiently in doubt at the present time that in most cases, it is advisable to stop at the simple LDA or—if ambitious—to attempt improvements other than series expansions in the density gradient (see below).

(c) Approach to E_{xc} via the Exchange-Correlation Hole. There exists an important exact expression, due to Harris and Jones,[26] Gunnarsson and Lundqvist,[20] and Langreth and Perdew[27] for $E_{xc}[n]$ in terms of a certain average pair correlation function $\bar{h}(\mathbf{r}_1,\mathbf{r}_2)$ of the interacting system. This expression sheds light on the a priori surprising success of the local density approximation and also allows some useful and relatively simple improvements.[20, 27, 27a, 28]

The function $\bar{h}(\mathbf{r}_1,\mathbf{r}_2)$ is defined as follows. Let $n(\mathbf{r})$ be the actual density distribution of a physical system of electrons in an external potential $v(\mathbf{r})$ and with Coulomb interaction $|\mathbf{r}_1-\mathbf{r}_2|^{-1}$. Now consider the interaction multiplied by a factor λ (≤ 1), and $v(\mathbf{r})$ changed to $v_\lambda(\mathbf{r})$ such that the density distribution $n(\mathbf{r})$ remains unchanged. Let $g_\lambda(\mathbf{r}_1,\mathrm{r}_2)$ be the pair correlation function of this modified system, defined by

$$n(\mathbf{r}_1)n(\mathbf{r}_2)g_\lambda(\mathbf{r}_1,\mathbf{r}_2) = \{\langle \mathsf{n}(\mathbf{r}_1)\mathsf{n}(\mathbf{r}_2)\rangle_\lambda - \delta(\mathbf{r}_1-\mathbf{r}_2)n(\mathbf{r}_1)\} \tag{62}$$

where $\mathsf{n}(\mathbf{r})$ is the density operator and $\langle\ \rangle_\lambda$ denotes expectation value in the state corresponding to λ. Plot of $g_\lambda(\mathbf{r}_1-\mathbf{r}_2)$ for a homogeneous system with $\lambda = 1$ and $r_s = 2$ is shown in Fig. 6. The Coulomb interaction energy in the state λ is given by

$$\langle U\rangle_\lambda = \frac{\lambda}{2}\int d\mathbf{r}_1\, d\mathbf{r}_2 \frac{1}{|\mathbf{r}_1-\mathbf{r}_2|} n(\mathbf{r}_1)g_\lambda(\mathbf{r}_1,\mathbf{r}_2)n(\mathbf{r}_2) \tag{63}$$

We shall find it convenient to define the function

$$h_\lambda(\mathbf{r}_1,\mathbf{r}_2) \equiv g_\lambda(\mathbf{r}_1,\mathbf{r}_2) - 1 \tag{64}$$

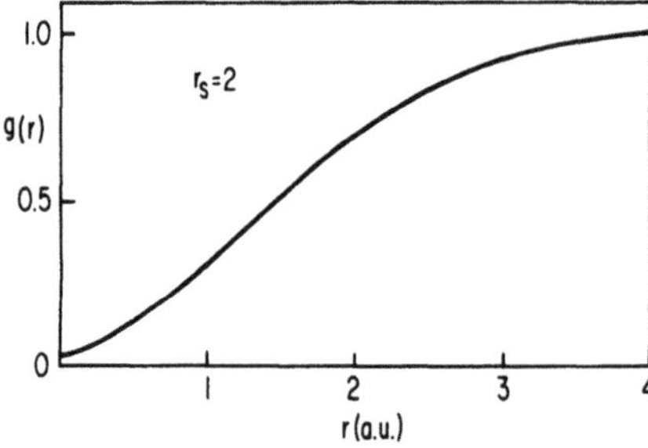

FIGURE 6. The pair correlation function of a homogeneous electron gas for $r_s = 2$. (After Ref. 11.)

which has the asymptotic behavior

$$\lim_{|\mathbf{r}_2-\mathbf{r}_1|\to\infty} h_\lambda(\mathbf{r}_1,\mathbf{r}_2) = 0 \tag{65}$$

It follows from equation (62) that h_λ is symmetric,

$$h_\lambda(\mathbf{r}_1,\mathbf{r}_2) = h_\lambda(\mathbf{r}_2,\mathbf{r}_1) \tag{66}$$

and satisfies the sum rules

$$\int d\mathbf{r}_2\, n(\mathbf{r}_2)\, h_\lambda(\mathbf{r}_1,\mathbf{r}_2) = -1 \tag{67a}$$

$$\int d\mathbf{r}_1\, n(\mathbf{r}_1)\, h_\lambda(\mathbf{r}_1,\mathbf{r}_2) = -1 \tag{67b}$$

The function

$$n_{xc,\lambda}(\mathbf{r}_1,\mathbf{r}_2) \equiv n(\mathbf{r}_2)h_\lambda(\mathbf{r}_1,\mathbf{r}_2) \tag{68}$$

which, when integrated over $\mathbf{r}_2$ gives -1, is called the exchange-correlation hole of an electron at $\mathbf{r}_1$. Regarded as function of $\mathbf{r}_2$, it represents the depletion of the average electron density, $n(\mathbf{r}_2)$, due to the presence of the electron at $\mathbf{r}_1$. The integral of this missing electron density has absolute value 1. The function $\bar{h}(\mathbf{r}_1,\mathbf{r}_2)$ needed for calculating E_{xc} is the average of $h_\lambda(\mathbf{r}_1,\mathbf{r}_2)$ over the coupling constant λ,

$$\bar{h}(\mathbf{r}_1,\mathbf{r}_2) \equiv \int_0^1 d\lambda\, h_\lambda(\mathbf{r}_1,\mathbf{r}_2) \tag{69}$$

It is of the same general character as $h_\lambda(\mathbf{r}_1,\mathbf{r}_2)$, and, of course, satisfies the same symmetry condition (66) and sum rules (67).

It can be shown[20,26,27] (Appendix C) that the exchange-correlation energy is given exactly by the expressions.

$$E_{xc}[n] = \frac{1}{2}\int d\mathbf{r}_1\, d\mathbf{r}_2 \frac{1}{|\mathbf{r}_1-\mathbf{r}_2|}\, n(\mathbf{r}_1)\bar{h}(\mathbf{r}_1,\mathbf{r}_2)n(\mathbf{r}_2) \tag{70}$$

or

$$E_{xc}[n] = \frac{1}{2}\int d\mathbf{r}_1\, d\mathbf{r}_2 \frac{1}{|\mathbf{r}_1-\mathbf{r}_2|}\, n(\mathbf{r}_1)\bar{n}_{xc}(\mathbf{r}_1,\mathbf{r}_2) \tag{71}$$

where

$$\bar{n}_{xc}(\mathbf{r}_1,\mathbf{r}_2) = n(\mathbf{r}_2)\bar{h}(\mathbf{r}_1,\mathbf{r}_2) \tag{72}$$

is the average exchange-correlation hole density satisfying the perfect screening sum rule

$$\int \bar{n}_{xc}(\mathbf{r}_1,\mathbf{r}_2)\, d\mathbf{r}_2 = -1 \tag{73}$$

A plot of $\bar{n}_{xc}$ for a uniform electron gas with $r_s = 2$ is shown in Fig. 7. We can also write E_{xc} in the suggestive form

$$E_{xc}[n] = -\tfrac{1}{2} \int d\mathbf{r}_1\, n(\mathbf{r}_1) \frac{1}{R(\mathbf{r}_1)} \tag{74}$$

where $R(\mathbf{r}_1)$ is an average range of the exchange-correlation hole $\bar{n}_{xc}(\mathbf{r}_1,\mathbf{r}_2)$ as function of $(\mathbf{r}_2 - \mathbf{r}_1)$ in the sense

$$\frac{1}{R(\mathbf{r}_1)} = \int \frac{1}{|\mathbf{r}_1 - \mathbf{r}_2|} [-\bar{n}_{xc}(\mathbf{r}_1,\mathbf{r}_2)]\, d\mathbf{r}_2 \tag{75}$$

Of course, for an interacting, inhomogeneous system the functions h and $\bar{n}_{xc}$ are not exactly known and approximations must be made. In this process the sum rules given in equation (67) and equation (73) provide a useful guide.

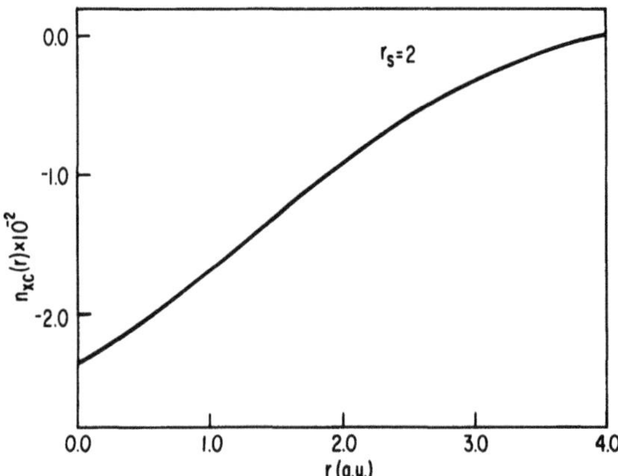

FIGURE 7. The λ-averaged exchange-correlation hole function $\bar{n}_{xc}(r)$, equation (72), for a uniform electron gas with $r_s = 2$. (After Ref. 11.)

The local density approximation corresponds to

$$\bar{n}_{xc}^{\mathrm{LD}}\,(\mathbf{r}_1,\mathbf{r}_2) \equiv n(\mathbf{r}_1)\bar{h}_0(|\mathbf{r}_1 - \mathbf{r}_2|;\, n(\mathbf{r}_1)) \tag{76}$$

where $\bar{h}_0(|\mathbf{r}_1 - \mathbf{r}_2|;\, n)$ is the hole function for an interacting uniform electron gas of density n. For given $\mathbf{r}_1$, by comparison with a uniform gas of density $n = n(\mathbf{r}_1)$, we see that (76) satisfies the sum rule (73). It does not, however, satisfy the symmetry condition (66). The corresponding effective range of this exchange-correlation hole is given by

$$\frac{1}{R^{\mathrm{LD}}(\mathbf{r}_1)} = \int \frac{1}{|\mathbf{r}_1 - \mathbf{r}_2|}\,[-\bar{n}_{xc}^{\mathrm{LD}}\,(\mathbf{r}_1,\mathbf{r}_2)]\, d\mathbf{r}_2 \tag{77a}$$

We see that provided only that the effective range of $\bar{n}_{xc}^{\mathrm{LD}}$ is similar to that of the actual $\bar{n}_{xc}$, the corresponding energy, E_{xc}^{LD}, equation (74), will be a good approximation to the correct E_{xc}. It is evident from equations (74), (75), and (76) that this E_{xc}^{LD} is just the LDA of equation (55) with

$$-\frac{1}{2}\frac{1}{R^{\mathrm{LD}}(\mathbf{r}_1)} = \epsilon_{xc}\,[n(\mathbf{r}_1)] \tag{77b}$$

Now except in rather extreme situations, such as the far tail of the electron distribution of a metal surface, $R(\mathbf{r}_1)$ is a relatively insensitive functional of the density $n(\mathbf{r})$ and hence we have a qualitative rationalization for the LDA version of the KS equations which treats $T[n]$ exactly and E_{xc} in the LDA.

The form for E_{xc}, equation (70), suggests some other simple approximations which should be more accurate than the LDA, equation (76). We shall limit ourselves to approximations based on the exchange-correlation hole functions $\bar{h}_0(\mathbf{r};n)$, which are directly obtainable from the pair correlation functions $g_\lambda(\mathbf{r};n)$ of a homogeneous electron gas with interaction $\lambda/|\mathbf{r}_1 - \mathbf{r}_2|$.

One such approximation, which has been quantitatively very successful in preliminary applications is due to O. Gunnarsson *et al.*[28] It consists of approximating the actual $\bar{h}(\mathbf{r}_1,\mathbf{r}_2)$ of the inhomogeneous system by the function*

$$\bar{h}^{\mathrm{WD}}(\mathbf{r}_1,\mathbf{r}_2) \equiv \bar{h}_0\,(\mathbf{r}_1,\mathbf{r}_2;\, \tilde{n}(\mathbf{r}_1)) \tag{78}$$

where, for every $\mathbf{r}_1$, $\tilde{n}(\mathbf{r}_1)$ is so chosen as to satisfy the sum rule, equation (73). This approximation, unlike the LDA, preserves the densities $n(\mathbf{r}_1)$ and $n(\mathbf{r}_2)$ in equation (70) and replaces the actual $\bar{h}\,(\mathbf{r}_1,\mathbf{r}_2)$ by an approximate $\bar{h}$ which,

*WD stands for weighted density.

while of course not correct, does at least satisfy the sum rule (73). Negative features are that for each $\mathbf{r}_1$, $\tilde{n}(\mathbf{r}_1)$ has to be found so that the sum rule (73) is satisfied; and that the approximate $\bar{h}$ of equation (78) violates the symmetry condition, equation (66), of the actual $\bar{h}$. We denote the corresponding exchange-correlation energy by $E_{xc}^{\mathrm{WD}}[n]$.

Since the LDA has been found to give good approximations to the density $n(\mathbf{r})$, one can take advantage of the stationary character of $E_v[n]$ to avoid recalculating $n(\mathbf{r})$ with the more complex E_{xc}^{WD} and take for the total energy

$$E^{\mathrm{WD}} = E_v^{\mathrm{LD}}[n^{\mathrm{LD}}] + \{E_{xc}^{\mathrm{WD}}[n^{\mathrm{LD}}] - E_{xc}^{\mathrm{LD}}[n^{\mathrm{LD}}]\} \tag{79}$$

2.3. Comparison with Other Theories

As its name implies, the central quantity of DFT is the particle density $n(\mathbf{r})$. In the following paragraphs we would like to make some remarks about the relationship of this theory to three other approaches to many-body systems, the density matrix approach,[29,29a] the Landau theory of Fermi liquids,[30,31] and Hartree and Hartree–Fock theories.[32,33]

2.3.1. Comparison with Density Matrix Theory. It is well known that the energy of any many electron system can be expressed in terms of the two-particle density matrix $n_2(\mathbf{r}_1,\mathbf{r}_2;\mathbf{r}_1',\mathbf{r}_2')$. First of all the one particle density matrix, $n_1(\mathbf{r}_1,\mathbf{r}_2')$, is defined as a partial trace of n_2,

$$n_1(\mathbf{r}_1,\mathbf{r}_1') \equiv \int d\mathbf{r}_2\, n_2(\mathbf{r}_1,\mathbf{r}_2;\, \mathbf{r}_1',\mathbf{r}_2) \tag{80}$$

and the particle density is given by

$$n(\mathbf{r}_1) = n_1(\mathbf{r}_1,\mathbf{r}_1) \tag{81}$$

The several parts of the energy [see equation (2)] are then given by

$$\langle \mathbf{T} \rangle = ½ \int [\nabla\nabla' n_1(\mathbf{r}_1,\mathbf{r}'_1)]_{\mathbf{r}_1=\mathbf{r}_1'}\, d\mathbf{r}_1 \tag{82}$$

$$\langle \mathbf{V} \rangle = \int v(\mathbf{r}_1) n(\mathbf{r}_1)\, d\mathbf{r}_1 \tag{83}$$

$$\langle \mathbf{U} \rangle = \frac{1}{2} \int \frac{1}{|\mathbf{r}_1 - \mathbf{r}_2|}\, n_2(\mathbf{r}_1,\mathbf{r}_2;\, \mathbf{r}_1,\mathbf{r}_2)\, d\mathbf{r}_1\, d\mathbf{r}_2 \tag{84}$$

For this reason much attention has been given to the as yet unsolved problem of determining the two-particle density matrix, n_2, without having to solve the full many-particle problem.[29]

This matrix must satisfy certain simple symmetry, positiveness, and normalization requirements. However, even when one restricts oneself to all those matrices $n_2(\mathbf{r}_1,\mathbf{r}_2;\mathbf{r}_1',\mathbf{r}_2')$ which satisfy these general requirements, only a set of relative measure zero can in fact be realized by a physical system in its ground state. For such a system is completely characterized by the number of particles N and the external potential $v(\mathbf{r})$ while n_2, even when obeying the general requirements, is a function of four independent variables.

On the other hand we have seen earlier that two physical ground-state systems with the same density distribution $n(\mathbf{r})$ must be identical. Since n_2 determines $n(\mathbf{r})$ by equations (80) and (81), we reach, in summary, the following comparative conclusions:

Two physical ground-state systems with identical two-particle density matrices are necessarily identical; however only a subset of relative measure zero of all matrices n_2 $(\mathbf{r}_1,\mathbf{r}_2;\mathbf{r}_1',\mathbf{r}_2')$, which satisfy certain general requirements, can be realized in actual physical systems. On the other hand density functional theory shows that two physical ground-state systems with the same density distribution $n(\mathbf{r})$ are identical; and that *all* "reasonable" density distributions, $n(\mathbf{r})$, can be realized by actual physical systems. Thus the set of all "reasonable" distributions, $n(\mathbf{r})$, is in one-to-one correspondence with all ground states of actual physical systems.

As an additional comparative comment we remark also that discussions of many-body systems in terms of two-particle density matrices apply equally to ground states and excited states whereas density functional theory is primarily a ground-state theory.*

2.3.2. Comparison with Landau–Fermi Liquid Theory. The central quantity of the Landau Fermi liquid theory (FLT)[30] is $\delta n(\mathbf{k},\mathbf{r},t)$, the distribution function of quasiparticles excited in a uniform Fermi liquid. The following are the chief points of comparison.

(a) DFT is primarily a ground state theory; FLT is silent on absolute ground state properties and is in its essence a theory of excitations.

(b) DFT is intrinsically a theory of spatially inhomogeneous systems, including strongly inhomogeneous systems†; FLT has important implications for strictly homogeneous systems (e.g., heat capacity, compressibility), and for certain small perturbations of homogeneous systems (e.g., long wavelength zero sound collective modes).

(c) DFT is primarily a static theory; FLT includes naturally low-frequency dynamic phenomena.

*As discussed in Sections 5 and 6, DFT can be extended to thermal ensembles and, in a limited way, to individual excited states.

†It reduces to previously known theories in the homogeneous limit; and it makes use of theories of homogeneous systems for treating systems whose density has a slow spatial variation, regardless of the amplitude of this variation.

(d) The connection between the central quantities in the two theories for a nearly uniform system is simply

$$n(\mathbf{r},t) = n_0 + \int d\mathbf{k}\ \delta n(\mathbf{k},\mathbf{r},t)$$

where $n(\mathbf{r},t)$ is the total time-dependent density of DFT (usually independent of t), n_0 is the uniform unperturbed density and δn was defined above.

(e) Very recently the concepts of FLT and DFT have been combined to derive a transport equation in the low-wave-number, low-frequency regime, for systems whose equilibrium density is spatially slowly varying.[31] (See also Section 6.2.2.)

2.3.3. Comparison with Hartree and Hartree–Fock Theories. To make the comparison of DFT with Hartree and Hartree–Fock theories (HT and HFT), we first split the exchange-correlation energy, E_{xc}, into its exchange and correlation parts. The energy functional given in equation (43) then takes the form

$$\begin{aligned} E_v[n'] = \int v(\mathbf{r})\, n'(\mathbf{r})\, d\mathbf{r} + \frac{1}{2}\int \frac{n'(\mathbf{r})n'(\mathbf{r}')}{|\mathbf{r}-\mathbf{r}'|}\, d\mathbf{r}\, d\mathbf{r}' \\ + T_s[n'] + E_x[n'] + E_c[n'] \end{aligned} \tag{85}$$

where $T_s[n]$ is the kinetic energy of a noninteracting system of density $n(\mathbf{r})$, $E_x[n]$ is the exchange energy (as determined by a complete HF calculation) and $E_c[n]$ is the correlation energy, i.e., $G[n] - T_s[n] - E_x[n]$. [See equation (42)].

The exact equations of HT and HFT are obtained from equation (85) by dropping $E_x + E_c$ or only E_c. (See Section 2.2.1.) Obviously the inclusion of any reasonable approximation for E_c will improve these equations.

Use of the *exact* $E_x[n]$ implies solution of the full HF equations which are difficult integrodifferential equations. Experience has shown that at a small cost in accuracy[32] (usually small compared to the cost of omitting the correlation energy) the exact E_x can be replaced by the local approximation,

$$E_x[n] = -\frac{2}{3}\left(\frac{9}{8\pi}\right)(3\pi^2)^{1/3}\int d\mathbf{r}\ n^{4/3}(\mathbf{r}) \tag{86}$$

which leads to what are called the Hartree–Fock–Slater equations

$$\left\{-\frac{1}{2}\nabla^2 + v(\mathbf{r}) + \int \frac{n(\mathbf{r}')}{|\mathbf{r}-\mathbf{r}'|}\, d\mathbf{r}' - \frac{1}{\pi}[3\pi^2 n(\mathbf{r})]^{1/3}\right\}\psi_i(\mathbf{r}) = \epsilon_i\psi_i(\mathbf{r}) \tag{87}$$

$$n(\mathbf{r}) = \sum |\psi_i(\mathbf{r})|^2 \tag{88}$$

but with the exchange term smaller by a factor of ⅔ than in the original work of Slater.*[33] These equations are ordinary differential equations, unlike the much more complex full HF equations.

One of the most useful forms of DFT have been the Kohn–Sham equations, equations (47)–(49), with $E_x + E_c$ treated in the local approximation. This gives rise to the effective exchange-correlation potential v_{xc} of equation (56) and Fig. 4.

The solution of these equations is only negligibly more difficult than the solution of the Hartree equations and much easier than of the full HF equations. Nevertheless, because this v_{xc} is a rather good representation of *both* exchange and correlation, the results are usually considerably more reliable than those of the full HF equations, in which correlation effects are omitted.

2.4. *Illustrations*

2.4.1. General Discussion. Since the middle 1960s, DFT has been extensively used to calculate the ground-state properties of a wide variety of systems. These include atoms, molecules, bulk solids, defects in solids, surfaces and interfaces of solids, adsorbates on surfaces, inversion layers at semiconductor surfaces, etc. Chapters 4 and 5 of this book deal with applications of DFT to some of these systems. In this section we shall merely give a few illustrations to indicate the scope and the accuracy of the method, when applied to one-component nonmagnetic systems.

A large fraction of all calculations have been made in the local density approximation (LDA) of the KS self-consistent equations (see Section 2.2.2). Here the only input is the form of the exchange-correlation energy per particle, $\epsilon_{xc}(n)$. This is a property of the uniform electron gas of density n, which is generally considered as accurately known. The corresponding contribution to the effective potential v_{eff} is $\mu_{xc} = d(n\epsilon_{xc})/dn$. An accurate analytic approximation to $\epsilon_{xc}(n)$ was given in equations (58)–(59). Here we present a

*Some controversy over this discrepancy exists in the literature of the 1960s. It led Slater and others to introduce a phenomenological multiplicative factor α to the original Slater exchange potential with values between ⅔ [corresponding to our equation (87)] and 1 (corresponding to Slater's original proposal). This was the origin of the widely used $X\alpha$ method. At this time there is general agreement that the correct local expression for the *exchange* potential is given in equation (87). However, somewhat larger values of α—different for different systems—can partially account for the effects of correlations.

similar approximation for μ_{xc}, regarded as function of r_s [defined in equation (26)]. One can write

$$\mu_{xc}(r_s) = \beta(r_s)\mu_x(r_s) \tag{89}$$

where μ_x is the exchange potential,

$$\mu_x(r_s) = -\left(\frac{9}{4\pi^2}\right)^{1/3} r_s^{-1} = -0.611 r_s^{-1} \tag{90}$$

and $\beta(r_s)$ can be accurately represented by the expression[20]

$$\beta(r_s) = 1 + 0.0545 r_s \log(1 + 11.4/r_s) \tag{91}$$

The enhancement factor β, describing the effect of correlation, varies from 1.0 to 1.25 as r_s varies from 0 to 6. Thus we see that for most practical densities the exchange portion of ϵ_{xc} dominates the correlation contribution.

A different approach, following a phenomenological philosophy, has been adopted by J. C. Slater and co-workers, the so-called $X\alpha$ method.[34–37] This takes as starting point an early proposal of Slater's[33] to represent the nonlocal Hartree–Fock exchange potential by the local potential

$$v_x^{\mathrm{Sl}} \equiv -\frac{3}{2}\left(\frac{9}{4\pi^2}\right)^{1/3} r_s^{-1} = -0.916 r_s^{-1} \tag{92}$$

Note that v_x^{Sl} is 3/2 times the μ_x of equation (90). This potential is then multiplied by a correction factor α, giving as effective exchange-correlation potential

$$v_{x\alpha} \equiv \alpha v_x^{\mathrm{Sl}} \tag{93}$$

The factor α is chosen so as to fit some experimental quantity such as the total energy of the system. In practice α is generally found to lie between 2/3 (pure exchange) and 0.75, in line with the a priori values of $\beta(r_s)$ of equation (91).* The $X\alpha$ method has been extensively used not only for ground states but also for excited states. For further details we refer the reader to the reviews in Refs. 34–37 and 1.

A number of calculations have made direct use of the stationary expres-

*The present authors generally prefer the use of a priori expression (89), which allows a clear comparison of the LDA with experiment, to the fitting procedure of the $X\alpha$ method.

TABLE 1. Two-Electron Atomic Systems[a]

	Energy (eV)	
Atom or ion	LDA[b]	Experiment
H^-	−14.4	−14.4
He	−77.8	−79.0
Li^+	−195.2	−198.1

[a]After Ref. 20.
[b]Local density approximation of the Kohn–Sham equations.

sion (19), with $G[n]$ represented either by its local approximation or by some gradient refinement of this approximation (see Section 2.1.2). However, because the single-particle kinetic energy $T_s[n]$ (which is a very large part of $G[n]$) is then treated only approximately, the results are in general less accurate and reliable than those obtained from the KS equations, in which $T_s[n]$ is treated exactly.

Finally some very good results have recently been obtained by combining the solutions of the KS equations with the nonlocal exchange-correlation energy of the so-called weighted density approximation of equation (78), and other nonlocal corrections.[38]

In the following subsections we present results of a few representative ground-state calculations for spin-compensated electronic systems.

2.4.2. Kohn–Sham Equations in Local Density Approximation. (a) Atoms. Some total energies of two-electron atoms are shown in Table 1. Errors are in the range of 1%–2%. Total ground-state energies of rare gas atoms are given in Table 2. The error for He is 3%, for the heavier atoms about 0.5%.

(b) Molecules. Various properties of diatomic molecules are given in Table 3. Errors of the binding energy are of the order of 20%, of the equilibrium spacing about 5%, and of the vibration frequency about 10%. These results are based on approximate solutions of the KS equations and accurate solutions may well give substantially different results. Preliminary indications are that

TABLE 2. Rare Gas Atoms in ^1S State[a]

	Energy (Ry)	
Atom	LDA	Experiment
He	−5.651	−5.807
Ne	−256.349	−257.880
Ar	−1051.709	−1055.21

[a]After Ref. 32.

TABLE 3. Diatomic Molecules[a]

Molecule	Binding Energy (eV)		Equilibrium spacing (a_0)		Vibration frequency (cm^{-1})	
	LDA	Experiment	LDA	Experiment	LDA	Experiment
N_2	7.8	9.9	2.16	2.07	2070	2359
CO	9.6	11.2	2.22	2.13	2090	2170

[a]After Ref. 39.

the LDA results are significantly better than those of complete Hartree–Fock calculations.

(c) Metals. Complete calculations, within the LDA, for all metals from $Z = 1$ (metallic H) to $Z = 49$ (In) have recently been published in a book by V. L. Moruzzi *et al.*[(40)] The results give an excellent view, with an accuracy of the order of 10%,* of all important physical properties of metals in equilibrium. This is true even of the transition metals in which, due to the tightly bound d electrons, the density $n(\mathbf{r})$ of the electrons participating in the binding is a rather rapidly varying function of position. By way of illustration, Table 4 gives results for the cohesive energy and Wigner–Seitz radius, r_s, of the $4d$ transition metals. (Results for the $3d$ transition metals, especially Mn, are less satisfactory.)

(d) Inversion Layers in Semiconductors. In metal–insulator–semiconductor (MIS) structures, application of an electric field normal (z direction) to the interfaces bends the bands of the semiconductor. If the bottom of the conduction band in a p-type material is bent below the Fermi level, the conduction

*Lattice parameters are usually much more accurate, typically within ~2%.

TABLE 4. Bulk Metals[a]

Metal	Cohesive energy (Ry)		Wigner–Seitz radius (a_0)[b]	
	LDA	Experiment	LDA	Experiment
Y	0.36	0.31	3.60	3.75
Zr	0.50	0.46	3.20	3.30
Nb	0.55	0.56	3.00	3.05
Mo	0.49	0.50	2.90	2.90
Tc	0.55	—	2.85	—
Ru	0.56	0.49	2.80	2.80
Rh	0.45	0.42	2.80	2.80
Pd	0.27	0.28	3.05	3.00

[a]After Ref. 40.
[b]To nearest $0.05a_0$.

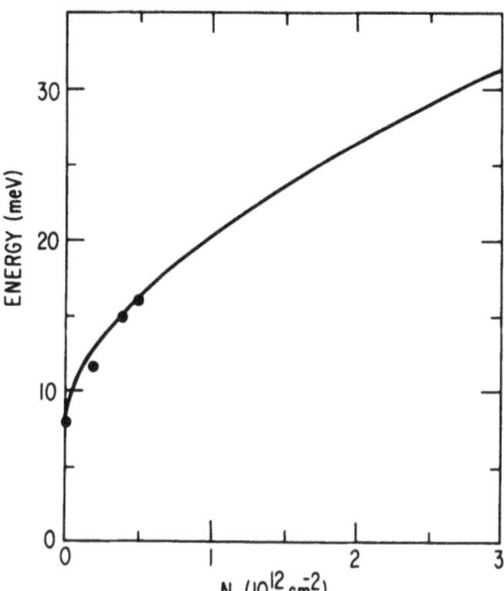

FIGURE 8. Density dependence of the separation of the $n = 0$ and 1 subbands in a (100) Si–SiO_2 inversion layer (solid curve). N_s is the number of electrons per unit area (after Ref. 43). The dots represent the results of infrared absorption measurements (after Ref. 44). (So-called "excitonlike" and "depolarization" corrections nearly cancel each other.)

band becomes partially populated by electrons whose motions are quantized in the z direction and free-particle-like along the interface. Such a quasi-two-dimensional system is called an inversion layer.[41]

Ando[42,43] has used DFT to investigate the dependence of energy level structure and the effective mass on the mean surface layer electron density. The position of the energy levels is found by solving the KS equations, which, for a Si-inversion layer in an (001) plane and the conduction band valleys along the (00 ± 1) axes can be written as

$$\left[\frac{k^2}{2m_{\parallel}} - \frac{1}{2m_{\perp}}\frac{d^2}{dz^2} + v_{\text{eff}}(z)\right]\psi_n(z)\, e^{i\mathbf{k}\cdot\mathbf{r}} = \epsilon_n(\mathbf{k})\psi_n(z)e^{i\mathbf{k}\cdot\mathbf{r}} \tag{94}$$

where $\mathbf{k}$ and $\mathbf{r}$ are the wave and position vectors in two dimensions, and $m_{\parallel}$ and $m_{\perp}$ are the electron masses parallel and perpendicular to the interface.

TABLE 5. Diatomic Rare Gas Molecules[a]

	Binding energy (10^{-6} erg)		Equilibrium spacing (Å)	
Molecule	LDA	Exact	LDA	Exact
Ne_2	56	63	2.99	3.03
Ar_2	175	195	3.63	3.70
Kr_2	248	273	3.89	3.95
NeAr	78.5	76–111	3.42	3.38–3.64

[a]After Ref. 45.

TABLE 6. Metal Surfaces[a]

Metal	Work function (eV)	
	Calculated	Experiment
Cs	2.64	1.81
Na	2.93	2.35
Al	3.64	4.25

[a]After Ref. 46.

The effective potential in equation (94) consists of four terms

$$v_{\mathrm{eff}}(z) = v_{\mathrm{depl}}(z) + v_{\mathrm{image}}(z) + v_{\mathrm{es}}(z) + v_{\mathrm{xc}}(z) \tag{95}$$

arising from the depletion layer charges, the image charges, the classical effects of the electronic charge density, and exchange and correlation effects. The electron density $n(z)$ is obtained by summing contributions from all valleys. Self-consistency is demanded.

In such a calculation one obtains not only total energy and electron density, but also the eigenvalues of the KS equations. These are taken by Ando to represent effective energy levels of the inversion layer electrons (see, however, the cautionary remarks in Section 2.2.1), and used to interpret infrared absorption measurements[44] corresponding to transition between states ψ_0 and ψ_1, of equation (94). Figure 8 shows the comparison between experimental results and LDA calculation, as function of the total electron density per unit surface area.

2.4.3. Direct Minimization of the Total Energy. Calculations, employing directly the minimal principle of Section 2.1 with a simple local form for E_{xc}, give strikingly good results shown in Table 5. Errors in the binding energies and equilibrium spacings are, respectively, of the order of 10% and 2%. This is all the more remarkable since much of the attractive force between the atoms is a van der Waals force and thus a pure correlation effect. Nevertheless, at least in the vicinity of the energy minimum, a local description of correlation energy leads to very accurate results.

In this early DFT work shown in Table 6, metals are represented by a half jellium model. The energy functional used is that of the LDA plus the leading gradient correction for the kinetic energy which is essential. More recent and more accurate calculations, employing various forms of $G[n]$ give significantly different results.[47]

3. Extension to Magnetic Systems

In Section 2 we have limited the discussion to nonmagnetic systems in an external potential $v(\mathbf{r})$ and developed a theory based on an energy functional $F[n(\mathbf{r})]$ depending on the particle density $n(\mathbf{r})$. We now turn to systems in magnetic fields $\mathbf{b}(\mathbf{r})$ in which the spin indices α of the one-electron operators $\psi_\alpha(\mathbf{r})$ and the magnetization density $\mathbf{m}(\mathbf{r})$, become, of course, relevant. The generalization of the nonmagnetic theory to systems in which there is also a magnetic term of the form $-\int \mathbf{b}(\mathbf{r})\cdot\mathbf{m}(\mathbf{r})d\mathbf{r}$ is conceptually straightforward[6,19,48,48a] and will be rapidly developed and illustrated in the following sections.

3.1. Variational Formulation

3.1.1. Formal Theory. The Density and Magnetization Density as Basic Variables. As in Section 2, we write the Hamiltonian in the form

$$\mathsf{H} = \mathsf{T} + \mathsf{V} + \mathsf{U} \tag{96}$$

where now the one-particle operator includes the interaction of the magnetic field $\mathbf{b}(\mathbf{r})$ with the electron spin:*

$$\mathsf{V} = \int [v(\mathbf{r})\mathsf{n}(\mathbf{r}) - \mathbf{b}(\mathbf{r})\cdot\mathsf{m}(\mathbf{r})]\, d\mathbf{r} \tag{97}$$

where

$$\mathsf{n}(\mathbf{r}) = \psi_\alpha^*(\mathbf{r})\psi_\alpha(\mathbf{r}) \tag{98}$$

and

$$\mathsf{m}(\mathbf{r}) = -\mu_0 \sum_{\alpha,\beta} \boldsymbol{\sigma}_{\alpha\beta}\psi_\alpha^*(\mathbf{r})\psi_\beta(\mathbf{r})$$

here μ_0 is the Bohr magneton, $e\hbar/2mc$, and $\sigma_{\alpha\beta}$ is the Pauli spin matrix vector. For compactness of notation we shall write

$$w(\mathbf{r}) \equiv \{v(\mathbf{r}), \mathbf{b}(\mathbf{r})\} \tag{99}$$

*The interaction with the electronic currents, leading to diamagnetism, is neglected in this formulation, as are the magnetic interactions between electrons. The relevance of this partial theory must therefore be checked for each given magnetic system.

$$\nu(\mathbf{r}) \equiv \{n(\mathbf{r}), \mathbf{m}(\mathbf{r})\} \tag{100}$$

where all quantities are c-numbers.

In complete analogy with the case of nonmagnetic systems treated in Section 2.1.1 one can show that $\nu(\mathbf{r})$ uniquely determines $w(\mathbf{r})$. The appropriate energy functional is given by

$$E_w[\nu(\mathbf{r})] \equiv \int d\mathbf{r}\{v(\mathbf{r})n(\mathbf{r}) - \mathbf{b}(\mathbf{r}) \cdot \mathbf{m}(\mathbf{r})\} + \frac{1}{2}\int d\mathbf{r}\, d\mathbf{r}' \frac{n(\mathbf{r})n(\mathbf{r}')}{|\mathbf{r}-\mathbf{r}'|} + G[\nu(\mathbf{r})] \tag{101}$$

Here $G[\nu(\mathbf{r})]$ is a universal functional of $\nu(\mathbf{r})$ given by

$$G[\nu(\mathbf{r})] = (\Psi, (\mathbf{T} + \mathbf{U})\Psi) - \frac{1}{2}\int d\mathbf{r}\, d\mathbf{r}' \frac{n(\mathbf{r})n(\mathbf{r}')}{|\mathbf{r}-\mathbf{r}'|} \tag{102}$$

Clearly because of the uncoupling of spin and orbit, $G[\nu(\mathbf{r})]$ is invariant under an $\mathbf{r}$-independent rotation of the magnetization field $\mathbf{m}(\mathbf{r})$. For given $w(\mathbf{r})$, the functional $E_w[\nu]$ attains its minimum for the correct $\nu(\mathbf{r})$.

3.1.2. Approximations for G[n,m]; Local Spin Density Theory. As in Section 2, it is convenient to separate off the kinetic energy of a noninteracting system of the given $\nu(\mathbf{r})$, plus a remainder—the exchange-correlation energy:

$$G[\nu] = T_s[\nu] + E_{xc}[\nu] \tag{103}$$

For a gas of slowly varying $\nu(\mathbf{r})$ one has expansions

$$T_s[\nu] = \int d\mathbf{r}\, n(\mathbf{r})t_s[\nu(\mathbf{r})] + \cdots \tag{104}$$

$$E_{xc}[\nu] = \int d\mathbf{r}\, n(\mathbf{r})\epsilon_{xc}[\nu(\mathbf{r})] + \cdots \tag{105}$$

where t_s and ϵ_{xc} refer to uniform systems and the omitted terms are gradient corrections. This is the so-called "local spin density approximation (LSDA)".

We now introduce the parameter $\boldsymbol{\zeta}(\mathbf{r})$ describing the degree of local magnetization

$$\boldsymbol{\zeta}(\mathbf{r}) = \frac{\mathbf{m}(\mathbf{r})}{\bar{m}(\mathbf{r})} \tag{106}$$

where $\bar{m}(\mathbf{r})$ is the maximum magnetization consistent with $n(\mathbf{r})$,

$$\bar{m}(\mathbf{r}) \equiv -\mu_0 n(\mathbf{r}) \tag{107}$$

Then we can write t_s and ϵ_{xc} as functions of n and ζ (= $|\boldsymbol{\zeta}|$)

For t_s an elementary calculation gives

$$t_s(n,\zeta) = \frac{3}{10} k_F^2(n) \cdot \frac{(1+\zeta)^{5/3} + (1-\zeta)^{5/3}}{2} \tag{108}$$

where $k_F(n)$ is the Fermi momentum of a nonmagnetic electron gas of density n.

The local exchange and correlation energy ϵ_{xc} has been approximated[19] by an interpolation formula between the paramagnetic ($\zeta=0$) and ferromagnetic case ($\zeta = 1$):

$$\epsilon_{xc}(n,\zeta) = \epsilon_{xc}^P(n) + f(\zeta)\Delta\epsilon_{xc}(n)^* \tag{109}$$

where

$$\epsilon_{xc}^P(n) \equiv \epsilon_{xc}(n,\zeta=0) \tag{110}$$

$$\Delta\epsilon_{xc}(n) \equiv \epsilon_{xc}(n,\zeta=1) - \epsilon_{xc}(n,\zeta=0) \equiv \epsilon_{xc}^F(n) - \epsilon_{xc}^P(n) \tag{111}$$

and

$$f(0) = 0, \qquad f(1) = 1 \tag{112}$$

Several useful and accurate forms for the functions ϵ_{xc}, $\Delta\epsilon_{xc}$, and $f(\zeta)$ have been proposed of which we quote those used by Gunnarsson and Lundqvist[20]:

$$\epsilon_{xc}^i(n) = [\epsilon_x^i(r_s) - C^i G(r_s/r^i)] \text{ a.u.}, \qquad i = P, F \tag{113}$$

where

$$\epsilon_x^P(r_s) = -\frac{0.458}{r_s}, \qquad C^P = 0.0666, \qquad r^P = 11.4 \tag{114}$$

$$\epsilon_x^F(r_s) = -\frac{0.577}{r_s}, \qquad C^F = 0.0406, \qquad r^F = 15.9 \tag{115}$$

*The *exact* $\epsilon_{xc}(n,\zeta)$ requires a function $f(n,\zeta)$ depending on *both* n and ζ.

and the function $G(x)$ is given in equation (59); the function $f(\zeta)$ is given by[19]

$$f(\zeta) = [(1+\zeta)^{4/3} + (1-\zeta)^{4/3} - 2]/[2^{4/3}-2] \tag{116}$$

3.1.3. Weak Magnetization. For paramagnetic systems for which, in the absence of an external field $\mathbf{b}$, the ground state corresponds to $\mathbf{m} \equiv 0$, the functional $G[n,\mathbf{m}]$ must have the following expansion for small $\mathbf{m}(\mathbf{r})$:

$$G[n,\mathbf{m}] = G[n] + \frac{1}{2}\sum_{ij} \int d\mathbf{r}\, d\mathbf{r}'\, m_i(\mathbf{r})G_{ij}(\mathbf{r},\mathbf{r}';[n])m_j(\mathbf{r}') + \cdots \tag{117}$$

where

$$G_{ij}(\mathbf{r},\mathbf{r}';[n]) \equiv \left.\frac{\delta^2 G[n,\mathbf{m}]}{\delta m_i(\mathbf{r})\delta m_j(\mathbf{r}')}\right|_{\mathbf{m}=0} \tag{118}$$

In the local approximation, where $n(\mathbf{r})$ and $\mathbf{m}(\mathbf{r})$ are slowly varying, this becomes[6]

$$G[n,\mathbf{m}] = \int g_0(n(\mathbf{r}))\, d\mathbf{r} + {\textstyle\frac{1}{2}} \int \chi^{-1}(n(\mathbf{r}))m^2(\mathbf{r})\, d\mathbf{r} \tag{119}$$

where $\chi(n)$ is the magnetic susceptibility of a uniform system, defined by

$$m = \chi(n)b \tag{120}$$

Table 7 gives $\chi(n)$ according to von Barth and Hedin[19]; very similar results have been obtained by other authors.[49–51]

3.2. Self-Consistent Equations

As for nonmagnetic systems, one can derive from the stationary expression $E_w[\nu]$, equation (101), a set of self-consistent equations by treating the single-particle kinetic energy functional $T_s[\nu]$ exactly. For simplicity we restrict

TABLE 7. Static Spin Susceptibility of a Uniform Interacting Electron Gas

r_s	1	2	3	4	5
χ/χ_0^a	1.15	1.28	1.43	1.60	1.82

$^a\chi_0$ is the susceptibility of a noninteracting free electron gas.

ourselves here to the case where both $\mathbf{b}(\mathbf{r})$ and $\mathbf{m}(\mathbf{r})$ have only one nonvanishing component, say $b_z(\mathbf{r}) \equiv b(\mathbf{r})$ and $m_z(\mathbf{r}) \equiv m(\mathbf{r})$*. Then we obtain, exactly as in Section 2.2.1, the following self-consistent equations for spin up and down electrons:

$$\left\{ -\frac{1}{2}\nabla^2 + v(\mathbf{r}) + \int \frac{n(\mathbf{r}')}{|\mathbf{r}-\mathbf{r}'|} d\mathbf{r}' + v_{xc}(\mathbf{r}) - \sigma\mu_0 [b(\mathbf{r}) - w_{xc}(\mathbf{r})] \right\} \psi_{j\sigma}(\mathbf{r}) = \epsilon_{j\sigma}\, \psi_{j\sigma}(\mathbf{r}), \qquad \sigma = \pm 1 \tag{121}$$

where

$$v_{xc}(\mathbf{r}) = \frac{\delta E_{xc}[n,m]}{\delta n(\mathbf{r})} \tag{122}$$

$$w_{xc}(\mathbf{r}) = \frac{\delta E_{xc}[n,m]}{\delta m(\mathbf{r})} \tag{123}$$

and

$$n(\mathbf{r}) = \sum_{\sigma} n_\sigma(\mathbf{r}), \qquad m(\mathbf{r}) = -\mu_0 [n_{+1}(\mathbf{r}) - n_{-1}(\mathbf{r})] \tag{124}$$

$$n_\sigma(\mathbf{r}) = \sum_{\epsilon_{j\sigma} \leq \bar{\mu}} |\psi_{j\sigma}(\mathbf{r})|^2 \tag{125}$$

and $\bar{\mu}$ is the chemical potential adjusted so that the total number of occupied electron states is equal to the total number of electrons. If one makes the local spin density approximation for $E_{xc}[\nu]$, one obtains the following interpolation formulas[20]:

$$v_{xc} = \mu_x^P(r_s) \left[\beta(r_s) - \frac{1}{3}\, \delta(r_s)\gamma(r_s) \frac{\zeta^2}{1-\gamma^2\zeta^2} \right] \tag{126}$$

and

$$\mu_0 w_{xc} = \mu_x^P(r_s) \frac{1}{3}\, \delta(r_s) \frac{\zeta}{1-\gamma^2\zeta^2} \tag{127}$$

*The general case is treated in Refs. (19) and (20).

where

$$\mu_x^P = -0.611 r_s^{-1} \text{ a.u.}$$

$$\beta(r_s) = 1 + 0.0545 r_s \log\left(1 + \frac{11.4}{r_s}\right) \quad (128)$$

$$\delta(r_s) = 1 - 0.036 r_s - 1.36 r_s/(1 + 10 r_s)$$

$$\gamma = 0.297$$

Here ζ is an *algebraic* quantity,

$$\zeta = \frac{m(\mathbf{r})}{\bar{m}(\mathbf{r})} = \frac{n_{+1}(\mathbf{r}) - n_{-1}(\mathbf{r})}{n(\mathbf{r})} \quad (129)$$

The LSDA shares with the unrestricted HF method the flexibility of allowing different orbitals for different spins. It differs from it by the simplified local treatment of exchange and by the inclusion of correlation effects. In applications it is generally found to be both simpler and more accurate.

TABLE 8. Ionization Potentials[a]

	Ionization potential (eV)	
Atom	LSDA[b]	Experiment
H	13.4	13.6
He	24.5	24.6
Li	5.7	5.4
Be	9.1	9.3
B	8.8	8.3
C	12.1	11.3
N	15.3	14.5
O	14.2	13.6
F	18.4	17.4
Ne	22.6	21.6
Na	5.6	5.1
Ar	16.2	15.8
K	4.7	4.3

[a]After Ref. 20.
[b]Local spin density approximation.

TABLE 9. Spin Susceptibility of the Alkali Metals[a]

Metal	χ/χ_0^b	
	Variational theory	Experiment
Li	2.66	2.57
Na	1.62	1.65
K	1.79	1.70
Rb	1.78	1.72
Cs	2.20	1.76 or 2.24[c]

[a]After Ref. 54.
[b]χ_0 is the Pauli susceptibility of a free electron gas of the appropriate density.
[c]For Cs the experimental data are consistent with two different values of χ.

3.3. *Illustrations*

Local spin density functional theory has been applied with good success to atoms and molecules with finite spin, and hence finite magnetization density, and to magnetic solids. Following are a few illustrative results:

(a) Atoms. Table 8 gives the ionization potentials of some atoms which were calculated in the LSD approximation as the difference between the neutral and the singly ionized ground states. The average fractional error of ionization energies is 5%. It is interesting to note that errors are similar for the lighter and heavier atoms. *Total* energies are fractionally more accurate for heavier atoms, since the Hartree part of the energy (treated exactly in LSDA) becomes an increasingly large part of the total energy.

(b) Molecules. A calculation of the total energy of the H_2 molecules by Gunnarsson and Lundqvist[(20)] in the LSD approximation gave excellent results for all separation R from 1 to 4 atomic units. Errors are in the range of 0% to 5%.

Calculations of single-particle wave functions and energy levels for many other molecules have also been carried out by the $X\alpha$ method combined with the scattered wave method or other approximate methods for solving the single-particle Schrödinger equations,[(36,52)] but these do not generally include total energies. The practical feasibility of such calculations, even for molecules consisting of 20 or more atoms, is, however, established.

(c) Spin Susceptibility of Solids. Here we mention first of all calculations of the spin susceptibility of paramagnetic metals. This can be obtained by solving the self-consistent LSD equations in the presence of a small uniform magnetic field B and either computing the total magnetization, $M = \chi B$, or the change of total energy, $E = \frac{1}{2}\chi B^2$.

A very useful variational principle has been derived by Vosko, Perdew, and collaborators[(53–56)] which circumvents the need to solve the LSD equations in a magnetic field B and has given results in very good agreement (on the

TABLE 10. Saturation Magnetization of Ferromagnetic Elements

	Magnetization per atom (μ_0)	
Element	LSDA[a]	Experiment
Fe	2.15	2.22
Co	1.56	1.56
Ni	0.59	0.61

[a]Local spin density approximation.

order of 5%) with those of the exact solutions of the self-consistent equations.[57] This variation method also has the merit of giving the susceptibility χ in the physically suggestive form

$$\chi \geq \frac{\chi_s}{1 - I\chi_s} \tag{130}$$

where χ_s is the single-particle susceptibility calculated with the nonmagnetic, self-consistent v_{xc}, and I is an enhancement factor (the so-called Stoner factor) incorporating many-body magnetic effects.

Table 9 gives theoretical results for the alkali metals obtained by the variational method as compared with experimental data:

(d) Ferromagnetism. Spin density calculations of all metals with $3 < Z < 49$ (Li to In) correctly yield ferromagnetic ground states only for the elements Fe, Co, and Ni, (see Fig. 6 in Chapter 4) with magnetization in excellent agreement with experiment.[40] Results are given in Table 10.

The spin wave stiffness coefficient D in ferromagnets has been shown to be a property of the ferromagnetic ground state. It is determined by the small-q behavior of the static transverse magnetic susceptibility $\chi(q)$.[58] Calculations for Ni are in good agreement with experiment: $D_{\text{theory}} = 0.104$ a.u., and $D_{\text{exp}} = 0.113$.[59]

4. Extension to Multicomponent Systems

In the preceding sections we have discussed the ground-state properties of a one-component system. Density functional theory (DFT) has also been used to study the many interesting properties of multicomponent systems. Examples of multicomponent systems where DFT has been applied with success are (i) electron–hole drops in semiconductors,[60–71] and (ii) protons and neutrons in finite nuclei.[72–74]

In the present section we shall first generalize the HK theorem and the KS equations for a multicomponent system.[60,71] Subsequently, we shall discuss the calculations of the surface tension[60–71] and the surface charge on electron–hole drops[62,63,68,70,71]

4.1. Formal Theory

4.1.1. Variational Formulation. Consider an M-component system in the presence of external potentials $v_1(\mathbf{r})$, $v_2(\mathbf{r})$, . . ., $v_M(\mathbf{r})$, which couple, respectively, with the densities $n_1(\mathbf{r})$, $n_2(\mathbf{r})$, . . ., $n_M(\mathbf{r})$ of the different components. The Hamiltonian operator of the system is

$$\mathsf{H} = \mathsf{T} + \mathsf{U} + \mathsf{V} \tag{131}$$

where

$$\mathsf{T} = \sum_{\alpha=1}^{M} -\frac{1}{2m_\alpha} \int d\mathbf{r}\ \psi_\alpha^*(\mathbf{r})\nabla^2\psi_\alpha(\mathbf{r}) \tag{132}$$

$$\mathsf{U} = \frac{1}{2}\sum_{\sigma,\beta=1}^{M} \int d\mathbf{r}\ d\mathbf{r}'\ \psi_\alpha^*(\mathbf{r})\psi_\beta^*(\mathbf{r}')u_{\alpha\beta}(\mathbf{r}-\mathbf{r}')\psi_\beta(\mathbf{r}')\psi_\alpha(\mathbf{r}) \tag{133}$$

$$\mathsf{V} = \sum_{\alpha=1}^{M} \int d\mathbf{r}\ \psi_\alpha^*(\mathbf{r})\ v_\alpha(\mathbf{r})\psi_\alpha(\mathbf{r}) \tag{134}$$

and $u_{\alpha\beta}(\mathbf{r})$ is the interaction potential between particles of αth and βth components. The density operator for the αth component is given by

$$\mathsf{n}_\alpha(\mathbf{r}) = \psi_\alpha^*(\mathbf{r})\psi_\alpha(\mathbf{r}) \tag{135}$$

In analogy to the case of one-component magnetic systems, one finds that the set of densities $\{n_\alpha(\mathbf{r})\}$ uniquely determines the set of external potentials $\{v_\alpha(\mathbf{r})\}$. From this one shows again the existence of an energy functional $E[n_1(\mathbf{r}), n_2(\mathbf{r}), \ldots]$ which assumes its minimum, equal to the correct ground-state energy, for the correct densities $\{n_\alpha(\mathbf{r})\}$. This functional can be written as

$$\begin{aligned} E[n_1,n_2,\ldots] = {} & \sum_{\alpha=1}^{M} \int d\mathbf{r}\ v_\alpha(\mathbf{r})n_\alpha(\mathbf{r}) \\ & + \frac{1}{2}\sum_{\alpha=1}^{M}\sum_{\beta=1}^{M} \int d\mathbf{r}\ d\mathbf{r}' u_{\alpha\beta}(\mathbf{r}-\mathbf{r}')n_\alpha(\mathbf{r})n_\beta(\mathbf{r}') \\ & + G[n_1,n_2,\ldots] \end{aligned} \tag{136}$$

In equation (136) G is a universal functional that depends on the density distributions of the various components, and can be written as

$$G[n_1,n_2, \ldots] = \sum_{\alpha=1}^{M} T_{s,\alpha}[n_\alpha] + E_{xc}[n_1,n_2, \ldots] \tag{137}$$

where $T_{s,\alpha}$ is the noninteracting kinetic energy of the αth component and E_{xc} is the exchange-correlation energy of the system.

4.1.2. Self-Consistent Equations. Using the stationary property of the ground-state energy functional subject to the constancy of the number of particles in each component, we find

$$\begin{aligned} v_\alpha(\mathbf{r}) &+ \sum_{\beta=1}^{M} \int d\mathbf{r}' \, u_{\alpha\beta}(\mathbf{r}-\mathbf{r}') n_\beta(\mathbf{r}') \\ &+ \frac{\delta T_{s,\alpha}[n_\alpha]}{\delta n_\alpha(\mathbf{r})} + \frac{\delta E_{xc}[n_1,n_2, \ldots]}{\delta n_\alpha(\mathbf{r})} = \mu_\alpha \end{aligned} \tag{138}$$

where μ_α denotes the chemical potential of the αth species. From the above equation it is easy to identify the effective potential, v_α^{eff}, felt by a particle of the αth component:

$$\begin{aligned} v_\alpha^{\text{eff}}[\mathbf{r};n_1,n_2, \ldots] &= v_\alpha(\mathbf{r}) + \sum_{\beta=1}^{M} \int d\mathbf{r}' \, u_{\alpha\beta}(\mathbf{r}-\mathbf{r}') n_\beta(\mathbf{r}') \\ &+ v_{xc,\alpha}[n_1,n_2, \ldots] \end{aligned} \tag{139}$$

where

$$v_{xc,\alpha}[n_1,n_2, \ldots] = \frac{\delta E_{xc}[n_1,n_2, \ldots]}{\delta n_\alpha(\mathbf{r})} \tag{140}$$

The density distribution $n_\alpha(\mathbf{r})$ can be obtained from the Kohn–Sham equations for the αth component:

$$\left\{ -\frac{1}{2m_\alpha} \nabla^2 + v_\alpha^{\text{eff}}[\mathbf{r};n_1,n_2, \ldots] \right\} \psi_{\alpha,i}(\mathbf{r}) = \epsilon_{\alpha,i} \psi_{\alpha,i}(\mathbf{r}) \tag{141}$$

with

$$n_\alpha(\mathbf{r}) = \sum_i f_{\alpha,i} \, | \psi_{\alpha,i}(\mathbf{r}) |^2 \tag{142}$$

where $f_{\alpha,i}$ is the Fermi occupation factor (1 or 0) for the species α.

Equations (139), (141), and (142) constitute a set of self-consistent equations. From the knowledge of the eigenfunctions of equation (141) one can obtain the expression for the ground state energy, E:

$$E[n_1,n_2,\ldots] = \sum_{\alpha=1}^{M} \sum_{i} \epsilon_{\alpha,i} - \frac{1}{2} \sum_{\alpha=1}^{M} \sum_{\beta=1}^{M} \int d\mathbf{r}\, d\mathbf{r}'\, u_{\alpha\beta}(\mathbf{r}-\mathbf{r}') n_\alpha(\mathbf{r}) n_\beta(\mathbf{r}') + E_{xc}[n_1,n_2,\ldots] - \sum_{\alpha=1}^{M} \int d\mathbf{r}\, n_\alpha(\mathbf{r}) v_{xc,\alpha}[n_1,n_2,\ldots] \tag{143}$$

4.2. *Approximation for the Exchange-Correlation Energy* $E_{xc}[n_i,n_2,\ldots]$

To solve the Kohn–Sham equations one requires a useful approximation for the exchange-correlation energy functional. The simplest approximation is

$$E_{xc}[n_1,n_2,\ldots] = \int d\mathbf{r}\, e_{xc}(n_1(\mathbf{r}),n_2(\mathbf{r}),\ldots) \tag{144}$$

where $e_{xc}(n_1,n_2,\ldots)$ is the exchange-correlation energy per unit volume* of a homogeneous system whose components have the densities $n_1,n_2,\ldots$. Equation (144) is the local density approximation (LDA) for the exchange-correlation energy.

It is possible to go beyond LDA if one assumes that the densities in a multicomponent system are varying slowly. The, besides the leading-order term in equation (144), one obtains a series of terms in the gradients of the densities. The lowest-order gradient term is of the form

$$\sum_{\alpha=1}^{M} \sum_{\beta=1}^{M} \int d\mathbf{r}\, B_{\alpha\beta}(n_1,n_2,\ldots)\, \nabla n_\alpha(\mathbf{r}) \cdot \nabla n_\beta(\mathbf{r}) \tag{145}$$

for a system with isotropic interactions, $u_{\alpha\beta}(|\mathbf{r}-\mathbf{r}'|)$. As we will discuss in the next section, the coefficients $B_{\alpha\beta}$ have been calculated for the electron–hole liquid.[69]

4.3. *Illustration: Electron–Hole Droplet*

In thermal equilibrium, intrinsic semiconductors such as germanium and silicon are free of charge carriers at low temperatures. By means of a weak

*For one-component systems we used $\epsilon_{xc}(n)$, the exchange-correlation energy per *particle* [see equation (55)].

illumination of sufficient frequency, one can generate a low-density gas of electrons and holes which can bind in pairs as excitons. The number of photoexcited carriers can be controlled by the intensity of the incident light. Experiments in germanium and silicon have revealed that at high densities optically pumped electrons and holes undergo a phase transition.[75–79] The new phase of the system, called electron–hole liquid (EHL), consists of interpenetrating and interacting electron and hole degenerate Fermi liquids. In the spirit of the effective mass approximation (which is very accurate for these systems) carriers can be regarded as particles, characterized by the masses determined by their respective energy bands, and an effective charge which is $e/\sqrt{\kappa}$, where κ is the static dielectric constant of the semiconductor.[81]

Consider N electrons and N holes occupying a volume Ω at $T = 0$ K. The total energy of a homogeneous EHL consists of kinetic, exchange, and correlation contributions.[78–84] If the electrons reside in ν_e equivalent conduction bands, the mean density in each will be $\bar{n}/\nu_e$, where $\bar{n} = N/\Omega$. The mean kinetic energy of an electron is given by

$$t_e = \frac{3}{10}\frac{1}{m_{ed}}\left(3\pi^2\frac{\bar{n}}{\nu_e}\right)^{2/3} \tag{146}$$

where m_{ed} is its density of states mass. If the bands are ellipsoids, as in germanium and silicon, m_{ed} depends on the transverse (m_{et}) and longitudinal (m_{el}) electron masses:

$$m_{ed} = (m_{et}^2 m_{el})^{1/3} \tag{147}$$

The holes in germanium or silicon occupy light and heavy hole bands. For the model of uncoupled light and heavy hole bands, the kinetic energy per hole can be written as

$$t_h = \frac{3}{5}\frac{K_0^2}{2m_{hh}}\left[1 + \left(\frac{m_{lh}}{m_{hh}}\right)^{3/2}\right]^{-2/3} \tag{148}$$

where $K_0^3 \equiv 3\pi^2\bar{n}$, and m_{lh} and m_{hh} are the light and heavy hole masses, respectively.

The calculation of exchange energy of the EHL proceeds very much like that for the homogeneous electron gas. The exchange energy per pair, ϵ_x, is given by

$$\epsilon_x = -\frac{e^2}{\kappa}(\bar{n})^{1/3}\left[\frac{a(m_{et},m_{el})}{\nu_e^{1/3}} + b(m_{lh},m_{hh})\right] \tag{149}$$

where $a(m_{et},m_{el})$ and $b(m_{lh},m_{hh})$ are constants independent of the density. The effects of anisotropy of conduction bands and the coupling of light and heavy hole bands can be included in a and b.[83]

Various calculations of the ground-state energy of the EHL differ in the approximations for the correlation contribution. Most commonly used approximations are: (i) Random phase approximation (RPA),[82] (ii) Nozieres–Pines interpolation scheme,[83] (iii) Hubbard approximation,[84,85] and (iv) fully self-consistent approximation.[86,87] Calculations of the ground-state energy, equilibrium density, isothermal compressibility, as well as thermodynamic properties have been found to be in good agreement with experiments.[67,80]

The surface structure of electron–hole droplets has been studied with the variational procedure of HK and the self-consistent method of KS. In the former scheme, the ground-state energy is approximated by a sum of a local density term and the first gradient correction to it. The densities of electrons and holes are parametrized, and the parameters are determined by minimizing the ground-state energy. The merit of the KS self-consistent equations lies in the exact treatment of the kinetic energy, which leads to a considerable improvement over the variational approach.

Before we illustrate how the KS method is used to study the surface properties of EHL, let us outline the physical model. Since the radius of an electron–hole drop is much larger than the interparticle distance, it is reasonable to model the surface by a semi-infinite system, in which the densities of electron and holes vary only in the z direction. Because in this model there is no external potential, the position and orientation of the surface is indeterminate. This logical difficulty can be overcome by the introduction of an appropriate infinitesimal external potential.

In the local density approximation, equation (139) for the effective potential reads

$$v_{\alpha}^{\text{eff}}[z;n_e,n_h] = \frac{4\pi}{\kappa}\eta_\alpha \int_{-\infty}^{z} dz'\,(z-z')[n_e(z') - n_h(z')] - \bar{\mu}_\alpha(n_e,n_h) + \mu_{xc,\alpha}(n_e,n_h) \tag{150}$$

where $n_{e,h}$ are electron and hole densities, and $\eta_h = -\eta_e = 1$. In the above equation, the first term represents the electrostatic dipole barrier, the second term is the total chemical potential in the bulk, and the last one is the exchange-correlation contribution to the effective potential in the local density approximation, which was evaluated in an approximate manner.

From the eigenvalues of the KS equation, equation (141), and the densities, $n_e(z)$ and $n_h(z)$, one can obtain the ground-state energy of the inhomogeneous EHL from equation (143). The surface tension σ is then found by dividing the difference of the ground-state energies of inhomogeneous and homogeneous EHL, with the same number of particles, by the surface area. In ordinary germanium, the value of σ obtained from the self-consistent calculation is 3.7×10^{-4} erg/cm^2.[71] It agrees well with the latest experimental measurement (3×10^{-4} erg/cm^2).[88] Within the KS scheme, Rose and Shore have also carried out a calculation of surface tension of EHL in Ge by parametrizing the effective potential. These authors find a value of 2.5×10^{-4} erg/cm^2.[70] Calculations of surface tension have also been carried out in Ge under $\langle 111 \rangle$ uniaxial stress and Si under $\langle 100 \rangle$ stress.[71]

Attempts have been made to improve upon the local density approximation (LDA) for the exchange-correlation energy functional. If the density is varying slowly in space, the term beyond LDA consists of density gradients. Rasolt and Geldart have evaluated the coefficients B_{ij} [see equation (145)] for the different components of EHL in the high-density limit.[69] For a model system (single isotropic electron and hole bands),

$$B_{ij}(n_e,n_h) = \frac{C_{ij}(n_e,n_h)}{(n_e n_h)^{2/3}} \tag{151}$$

where n_e and n_h are the electron and hole densities, respectively. For the purpose of illustration, in the special case of $n_e = n_h$ we show in Fig. 9 the dependence of C_{ij} on $\gamma(= m_e/m_h)$, the ratio of electron to hole mass.

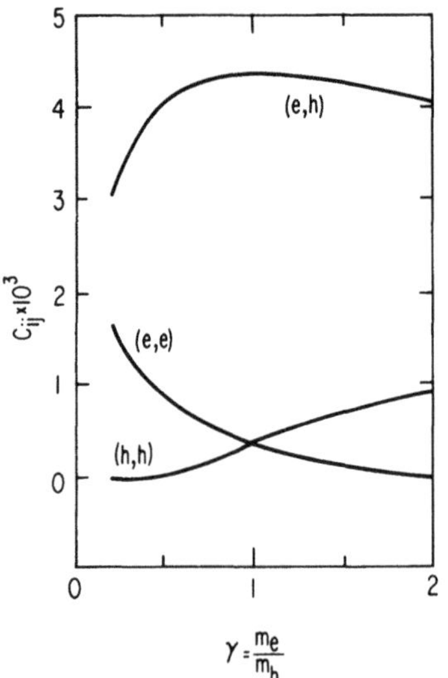

FIGURE 9. The dimensionless functions C_{ij} which determine the coefficients B_{ij} of the leading gradient corrections of an inhomogeneous electron–hole fluid, Equation (151) (after Ref. 69). The electron and hole masses are shown by m_e and m_h, respectively. The values shown are for the high density limit.

The gradient correction, equation (145) calculated with the values of B_{ij}, obtained in Ref. 22 has been estimated for the surface energy of the EHL in Ge.[70] Again, as in the case of the one-component systems discussed in Section 2.2.2., gradient corrections do not seem to lead to more satisfactory theoretical results.

Experimentally the electron–hole droplets in Ge are found to carry a negative charge of the order of $-100|e|$. This can be qualitatively understood as follows. If $\bar{\mu}_e$ and $\bar{\mu}_h$ represent the chemical potentials of electrons and holes in an infinite EHL relative to the internal electrostatic potential, then the difference of the work function is given by

$$\Phi_h - \Phi_e = (\Delta\rho - \bar{\mu}_h) - (-\Delta\rho - \bar{\mu}_e) = \bar{\mu}_e - \bar{\mu}_h + 2\Delta\rho \qquad (152)$$

where $\Delta\rho$ is the dipole potential barrier. If, for example, this quantity is negative, holes will evaporate preferentially and leave the drop negatively charged. Approximate quantitative calculations for Ge[62,70,71] agree in sign and order of magnitude with the experiments of Pokrovsky and Svistunova[89] and of Ugumori *et al.*[90] Density profiles of electrons and holes for electron–hole droplets in Ge are shown in Fig. 10. Under a sufficiently large $\langle 111 \rangle$ stress the droplet charge becomes positive.[89] This sign reversal has been correctly accounted for by the theory.

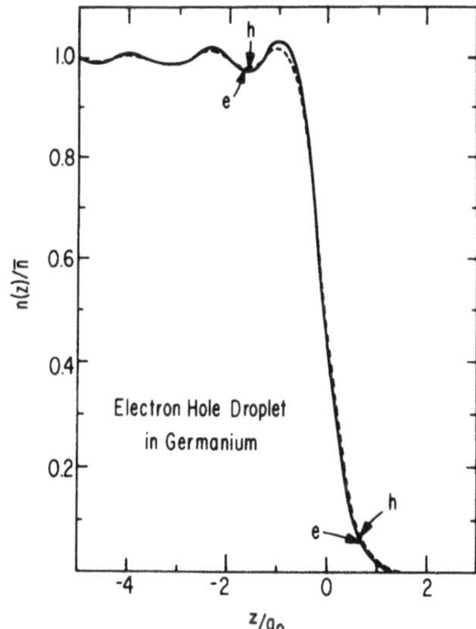

FIGURE 10. Electron and hole density profiles near the surface of an electron–hole droplet in Ge. a_0 is an excitonic Bohr radius (= 177 Å). $\bar{n}$ is the density of electrons and of holes in the interior. (After Ref. 71.)

5. Generalization of Finite Temperature Ensembles

5.1. Formal Theory

The generalization of the HK theorem for a given finite temperature states:

Theorem 1. In a grand canonical ensemble at a given temperature τ the density distribution $n(\mathbf{r})$ uniquely determines the quantity $v(\mathbf{r}) - \mu$, where v is the external potential and μ is the chemical potential.

Theorem 2. For given $v(\mathbf{r})$ and μ, there exists a functional of $n'(\mathbf{r})$,

$$\Omega_{v-\mu}[n'(\mathbf{r})] = \int (v(\mathbf{r}) - \mu)\, n'(\mathbf{r})\, d\mathbf{r} + F[n'(\mathbf{r})] \tag{153}$$

which has the following properties: It is an absolute minimum when $n'(\mathbf{r})$ is the correct $n(\mathbf{r})$ corresponding to $v(\mathbf{r})$, and the minimum value is equal to the grand potential. The functional F is a universal, τ-dependent functional of $n'(\mathbf{r})$ only.

We now demonstrate these theorems, following the paper of N. D. Mermin.[91] Consider a system at a given temperature, with $\beta = (k_B\tau)^{-1}$, and given $v(\mathbf{r})$ and μ. Its grand canonical density matrix operator is given by

$$\boldsymbol{\rho} \equiv e^{-\beta(\mathsf{H}-\mu\mathsf{N})} / \mathrm{Tr}\, e^{-\beta(\mathsf{H}-\mu\mathsf{N})} \tag{154}$$

where

$$\mathsf{H} = \mathsf{T} + \mathsf{V} + \mathsf{U} \tag{155}$$

and N is the number operator. The grand potential, which clearly depends only on $v(\mathbf{r}) - \mu$, is given by

$$\Omega_{v-\mu} = \mathrm{Tr}\, \rho\left(\mathsf{H} - \mu\mathsf{N} + \frac{1}{\beta}\log\rho\right) \tag{156}$$

We now define the following functional of an arbitrary, normalized density matrix ρ' taken in the $\mathbf{r}$ representation,

$$\Omega_{v-\mu}[\rho'] \equiv \mathrm{Tr}\, \rho'\left(\mathsf{H} - \mu\mathsf{N} + \frac{1}{\beta}\log\rho'\right) \tag{157}$$

Then, following the work of Gibbs and von Newmann, Mermin [91] has shown that, for $\rho' \neq \rho$

$$\Omega_{v-\mu}[\rho'] > \Omega_{v-\mu}[\rho] \tag{158}$$

i.e., the absolute minimum of the functional in equation (157) with respect to ρ' occurs at the correct ρ and is equal to the grand potential.

Now let $n(\mathbf{r})$ be the density corresponding to $v(\mathbf{r})-\mu$, and assume that there is a function $v'(\mathbf{r})-\mu'$, different from $v(\mathbf{r})-\mu$, giving rise to the same density $n(\mathbf{r})$. We shall show that this is not possible.

The density matrix operator associated with $v'(\mathbf{r})-\mu'$ is given by

$$\rho' = e^{-\beta(\mathrm{H}'-\mu'\mathrm{N})}/\mathrm{Tr}\, e^{-\beta(\mathrm{H}'-\mu'\mathrm{N})} \tag{159}$$

where

$$\mathrm{H}' = \mathrm{H} + \mathrm{V}' - \mathrm{V} \tag{160}$$

Then

$$\begin{aligned} \Omega_{v'-\mu'}[\rho'] &= \mathrm{Tr}\,\rho'(\mathrm{H}' - \mu'\mathrm{N} + \frac{1}{\beta}\log\rho') \\ &= \int \{(v'(\mathbf{r}) - \mu') - (v(\mathbf{r}) - \mu)\}n(\mathbf{r})\,d\mathbf{r} \\ &\quad + \mathrm{Tr}\,\rho'(\mathrm{H} - \mu\mathrm{N} + \frac{1}{\beta}\log\rho') \\ &= \int \{[v'(\mathbf{r}) - \mu'] - [v(\mathbf{r}) - \mu]\}n(\mathbf{r})\,d\mathbf{r} \\ &\quad + \Omega_{v-\mu}[\rho'] \\ &> \int \{[v'(\mathbf{r}) - \mu'] - [v(\mathbf{r}) - \mu]\}n(\mathbf{r})\,d\mathbf{r} \\ &\quad + \Omega_{v-\mu}[\rho] \end{aligned} \tag{161}$$

where in the last step we have used the inequality in equation (158). Now, interchanging primed and unprimed quantities, but recalling that, by assumption, $n'(\mathbf{r}) = n(\mathbf{r})$, we obtain

$$\begin{aligned} \Omega_{v-\mu}[\rho] &> \int \{(v(\mathbf{r}) - \mu) - (v'(\mathbf{r}) - \mu')\}\, n(\mathbf{r})\,d\mathbf{r} \\ &\quad + \Omega_{v'-\mu'}[\rho'] \end{aligned} \tag{162}$$

Adding equations (161) and (162) gives the self-contradictory result

$$\Omega_{v-\mu}[\rho] + \Omega_{v'-\mu'}[\rho'] > \Omega_{v-\mu}[\rho] + \Omega_{v'-\mu'}[\rho'] \tag{163}$$

showing that the assumption $n'(\mathbf{r}) = n(\mathbf{r})$ was false. This proves theorem (1) above.

We now turn to Theorem 2. Referring to the definition of the grand potential in equation (157), and recalling that $n'(\mathbf{r})$ uniquely determines $v'(\mathbf{r}) - \mu'$ and ρ', we define

$$\begin{aligned} F[n'(\mathbf{r})] &= \operatorname{Tr} \rho' \left[\mathsf{T} + \mathsf{U} - \frac{1}{\beta} \log \rho' \right] \\ &= T[n'(\mathbf{r})] + U[n'(\mathbf{r})] - \tau S[n'(\mathbf{r})] \end{aligned} \tag{164}$$

where T, U, and S are, respectively, the kinetic energy, interaction energy, and entropy—all functionals of $n'(\mathbf{r})$ only (at a given temperature). Thus $\Omega_{v-\mu}[\rho']$ in fact depends only on $v-\mu$ and n', and can be written as

$$\Omega_{v-\mu}[n'(\mathbf{r})] \equiv \int [v(\mathbf{r}) - \mu]\, n'(\mathbf{r})\, d\mathbf{r} + F[n'] \tag{165}$$

The inequality in equation (158) can therefore be written as

$$\Omega_{v-\mu}[n'(\mathbf{r})] > \Omega_{v-\mu}[n(\mathbf{r})] \tag{166}$$

where the right-hand side is the correct grand potential corresponding to $v(\mathbf{r}) - \mu$. This proves Theorem 2.

Let us note that these theorems are free of two limitations of the corresponding ground-state theorems (Section 2.1.1): no assumption of nondegeneracy is required; and, for fixed τ, the function $v(\mathbf{r}) - \mu$ is uniquely determined by $n(\mathbf{r})$ (not only to within an additive constant).*

It is now straightforward to derive self-consistent equations for a given, τ analogous to the KS equations at $\tau = 0$ (Section 2.2.1).[6] We define $G_s[n'(\mathbf{r})]$ as the Helmholtz free energy of noninteracting electrons with density distribution $n(\mathbf{r})$. Thus, denoting the kinetic energy and entropy of such a system by T_s and S_s,

$$G_s[n(\mathbf{r})] = T_s[n(\mathbf{r})] - \tau S_s[n(\mathbf{r})] \tag{167}$$

*The footnote 4 in Ref. 91 is in error: $n(\mathbf{r})$ determines only $v(\mathbf{r}) - \mu$ and not $v(\mathbf{r})$ and μ separately. Also, in equation (9) in Ref. 91, $v(\mathbf{r})$ should be replaced by $v(\mathbf{r}) - \mu$.

Then we write

$$\Omega_{v-\mu}[n'(\mathbf{r})] \equiv \int (v(\mathbf{r})-\mu)n'(\mathbf{r})\,d\mathbf{r} + G_s[n'(\mathbf{r})] + \frac{1}{2}\int \frac{n'(\mathbf{r})n'(\mathbf{r}')}{|\mathbf{r}-\mathbf{r}'|}\,d\mathbf{r}\,d\mathbf{r}' + F_{xc}[n'(\mathbf{r})] \tag{168}$$

which defines $F_{xc}[n'(\mathbf{r})]$ as the exchange-correlation contribution to the free energy,

$$F[n'(\mathbf{r})] = \frac{1}{2}\int \frac{n'(\mathbf{r})n'(\mathbf{r}')}{|\mathbf{r}-\mathbf{r}'|}\,d\mathbf{r}\,d\mathbf{r}' + G_s[n'(\mathbf{r})] + F_{xc}[n'(\mathbf{r})] \tag{169}$$

The stationary property of equation (168) with respect to $n'(\mathbf{r})$ now results in the equation

$$0 = (\phi(\mathbf{r}) - \mu) + \frac{\delta G_s[n']}{\delta n'(\mathbf{r})} + v_{xc}[n'] \tag{170}$$

where

$$\phi(\mathbf{r}) \equiv v(\mathbf{r}) + \int \frac{n'(\mathbf{r}')}{|\mathbf{r}-\mathbf{r}'|}\,d\mathbf{r}' \tag{171}$$

is the total electrostatic potential, and

$$v_{xc}[n'] \equiv \frac{\delta F_{xc}[n']}{\delta n'(\mathbf{r})} \tag{172}$$

is the exchange-correlation contribution to the total effective potential

$$v_{\text{eff}}(\mathbf{r}) \equiv v(\mathbf{r}) + \int \frac{n(\mathbf{r}')}{|\mathbf{r}-\mathbf{r}'|}\,d\mathbf{r}' + v_{xc}[n'] \tag{173}$$

The equation (170) has the same form as the stationary condition for a noninteracting gas of electrons in an external potential $v(\mathbf{r})$, namely,

$$0 = (v(\mathbf{r})-\mu) + \frac{\delta G_s[n']}{\delta n'(\mathbf{r})} \tag{174}$$

with the replacement

$$v(\mathbf{r}) \rightarrow v_{\text{eff}}(\mathbf{r}) = \phi(\mathbf{r}) + v_{xc}[n'] \tag{175}$$

Since the solution of the noninteracting problem is known in terms of the eigenfunctions and eigenvalues of the single-particle Hamiltonian, we can write down the solution of equation (170) in terms of the following self-consistent equations:

$$[-\tfrac{1}{2}\nabla^2 + v_{\text{eff}}(\mathbf{r}) - \mu]\psi_i = \epsilon_i\psi_i \tag{176}$$

$$n(\mathbf{r}) = \sum_i |\psi_i(\mathbf{r})|^2 f(\epsilon_i - \mu) \tag{177}$$

$$v_{\text{eff}}(\mathbf{r}) = v(\mathbf{r}) \int \frac{n(\mathbf{r}')}{|\mathbf{r}-\mathbf{r}'|}\, d\mathbf{r}' + \frac{\delta F_{xc}[n]}{\delta n(\mathbf{r})} \tag{178}$$

where $f(\epsilon_i - \mu)$ if the Fermi function

$$f(\epsilon_i - \mu) \equiv 1/[1 + e^{\beta(\epsilon_i - \mu)}] \tag{179}$$

The grand potential can then in principle be calculated from equation (168). The single-particle quantity G_s, defined in (167), presents no difficulty. The kinetic energy part is given by

$$T_s = \sum_i \epsilon_i f(\epsilon_i - \mu) - \int v_{\text{eff}}(\mathbf{r}) n(\mathbf{r})\, d\mathbf{r} \tag{180}$$

and the entropy by

$$S_s = -k_B \sum_i \{f_i \ln f_i + (1 - f_i) \ln (1 - f_i)\} \tag{181}$$

The difficult many-body quantity, $F_{xc}[n]$, analogous to $E_{xc}[n']$ for the ground state, can be approximated by the LDA,

$$F_{xc}[n(\mathbf{r})] = \int n(\mathbf{r}) f_{xc}(n(\mathbf{r}))\, d\mathbf{r} \tag{182}$$

where $f_{xc}(n)$ is the exchange-correlation contribution per particle to the free energy of a homogeneous electron gas of density n at temperature τ. At low temperatures

$$f_{xc}(n) = \epsilon_{xc}(n) - \frac{\gamma_{xc}(n)}{n}\frac{\tau^2}{2} + \cdots \tag{183}$$

where γ_{xc} is the exchange-correlation contribution to the low-temperature specific heat coefficient per unit volume,

$$\gamma(n) = \lim_{\tau\to 0}\frac{C_v(n)}{\tau} \tag{184}$$

Many-body calculations of $\gamma(n)$ are reported in Ref. 92. To our knowledge calculations of $f_{xc}(n)$ for higher temperatures have not yet been made and would be interesting for a discussion of partially degenerate plasmas. In the high-temperature limit

$$\lim_{\tau\to 0} f_{xc}(n) = 0 \tag{185}$$

5.2. *Illustration: Metal–Insulator Transition*

Ghazali and Hugon[93] have used the finite-temperature DFT to study the metal–insulator transition in doped semiconductors. They used a model of a semiconductor with a single isotropic conduction band minimum at $\mathbf{k} = 0$ with effective mass m^* and dielectric constant κ. The effective potential was written as

$$v_{\rm eff}(\mathbf{r}) = -\frac{e^2}{\kappa}\sum_j \frac{1}{|\mathbf{r}-\mathbf{R}_j|} + \frac{e^2}{\kappa}\int d\mathbf{r}'\frac{n(\mathbf{r}')}{|\mathbf{r}-\mathbf{r}'|} + v_{xc}(n(\mathbf{r})) \tag{186}$$

where the first two terms arise from the electron–donor and electron–electron interactions and v_{xc} was approximated by an appropriate zero-temperature form. Equations (176), (177), and (186) were solved self-consistently in the cellular approximation.

Figure 11 shows the dependence of the free carrier density, n_f, on the impurity concentration, n_I, at various reduced temperatures, Θ. At low values of n_I there is an energy gap between localized states and extended impurity band states. At $\Theta = 0$, the system is insulating for $n_I \leq 0.28$ and becomes metallic for $n_I > 0.28$, corresponding to the closing of the gap. At $\Theta > 0$, the gap closes at higher values of n_I. For $n_I > 0.5$ the system is metallic at all temperatures.

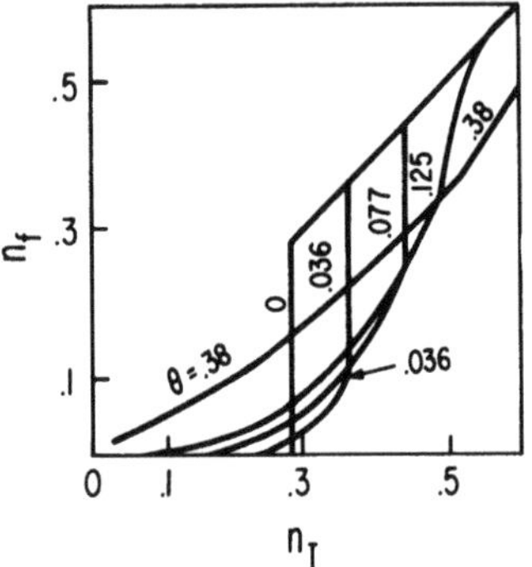

FIGURE 11. Free carrier density, n_f, in a doped semiconductor, as a function of donor concentration, n_I, for various temperatures, Θ. Densities are measured in units of $n_0 = (\pi/3)\,(4a_0)^{-3}$, where a_0 is the effective Bohr radius ($= \kappa\hbar^2/m^*e^2$) and the temperature is measured in units of effective Rydberg ($= e^2/2\kappa a_0$).

6. Excited States and Transport Coefficients*

6.1. Excited States

In the previous sections we have described DFT for ground states and thermal ensembles. In the present section we shall deal with individual or finite groups of excited states.

We begin with some strict formal remarks. For simplicity we consider systems which have a nondegenerate ground state in an external potential $v(\mathbf{r})$. We have seen that for such systems the density distribution $n(\mathbf{r})$ of the ground state determines the external potential $v(\mathbf{r})$, and thus the total Hamiltonian, as well as, of course, the total number of particles N. Thus implicitly all excited states Ψ_j and energies E_j are unique functionals of $n(\mathbf{r})$.

6.1.1. Green's Function Theory. In many systems so-called excited states are not really strict eigenstates of the Hamiltonian but rather more or less well defined resonances in a continuum background. Examples are Auger-unstable excited states in atoms and electron–hole pairs in metals. A convenient formalism to deal with both strict excited states and resonances is Green's function theory.

The m-particle ground-state Green's function is defined as

$$\begin{aligned} G_n(1, \ldots, m; 1', \ldots, m') \\ = (-i)^m \langle \Psi \,| T\, \psi(1) \ldots \psi(m)\psi^*(1') \ldots \psi^*(m')|\, \Psi\rangle \quad (187) \end{aligned}$$

where the symbol "1" stands for $(\mathbf{r}_1, t_1, \alpha_1)$, etc; Ψ is the ground state, and T is the time ordering operator. As the definition makes evident, the Green's functions are all functionals of $n(\mathbf{r})$.

*This entire section is limited to one-component systems.

As example, consider the one-particle Green's function which, when Fourier transformed with respect to $t'-t$, has the form[17,94]

$$G_{\alpha\beta}(\mathbf{r},\mathbf{r}';E) = \sum_{l} \frac{A_{\alpha\beta}(\mathbf{r},\mathbf{r}';E_l)}{E-E_l} \tag{188}$$

where E_l are all the $N \pm 1$ particle states connected to the ground state by $\psi^*(\mathbf{r})$ or $\psi(\mathbf{r})$ and A is a spectral function. It has well-defined poles at isolated eigenvalues E_l, and resonance peaks near the energies E of long-lived quasieigenstates. Similarly, the two-particle Green's function, when appropriately energy-analyzed, will exhibit poles or resonance peaks at the eigenvalues or resonances of N, $N+2$, and $N-2$ particle states.

How can DFT be used to actually calculate Green's functions for inhomogeneous systems? We take the one-particle Green's function G as the simplest example, discussed in Ref. 17. This function satisfies the Dyson equation

$$(-\tfrac{1}{2}\nabla^2 - E)G(\mathbf{r},\mathbf{r}';E) + \int d\mathbf{r}''\, \Sigma(\mathbf{r},\mathbf{r}''; E)G(\mathbf{r}'',\mathbf{r}'; E) = \delta(\mathbf{r}-\mathbf{r}') \tag{189}$$

where Σ is the mass operator and spin indices have been suppressed. Now, for a system of uniform density n, Σ has the form

$$\Sigma(\mathbf{r},\mathbf{r}'';E) = \Sigma_u(|\mathbf{r}-\mathbf{r}''|; E; n) \tag{190}$$

and is a *short-range* function of $|\mathbf{r}-\mathbf{r}''|$. Therefore, for a system of slowly varying density, one may use the following approximation:

$$\Sigma(\mathbf{r},\mathbf{r}'';E) = \Sigma_u(|\mathbf{r}-\mathbf{r}''|, E - \phi(\mathbf{r}_0); n(\mathbf{r}_0)) \tag{191}$$

where $\mathbf{r}_0 \equiv (\mathbf{r} + \mathbf{r}'')/2$. This is an LDA for Σ. As in previous sections we can use available calculations for the uniform system, in this case of $\Sigma(\mathbf{k},E;n)$, the spatial Fourier transform of $\Sigma_u(\mathbf{r}-\mathbf{r}'';E;n)$. Eigenvalues of the $(N \pm 1)$-particle system are given by

$$(-\tfrac{1}{2}\nabla^2 - E)\,\chi(\mathbf{r}) + \int d\mathbf{r}''\, \Sigma_u(|\mathbf{r}-\mathbf{r}''|, E-\phi(\mathbf{r}_0); n(\mathbf{r}_0))\,\chi(\mathbf{r}'') = 0 \tag{192}$$

For further details see Ref. 17.

6.1.2. Excited State Functionals. In Sections 2 and 3 we established the existence of an energy functional, $E[\nu(\mathbf{r})]\,\{\nu(\mathbf{r}) \equiv [n(\mathbf{r}), m(\mathbf{r})]\}$ whose minimum was reached for the correct $\nu(\mathbf{r})$ and equaled the ground-state energy. We also showed how the minimization process can be carried out by solving self-consistent equations. In this section we show that for systems whose symmetry admits specification of quantum numbers λ, these considerations can be partially, but only *very partially*, generalized to the lowest energy state associated with given λ.

(a) The Variational Principle. First of all one can show, precisely as in Sections 2 and 3, that if $\nu_\lambda(\mathbf{r})$ corresponds to the lowest state with quantum numbers λ (the λ-ground state) in the external potential $w_\lambda(\mathbf{r})$ $\{\equiv [v_\lambda(\mathbf{r}), b_\lambda(\mathbf{r})]\}$ there exists no other potential $w'_\lambda(\mathbf{r})$ [except $v'(\mathbf{r}) = v(\mathbf{r}) + \text{const}$] leading to the same λ and $\nu_\lambda(\mathbf{r})$. Thus, for given λ, $\nu_\lambda(\mathbf{r})$ completely determines the external potential and hence the λ-ground state and all other electronic properties.

Note, however, that whereas for the unrestricted ground state all "reasonable" functions $\nu(\mathbf{r})$ could be attained with suitable potentials $w(\mathbf{r})$, the restriction to quantum numbers λ limits the class of admissible $w_\lambda(\mathbf{r})$ and $\nu_\lambda(\mathbf{r})$ drastically. For example, the specification $L = 0$ and $S = 0$ requires $v(\mathbf{r})$ and $n(\mathbf{r})$ to be spherically symmetric and $\mathbf{b}(\mathbf{r})$ and $\mathbf{m}(\mathbf{r})$ to be zero. To the best of our knowledge, there exists no general theory which determines the functional forms of $\nu_\lambda(\mathbf{r})$ compatible with a given set of quantum numbers λ. As a working hypothesis one may postulate that the set of densities $\nu_\lambda(\mathbf{r})$ for an interacting system is the same as for a noninteracting system and the same λ. We shall parametrize the potentials and corresponding densities compatible with λ by a set of variables, $p \equiv \{p_1, p_2, \ldots\}$, which for convenience we take to be denumerable. In what follows a *fixed* set λ will be assumed but, in general, not explictly written.

The stationary energy functional for a given potential $w(\mathbf{r};p)$ can be written as

$$E_{w(\mathbf{r};p)}[\nu(\mathbf{r};p')] = \int [v(\mathbf{r};p')n(\mathbf{r};p) - \mathbf{b}(\mathbf{r};p) \cdot \mathbf{m}(\mathbf{r};p')]d\mathbf{r} + F[\nu(\mathbf{r};p')] \quad (193)$$

In fact the left-hand side can be expressed as $E_p(p')$ and the F-functional as $F(p')$. It is given by

$$F(p') = (\Psi_p', (\mathsf{T} + \mathsf{U})\Psi_p') \quad (194)$$

Note that $F[\nu]$, for given λ, is in general *not* the same functional of ν as for the unrestricted ground state. It should, more completely, be denoted by $F_\lambda[\nu]$. Thus the same density $\nu_\lambda(\mathbf{r})$ can in general be associated not only with quantum numbers λ and the potential $w_\lambda(\mathbf{r})$, but also with the absolute ground state in a different potential $w(\mathbf{r})$. The two functionals F have different values for these two cases, unless λ happen to be the quantum numbers of the absolute ground state corresponding to $\nu_\lambda(\mathbf{r})$.

For given p, and hence given $w(\mathbf{r};p)$, the expression given in equation (193) is stationary with respect to variation of p' and attains its minimum, the desired λ-ground state energy, when $p' = p$, i.e., for the density $\nu(\mathbf{r};p)$. As before it is practical to separate out from $F[\nu]$ the classical Coulomb interaction and write

$$F[\nu] = \frac{1}{2}\int \frac{n(\mathbf{r})n(\mathbf{r}')}{|\mathbf{r}-\mathbf{r}'|}\, d\mathbf{r}\, d\mathbf{r}' + G[\nu] \tag{195}$$

The main difficulty in applying equations (193) and (195) directly for a determination of ν and E is our lack of knowledge of the functional $G[\nu]$ for given λ. To our knowledge no actual calculation for excited states has been carried out by direct minimization of $E_w[\nu]$.

(b) One-Particle Equations. To develop one-particle equations, as in Section 2.2 we first define $T_s[\nu(\mathbf{r};p)]$, the kinetic energy of a noninteracting system of the given density $\nu(\mathbf{r};p)$. Thus we must take a single particle potential, $w_s(\mathbf{r})$, then solve the equations

$$\{-\tfrac{1}{2}\nabla^2 + v_s(\mathbf{r};p) - \sigma\mu_0 b_s(\mathbf{r};p)\}\psi_{j\sigma} = \epsilon_{j\sigma}\psi_{j\sigma} \tag{196}$$

and form the noninteracting λ-ground state. If the density $\nu_s(\mathbf{r})$ is equal to the given density $\nu(\mathbf{r};p)$, this is the right w and will be denoted by $w_s(\mathbf{r},p)$. Call the corresponding single-particle energy $E_s[\nu(\mathbf{r};p)]$. Then T_s is defined as

$$\begin{aligned} T_s[\nu(\mathbf{r};p)] = E_s\,[\nu(\mathbf{r};p)] - \textstyle\int [v_s(\mathbf{r};p)n(\mathbf{r},p) \\ - \mathbf{b}_s(\mathbf{r};p)\cdot\mathbf{m}\,(\mathbf{r};p)]\, d\mathbf{r} \end{aligned} \tag{197}$$

We now define $E_{xc}[\nu]$ (more completely $E_{xc;\lambda}[\nu]$) by

$$G[\nu] = T_s[\nu] + E_{xc}[\nu] \tag{198}$$

so that the stationary energy functional in equation (193) becomes

$$\begin{aligned} E_{w(\mathbf{r},p)}[\nu(\mathbf{r};p')] = \int [v(\mathbf{r};p)n(\mathbf{r};p') - \mathbf{b}(\mathbf{r};p)\cdot\mathbf{m}(\mathbf{r};p')]\, d\mathbf{r} \\ + T_s[\nu(\mathbf{r};p')] + \frac{1}{2}\int \frac{n(\mathbf{r};p')n(\mathbf{r}';p')}{|\mathbf{r}-\mathbf{r}'|}\, d\mathbf{r}\, d\mathbf{r}' \\ + E_{xc}[\nu(\mathbf{r};p')] \end{aligned} \tag{199}$$

Our previous remarks concerning the lack of knowledge of $F_\lambda[\nu]$ apply also to $E_{xc,\lambda}[\nu]$. In particular, $E_{xc;\lambda}[\nu]$ is, for given ν, not the same as the $E_{xc}[\nu]$ for the same ν and the unrestricted ground state.

We now vary, in the stationary expression (199), the density $\nu(\mathbf{r};p')$ near $p' = p$ by varying the single-particle potential $w_s(\mathbf{r};p')$. Note that, from equation (197) and first-order perturbation theory,

$$\delta T_s[\nu(\mathbf{r};p)] = -v_s(\mathbf{r};p)\delta n(\mathbf{r};p) + \mathbf{b}_s(\mathbf{r};p)\cdot\delta\mathbf{m}(\mathbf{r};p) \tag{200}$$

Therefore

$$\begin{aligned} 0 = \delta E_{w(\mathbf{r};p)}[\nu(\mathbf{r};p)] &= \int \{[v(\mathbf{r};p) - v_s(\mathbf{r};p)]\, \delta n(\mathbf{r};p) \\ &- [\mathbf{b}(\mathbf{r};p) - \mathbf{b}_s(\mathbf{r};p)]\cdot\delta\mathbf{m}(\mathbf{r};p)\}\, d\mathbf{r} + \int \delta n(\mathbf{r};p)\, d\mathbf{r} \int \frac{n(\mathbf{r}';p)}{|\mathbf{r}-\mathbf{r}'|}\, d\mathbf{r}' \\ &+ \sum_j \frac{\partial E_{xc}[\nu(\mathbf{r};p)]}{\partial p_j}\, \delta p_j \end{aligned} \tag{201}$$

or

$$\begin{aligned} 0 = &\int [\phi(\mathbf{r};p) - v_s(\mathbf{r};p)] \frac{\partial n(\mathbf{r};p)}{\partial p_j}\, d\mathbf{r} - \int [\mathbf{b}(\mathbf{r};p) \\ &- \mathbf{b}_s(\mathbf{r};p)]\cdot \frac{\partial \mathbf{m}(\mathbf{r};p)}{\partial p_j}\, d\mathbf{r} + \frac{\partial E_{xc}[\nu(\mathbf{r};p)]}{\partial p_j} \end{aligned} \tag{202}$$

where $\phi(\mathbf{r};p)$ is the total electrostatic potential,

$$\phi(\mathbf{r};p) = v(\mathbf{r};p) + \int \frac{n(\mathbf{r}';p)}{|\mathbf{r}-\mathbf{r}'|}\, d\mathbf{r}' \tag{203}$$

The main difference, compared to the discussion of the absolute ground state, is that one cannot take a variation, $\delta\nu(\mathbf{r})$ at a given point $\mathbf{r}$, since, if $\nu(\mathbf{r})$ is compatible with λ, $\nu(\mathbf{r}) + \delta\nu(\mathbf{r})$ is not. Therefore we *cannot* conclude, as we did earlier, that v_s has the form

$$v_s(\mathbf{r};p) = \phi(\mathbf{r};p) + v_{xc}(\mathbf{r};p) \tag{204}$$

$$\mathbf{b}_s(\mathbf{r};p) = \mathbf{b}(\mathbf{r};p) + \mathbf{b}_{xc}(\mathbf{r};p) \tag{205}$$

where v_{xc} and $\mathbf{b}_{xc}$ are obtained from E_{xc} by making variations of n and $\mathbf{m}$ at a given point $\mathbf{r}$.

The nature of the difficulties can be exemplified by taking for E_{xc} a λ-independent local approximation (a logically incorrect step):

$$E_{xc} = \int \epsilon_{xc}(n(\mathbf{r}), \mathbf{m}(\mathbf{r}))n(\mathbf{r})\, d\mathbf{r} \tag{206}$$

Then, in equation (201),

$$\sum_j \frac{\partial E_{xc}[\nu(\mathbf{r};p)]}{\partial p_j}\delta p_j = \int \frac{\partial[\epsilon_{xc}(n(\mathbf{r};p), \mathbf{m}(\mathbf{r};p))n(\mathbf{r};p)]}{\partial n(\mathbf{r};p)}\delta n(\mathbf{r};p) + \sum_i \int \frac{\partial[\epsilon_{xc}(n(\mathbf{r};p), \mathbf{m}(\mathbf{r};p))n(\mathbf{r};p)]}{\partial m_i(\mathbf{r};p)}\delta m_i(\mathbf{r};p) \tag{207}$$

Consequently, equation (201) could be satisfied by

$$v_s(\mathbf{r};p) = \phi(\mathbf{r};p) + \frac{\partial[\epsilon_{xc}(n(\mathbf{r};p), \mathbf{m}(\mathbf{r};p))n(\mathbf{r};p)]}{\partial n(\mathbf{r};p)} \tag{208}$$

$$b_{s,i}(\mathbf{r};p) = b_i(\mathbf{r};p) + \frac{\partial[\epsilon_{xc}(n(\mathbf{r};p), \mathbf{m}(\mathbf{r};p))n(\mathbf{r};p)]}{\partial m_i(\mathbf{r};p)} \tag{209}$$

Since, in line with earlier considerations, there is a unique w_s reproducing the given $\nu(\mathbf{r};p)$, this would seem to be the unique effective single-particle potential. However, in general, the above expressions for v_s and b_s are not compatible with the given λ. [As an example, consider a single-particle p-state, whose density has the form $z^2Q(r)$.]

*We conclude that self-consistent equations compatible with given quantum numbers λ do not, in general, exist.**

What *can* logically be done is the following: a trial $w_s(\mathbf{r})$ of the proper symmetry is chosen, giving rise to single-particle states by means of the Schrödinger equation (196), from which the noninteracting λ-ground state is formed. The corresponding density, $\nu(\mathbf{r})$, is calculated, as is $T_s[\nu]$, from equation (197). Finally, if some reasonable approximation for $E_{xc}[\nu]$ is known the energy is evaluated from equation (199), $w_s(\mathbf{r})$ is then varied to minimize the total energy. In practice, faute de mieux, the local density approximation for $E_{xc}[\nu]$, equation (105), appropriate to the absolute ground state is generally used.

*This conclusion is contrary to that of Ref. 20.

This procedure, which takes the place of solving self-consistent equations, strictly respects the imposed quantum numbers. The use of the LDA form of $E_{xc}[\nu]$, appropriate to the absolute ground state rather than to λ, needs to be critically examined in each case; it is the weak link of this procedure.

6.2. Transport Coefficients

In this section we shall consider the linear response of an inhomogeneous Fermi system, of density distribution $n(\mathbf{r})$ in its ground state, to an arbitrary electric field $E(\mathbf{r},t)$.

6.2.1. Formal Green's Function Theory. The response function which we shall study is the current-field response function, the conductivity tensor σ_{ij}, defined by the equation

$$j_i(\mathbf{r},t) = \int_{-\infty}^{t} dt' \int d\mathbf{r}' \sum_j \sigma_{ij}(\mathbf{r},\mathbf{r}'; t-t')\, E_j(\mathbf{r}',t') \tag{210}$$

where j_i is the current density per unit volume. Fourier analyzing with respect to time by writing

$$j_i(\mathbf{r},t) = \frac{1}{2\pi}\int j_i(\mathbf{r},\omega)\, e^{-i\omega t}\, d\omega, \qquad \text{etc.} \tag{211}$$

results in

$$j_i(\mathbf{r},\omega) = \int \sigma_{ij}(\mathbf{r},\mathbf{r}';\omega) E_j(\mathbf{r}',\omega)\, d\mathbf{r}' \tag{212}$$

Now it is well known from the fluctuation–dissipation theorem that the conductivity tensor is given by the current–current correlation function in the ground state[95]

$$\sigma_{ij}(\mathbf{r},\mathbf{r}';\omega) = -\frac{1}{i\omega}\delta_{ij} - \frac{1}{\omega}\int_0^{\infty} dt\, e^{-i\omega t}\, \langle \mathrm{j}_i(\mathbf{r},\mathrm{o})\, \mathrm{j}_j(\mathbf{r}',t)\rangle \tag{213}$$

where the current operator $\mathbf{j}_i(\mathbf{r},t)$ is given by

$$\mathrm{j}_i(\mathbf{r},t) = -\psi^*(\mathbf{r},t)\frac{1}{i}\frac{\partial}{\partial x_i}\psi(\mathbf{r},t) \tag{214}$$

The conductivity tensor is thus the ground-state expectation value of an operator quadratic in both ψ and a ψ^* and can be expressed in terms of the two-particle Green's function, a property of the ground state Ψ. Since the latter is, by virtue of the HK theory, a functional of the unperturbed ground-state density $n(\mathbf{r})$, we see that $\sigma_{ij}(\mathbf{r},r';\omega)$ is a functional of the density $n(\mathbf{r})$.

From the current response, one can directly obtain the density response using the continuity equation

$$n_1(\mathbf{r},\omega) = -\frac{1}{\omega}\,\mathrm{div}\, j(\mathbf{r},\omega) \tag{215}$$

For arbitrary $n(\mathbf{r})$ and $\mathbf{E}(\mathbf{r},t)$ one can only state the following general properties of σ_{ij}:

$$\sigma_{ij}\,(\mathbf{r},\mathbf{r}';t) = \sigma_{ji}(\mathbf{r}',\mathbf{r};-t) \tag{216}$$

$$\sigma_{ij}\,(\mathbf{r},\mathbf{r}';t) = 0, \qquad t < 0 \tag{217}$$

$$\sigma_{ij}\,(\mathbf{r},\mathbf{r}';\omega) \qquad \text{is analytic in the upper half } \omega \text{ plane} \tag{218}$$

However, in certain limiting cases, the functional form of σ_{ij} can be expressed in terms of the functional form of σ_{ij} for homogeneous systems. Thus for large ω, the local free electron gas behavior is approached

$$\omega \to \infty : \sigma_{ij}(\mathbf{r},\mathbf{r}';\omega) \to -\frac{n(\mathbf{r})}{i\omega}\,\delta\,(\mathbf{r}-\mathbf{r}')\delta_{ij} \tag{219}$$

When the field E is highly oscillating in space, with characteristic wave number $q >> k_F$, its effect on the electrons, which move with a velocity of order v_F, is similar to that of a high-frequency field with $\omega_{\mathrm{eff}} \sim v_F q$. This case is further discussed by Roulet and Nozieres.[96]

Here we discuss in some detail a long-wavelength, low-frequency limit.

6.2.2. Landau Theory for Weakly Inhomogeneous Systems. The Landau theory of Fermi liquids was developed by Landau for *homogeneous* Fermi liquids subject to long-wavelength low-frequency external perturbations.[30] Here following Chakravarty *et al.*[31] we extend it to systems whose unperturbed density varies on a length scale l large compared to the local p_F^{-1}. We shall be led to a local density functional formulation.

For a homogeneous electron system, subject to an external scalar potential $v_1(\mathbf{r},t)$ the Landau transport equation for the quasiparticle distribution function $n_1(\mathbf{p},\mathbf{r},t)$ is obtained from the classical Hamilton's equation

$$\dot{\mathbf{p}} = -\nabla_{\mathbf{r}} H, \qquad \dot{\mathbf{r}} = \nabla_{\mathbf{p}} H \tag{220}$$

using the Hamiltonian

$$H(\mathbf{p},\mathbf{r},t) = \epsilon(\mathbf{p}) + \phi_1(\mathbf{r},t) + \sum_{\mathbf{p}'} f_{\mathbf{p},\mathbf{p}'}\, n_1(\mathbf{r},\mathbf{p}',t) \tag{221}$$

Here spin indices have been suppressed; $\epsilon(\mathbf{p})$ is the one-particle energy

$$\epsilon(\mathbf{p}) \equiv \frac{\mathbf{p}^2}{2m^*} - \frac{\mathbf{p}_F^2}{2m^*} \tag{222}$$

where m^* is the effective mass parameter; $\phi_1(\mathbf{r},t)$ is the total change of electrostatic potential,

$$\phi_1(\mathbf{r},t) \equiv v_1(\mathbf{r},t) + \int \frac{n_1(\mathbf{r}',t)}{|\mathbf{r}-\mathbf{r}'|}\, d\mathbf{r}' \tag{223}$$

where

$$n_1(\mathbf{r},t) \equiv \int n_1(\mathbf{r},\mathbf{p},t)\, d\mathbf{p} \tag{224}$$

the last term in equation (221) arises from short-range quasiparticle interactions, $f_{\mathbf{p},\mathbf{p}'}$ being the interaction function usually characterized by the coefficients A_l and B_l of its spherical harmonic expansion.

The generalization for a weakly inhomogeneous system is physically evident. The quasiparticle distribution function $n_1(\mathbf{p},\mathbf{r},t)$ maintains its meaning, as does the short-range function $f_{\mathbf{p},\mathbf{p}'}$ which, however, now depends on the local density. The term in ϕ_1 remains unchanged. Finally $\epsilon(\mathbf{p})$ must be replaced by

$$\epsilon(\mathbf{p},\mathbf{r},t) \equiv \frac{p^2}{2m^*(\mathbf{r})} - \frac{p_F^2(\mathbf{r})}{2m^*(\mathbf{r})} \tag{225}$$

This expression has been derived in the theory of single-particle excitations of a weakly inhomogeneous electron gas.[17] It can also be understood as follows. The local wave number of a quasiparticle of vanishing excitation energy is equal to

$$p_F(\mathbf{r}) \equiv [3\pi^2\, n_0(\mathbf{r})]^{1/3} \tag{226}$$

where $n_0(\mathbf{r})$ is the local unperturbed density. In a system at equilibrium, quasiparticles of vanishing excitation energy have everywhere the local Fermi momentum. Furthermore, at a given point $\mathbf{r}$, the momentum dependence of ϵ for small $p-p_F(\mathbf{r})$ must be given by the local value of $m^*(\mathbf{r})$. Both of these requirements are satisfied by the expression in equation (225). The total Hamiltonian is then

$$\begin{aligned} H(\mathbf{p},\mathbf{r},t) &= \epsilon(\mathbf{p},\mathbf{r},t) + \phi_1(\mathbf{r},t) \\ &\quad + \sum_{\mathbf{p}'} f_{\mathbf{p},\mathbf{p}'}\,[n_0(\mathbf{r})] \cdot n_1(\mathbf{p}',\mathrm{r},t) \end{aligned} \tag{227}$$

This H governs the collisionless transport equation

$$\frac{\partial n}{\partial t} + \dot{\mathbf{p}} \cdot \nabla_{\mathbf{p}} n + \dot{\mathbf{r}} \cdot \nabla_{\mathbf{r}} n = 0 \tag{228}$$

where $\dot{\mathbf{p}}$ and $\dot{\mathbf{r}}$ are given by Hamilton's equations (220). We now write

$$n(\mathbf{p},\mathbf{r},t) = n_0(\mathbf{p},\mathbf{r}) + n_1(\mathbf{p},\mathbf{r},t) \tag{229}$$

where

$$\begin{aligned} n_0(\mathbf{p},\mathbf{r}) = 1, &\qquad p \leq p_F(\mathbf{r}) \\ 0, &\qquad p > p_F(\mathbf{r}) \end{aligned} \tag{230}$$

and linearize equation (228), treating v_1, and n_1 as small:

$$\frac{\partial n_1}{\partial t} + \nabla_{\mathbf{p}}\epsilon_0 \cdot \nabla_{\mathbf{r}} n_1 - \nabla_{\mathbf{r}}\epsilon_0 \cdot \nabla_{\mathbf{p}} n_1 = S_1(\mathbf{p},\mathbf{r},t) \tag{231}$$

where

$$\begin{aligned} S_1(\mathbf{p},\mathbf{r},t) &= -\nabla_{\mathbf{P}} \{\sum_{\dot{\mathbf{p}}} f_{\mathbf{p},\mathbf{p}'}(n_0(\mathbf{r}))n_1(\mathbf{p}',\mathbf{r},t)\} \cdot \nabla_{\mathbf{r}} n_0 \\ &\quad + \nabla_{\mathbf{r}}\,(\phi_1 + \sum_{\mathbf{p}'} f_{\mathbf{p},\mathbf{p}'}\,(n_0(\mathbf{r}))n_1(\mathbf{p}',\mathbf{r},t) \cdot \nabla_{\mathbf{p}} n_0 \end{aligned} \tag{232}$$

Equation (231) has the form of a classical transport equation with a source $S_1(\mathbf{p},\mathbf{r},t)$, which must be determined self-consistently.

Considering S_1 as *given* we first define the Green's function

$$G(\mathbf{p},\mathbf{r};\,\mathbf{p}',\mathbf{r}';\,t-t') = \delta(\mathbf{p} - \mathrm{P}(\mathbf{p}',\mathbf{r}',t))\,\delta(\mathbf{r}-\mathbf{Q}(\mathbf{p}',\mathbf{r}',t))\theta(t-t') \tag{233}$$

which describes the motion of a particle, moving under the action of the Hamiltonian $\epsilon(\mathbf{P},\mathbf{Q})$, which was created at time t' at the point $\mathbf{r}'$ with momentum $\mathbf{p}'$. Here the functions $\mathbf{P}$ and $\mathbf{Q}$ are determined by Hamilton's equation

$$\dot{\mathbf{p}} = -\nabla_{\mathbf{Q}}\,\epsilon(\mathbf{P},\mathbf{Q}), \qquad \dot{Q} = \nabla_{\mathbf{P}}\,\epsilon(\mathbf{P},\mathbf{Q}) \tag{234}$$

and the initial conditions

$$\mathbf{P}(t') = \mathbf{p}', \qquad \mathbf{Q}(t') = \mathbf{r}' \tag{235}$$

$$\theta(t-t') \equiv \begin{cases} 1, & t > t' \\ 0, & t < t' \end{cases} \tag{236}$$

G can be obtained from the *classical trajectories* corresponding to $\epsilon(\mathbf{p},\mathbf{r})$. The formal solution of equation (231) is then given by

$$n_1(\mathbf{p},\mathbf{r},t) = \tilde{n}_1(\mathbf{p},\mathbf{r},t) + \int G(\mathbf{p},\mathbf{r};\,\mathbf{p}',\mathbf{r}';t-t')\,S_1(\mathbf{p}',\mathbf{r}',t')\,d\mathbf{p}'\,d\mathbf{r}'\,dt' \tag{237}$$

where $\tilde{n}_1$ is an appropriate solution of the homogeneous transport equation. If we now consider a single time Fourier component of v_1, and the corresponding change of n_1,

$$\begin{aligned} v_1(\mathbf{r},t) &= v_1(\mathbf{r})\,e^{-i(\omega-i\eta)t} \\ n_1(\mathbf{r},\mathbf{p},t) &= n_1(\mathbf{r},\mathbf{p})\,e^{-i(\omega-i\eta)t} \end{aligned} \tag{238}$$

where $\eta = +0$, then equation (237) reduces to

$$n_1(\mathbf{p},\mathbf{r},\omega) = \delta\,(\epsilon(\mathbf{p},\mathbf{r}))\,[S_1(\mathbf{p},\mathbf{r},;\omega) + i\omega \int S_1(\mathbf{p}',\mathbf{r}',\omega)\,G(\mathbf{p},\mathbf{r};\mathbf{p}',\mathbf{r}';\omega)\,d\mathbf{p}'\,d\mathbf{r}'] \tag{239}$$

$\tilde{n}_1$ has been omitted since (except exactly at resonance) the homogeneous equation has no solution.

Equation (239) and the Fourier transform of equation (232) must be solved self-consistently. Thus, we see that under the stated conditions the density response of an inhomogeneous system to an external potential can be determined with the aid of a Green's function calculable from simple classical trajectories.

In summary, we have seen that for an inhomogeneous system, whose density is slowly varying, a Landau theory of Fermi liquids can be derived in which the Landau parameters m^* and $f_{\mathbf{p},\mathbf{p}'}$ are dependent on the local density $n(\mathbf{r})$.

Acknowledgments

This work was supported in part by the National Science Foundation, the Office of Naval Research, and the Department of Energy. We wish to thank Dr. R. K. Kalia for extensive help in the preparation of the manuscript and Mrs. S. Horyza for her good humor during the typing of this text. We also wish to express our admiration to the editors, Professors S. Lundqvist and N. H. March for their extraordinary patience.

Appendix A. Detailed Derivation of the Self-Consistent Equations

Our problem is to minimize equation (43) with respect to n'. Let $v'(\mathbf{r})$ be a trial single-particle potential which gives rise to an electron density n' in the following sense. We solve the Schrödinger equation

$$[-\tfrac{1}{2}\nabla^2 + v'(\mathbf{r})]\psi_i'(\mathbf{r}) = \epsilon_i'\psi_i'(\mathbf{r}) \tag{A.1}$$

and take

$$n'(\mathbf{r}) \equiv \sum_{i=1}^{N} |\psi_i'(\mathbf{r})|^2 \tag{A.2}$$

where the sum runs over the *lowest* N eigenvalues. Then clearly, from its definition given in equation (50)

$$T_s[n'] = \sum_i \epsilon_i' - \int v'(\mathbf{r})n'(\mathbf{r})\,d\mathbf{r} \tag{A.3}$$

Thus we need to minimize the quantity

$$E_v[n'] \equiv \sum_i \epsilon_i' - \int v'(\mathbf{r})n'(\mathbf{r})\, d\mathbf{r} + \int v(\mathbf{r})n'(\mathbf{r})\, d\mathbf{r} + \frac{1}{2}\int \frac{n'(\mathbf{r})n'(\mathbf{r}')}{|\mathbf{r}-\mathbf{r}'|}\, d\mathbf{r}\, d\mathbf{r}' + E_{xc}[n'] \tag{A.4}$$

with respect to $v'(\mathbf{r})$, with $n'(\mathbf{r})$ regarded as functional of $v'(\mathbf{r})$, or equivalently with respect to $n'(\mathbf{r})$, with $v'(\mathbf{r})$ regarded as a functional of $n'(\mathbf{r})$.

Taking the variation with respect to $n'(\mathbf{r})$ gives*

$$\delta E_v[n'] = \delta n'(\mathbf{r}) \left\{ \int \frac{\delta v'(\mathbf{r}'')}{\delta n'(\mathbf{r})} \cdot n'(\mathbf{r}'')\, d\mathbf{r}'' - v'(\mathbf{r}) - \int \frac{\delta v'(\mathbf{r}'')}{\delta n'(\mathbf{r})} n'(\mathbf{r}'')\, d\mathbf{r}'' + v(\mathbf{r}) + \int \frac{n'(\mathbf{r}')}{|\mathbf{r}-\mathbf{r}'|}\, d\mathbf{r}' + \frac{\delta E_{xc}[n']}{\delta n'(\mathbf{r})} \right\} \tag{A.5}$$

Thus we see that the $v'(\mathbf{r})$ which minimizes $E_v[n']$ must satisfy the self-consistent condition

$$v'(\mathbf{r}) = v(\mathbf{r}) + \int \frac{n'(\mathbf{r}')}{|\mathbf{r}-\mathbf{r}'|}\, d\mathbf{r}' + \frac{\delta E_{xc}[n']}{\delta n'(\mathbf{r})} + \text{const} = v_{\text{eff}} + \text{const} \tag{A.6}$$

where $n'(\mathbf{r})$ is given by (A.2) and (A.1). This justifies the self-consistent equations (47)–(49).

Appendix B. Koopmans' Theorem for Self-Consistent Equations

In this appendix we consider an extensive system in which the eigenfunctions, ψ_i, at the maximum occupied single-particle energy, ϵ_F, are all extended. We shall show that the difference in energy between the N- and $(N + m)$-particle states $(m << N)$ is $m\epsilon_F$, to first order in m.

Since the expression (43) is stationary with respect to particle-conserving changes, δn, we must have—for the correct ground state n—

$$\frac{\delta E_v[n]}{\delta n(\mathbf{r})} = \mu \tag{B.1}$$

*See footnote following equation (46).

a constant independent of $\mathbf{r}$ to be identified with the chemical potential. Let us take

$$\delta n(\mathbf{r}) = \sum_{i=N+1}^{N+m} |\psi_i(\mathbf{r})|^2 \tag{B.2}$$

with $m << N$. (Since all ψ_i are extended this is small everywhere). Then clearly we have

$$\begin{aligned} \delta E_v[n(\mathbf{r})] &= T_s[n(\mathbf{r}) + \delta n(\mathbf{r})] - T_s[n(\mathbf{r})] \\ &\quad + \int d\mathbf{r}\ \delta n(\mathbf{r}) \left[v(\mathbf{r}) + \int \frac{n(\mathbf{r}')}{|\mathbf{r}-\mathbf{r}'|}\, d\mathbf{r}' + \frac{\delta E_{xc}}{\delta n(\mathbf{r})} \right] \\ &= \sum_{i=N+1}^{N+m} \epsilon_1 = m\epsilon_F \end{aligned} \tag{B.3}$$

Thus, noting that by (B.2)

$$\int \delta n(\mathbf{r})\, d\mathbf{r} = m \tag{B.4}$$

we obtain from (B.1), the required result

$$\mu = \epsilon_F \tag{B.5}$$

Appendix C. $E_{xc}[n]$ In Terms of the Pair Correlation Function

This appendix contains the derivation of the important equation (70) of the text. We follow closely the work of Langreth and Perdew.[27]

We consider a system of N electrons and write the Hamiltonian as

$$\mathsf{H} = \mathsf{T} + \mathsf{H}' \tag{C.1}$$

where T is the kinetic energy and H' contains the interaction with an external potential as well as the mutual interaction. We consider this system not only when these interactions have their actual physical values but also, when the Coulomb repulsion has been reduced by a factor $\lambda(\leq 1)$ and at the same time the external potential $v(\mathbf{r})$ has been changed to $v_\lambda(\mathbf{r})$ in such a way that the density $n(\mathbf{r})$ is independent of λ. Thus

$$
\begin{aligned}
\mathrm{H}'_\lambda = {} & \frac{1}{2}\int d\mathbf{r}\, d\mathbf{r}' \frac{\lambda}{|\mathbf{r}-\mathbf{r}'|}\, \mathrm{n}(\mathbf{r})[\mathrm{n}(\mathbf{r}') - \delta(\mathbf{r}-\mathbf{r}')] \\
& + \int d\mathbf{r}\, v_\lambda(\mathbf{r})\mathrm{n}(\mathbf{r})
\end{aligned}
\tag{C.2}
$$

We denote the energy corresponding to λ by E_λ. Then, by first-order perturbation theory,

$$
\begin{aligned}
\frac{dE_\lambda}{d\lambda} = \left\langle \frac{\partial H'_\lambda}{d\lambda} \right\rangle = {} & \frac{1}{2}\int d\mathbf{r}\, d\mathbf{r}' \frac{1}{|\mathbf{r}-\mathbf{r}'|} \langle \mathrm{n}(\mathbf{r})\, [\mathrm{n}(\mathbf{r}') - \delta(\mathbf{r}-\mathbf{r}')]\rangle_\lambda \\
& + \frac{\partial}{\partial\lambda}\int d\mathbf{r}\, v_\lambda(\mathbf{r}) n(\mathbf{r})
\end{aligned}
\tag{C.3}
$$

When this is integrated over λ we obtain

$$
\begin{aligned}
E \equiv E_1 = E_0 + & \int_0^1 \frac{dE_\lambda}{d\lambda}\, d\lambda \\
= T_s[n] + & \int v(\mathbf{r}) n(\mathbf{r})\, d\mathbf{r} + \frac{1}{2}\int \frac{n(\mathbf{r})n(\mathbf{r}')}{|\mathbf{r}-\mathbf{r}'|}\, d\mathbf{r}\, d\mathbf{r}' + E_{xc}
\end{aligned}
\tag{C.4}
$$

here $T_s[n]$ ($=E_0$) is the kinetic energy of a noninteracting system with density n, as previously defined in equation (39), and

$$
\begin{aligned}
E_{xc} \equiv \frac{1}{2}\int_0^1 d\lambda \int d\mathbf{r}\, d\mathbf{r}' \frac{1}{|\mathbf{r}-\mathbf{r}'|} & \{\langle[\mathrm{n}(\mathbf{r}) - \langle \mathrm{n}(\mathbf{r})\rangle][\mathrm{n}(\mathbf{r}') \\
& - \langle \mathrm{n}(\mathbf{r}')\rangle]\rangle_\lambda - \langle \mathrm{n}(\mathbf{r})\rangle \delta(\mathbf{r}-\mathbf{r}')\}
\end{aligned}
\tag{C.5}
$$

This may be written as

$$
\begin{aligned}
E_{xc} & = \int d\mathbf{r}\, d\mathbf{r}' \frac{n(\mathbf{r})n(\mathbf{r}')}{|\mathbf{r}-\mathbf{r}'|} \int_0^1 d\lambda\, [g_\lambda(\mathbf{r},\mathbf{r}') - 1] \\
& = \int d\mathbf{r}\, d\mathbf{r}' \frac{n(\mathbf{r})\bar{h}(\mathbf{r},\mathbf{r}')n(\mathbf{r}')}{|\mathbf{r}-\mathbf{r}'|}
\end{aligned}
\tag{C.6}
$$

where $g_\lambda(\mathbf{r},\mathbf{r}')$ is the pair correlation function corresponding to strength λ, defined in equation (62), and

$$
\bar{h}(\mathbf{r},\mathbf{r}') = \int_0^1 d\lambda\, [g_\lambda(\mathbf{r},\mathbf{r}') - 1] \tag{C.7}
$$

References

1. A. K. Rajagopal, in *Advances in Chemical Physics*, I. Prigogine and S. A. Rice, editors, Vol. 41, pp. 59–193 (Wiley, New York, 1980).
2. L. H. Thomas, *Proc. Camb. Phil. Soc.* **23,** 542–547 (1927).
3. E. Fermi, *Rend. Acad. Naz. Lincei* **6,** 602–607 (1927).
4. N. H. March, *Self-Consistent Fields in Atoms* (Pergamon Press, New York, 1975).
5. P. Hohenberg and W. Kohn, *Phys. Rev.* **136,** B864–871 (1964).
6. W. Kohn and L. J. Sham, *Phys. Rev. A* **140,** 1133–1138 (1965)
7. E. P. Wigner, *Trans. Faraday Soc.* **34,** 678–685 (1938).
8. M. Gell-Mann and K. A. Brueckner, *Phys. Rev.* **106,** 364–368 (1957); L. Onsager, L. Mittag, and M. J. Stephen, *Ann. Phys. (Leipzig)* **18,** 71–77 (1966).
9. P. Nozieres and D. Pines, *Phys. Rev.* **111,** 442–454 (1958).
10. T. Gaskell, *Proc. Phys. Soc.* **77,** 1182–1192 (1961).
11. P. Vashishta and K. S. Singwi, *Phys. Rev. B* **6,** 875–887 (1972); (E) *Phys. Rev. B* **6,** 4883–4883 (1972).
12. K. S. Singwi, A. Sjölander, M. P. Tosi, and R. H. Land, *Phys. Rev. B* **1,** 1044–1053 (1970).
13. K. S. Singwi, M. P. Tosi, R. H. Land, and A. Sjölander, *Phys. Rev.* **176,** 589–599 (1968).
14. L. Kleinman, *Phys. Rev.* B **10,** 2221–2225 (1974); (E) *Phys. Rev. B* **12,** 3512 (1975).
15. C. F. von Weizsäcker, *Z. Phys.* **96,** 431–458 (1935)
16. D. A. Kirzhnits, *Zh. Eksp. Teor. Fiz.* **32,** 115–123 (1957) [English Transl.: *Sov. Phys.—JETP* **5,** 64–71 (1957)].
17. L. J. Sham and W. Kohn, *Phys. Rev.* **145,** 561–567 (1966).
18. T. C. Koopmans, *Physica* **1,** 104–113 (1933).
19. V. von Barth and L. Hedin, *J. Phys. C* **5,** 1629–1642 (1972).
20. O. Gunnarsson and B. I. Lundqvist, *Phys. Rev. B* **13,** 4274–4298 (1976).
21. S. K. Ma and K. Brueckner, *Phys. Rev.* **165,** 18–31 (1968).
22. D. J. W. Geldart and M. Rasolt, *Phys. Rev. B* **13,** 1477–1488 (1976).
23. A. K. Gupta and K. S. Singwi, *Phys. Rev. B* **15,** 1801–1810 (1977).
24. K. H. Lau and W. Kohn, *J. Phys. Chem. Solids* **37,** 99–104 (1976).
25. J. P. Perdew, D. C. Langreth, and V. Sahni, *Phys. Rev. Lett.* **38,** 1030–1033 (1977).
26. J. Harris and R. O. Jones, *J. Phys. F* **4,** 1170–1186 (1974).
27. D. C. Langreth and J. P. Perdew, *Phys. Rev. B* **15,** 2884–2901 (1977); (27a) O. Gunnarsson, in *Electrons in Disordered Metals and at Metallic Surfaces*, P. Phariseau, B. L. Gyorffy, and L. Schiene, editors, pp. 1–53 (Plenum Press, New York, 1979).
28. O. Gunnarsson, M. Jonson, and B. I. Lundquist, *Solid State Commun.* **24,** 765–768 (1977).
29. A. J. Coleman, *Rev. Mod. Phys.* **35,** 668–687 (1963); (29a) M. Levy, *Proc. Natl. Acad. Sci. USA* **76,** 6062–6065 (1979).
30. L. D. Landau, *Sov. Phys. JETP* **3,** 920–925 (1956).
31. S. Chakravarty, M. Fogel, and W. Kohn, *Phys. Rev. Lett.* **43,** 775–778 (1979).
32. B. Y. Tong and L. J. Sham, *Phys. Rev.* **144,** 1–4 (1966).
33. J. C. Slater, *Phys. Rev.* **81,** 385–390 (1951).
34. J. C. Slater in *Advances in Quantum Chemistry*, P. O. Lowdin, ed., Vol. 6, pp. 1–92, (Academic Press, New York, 1972)
35. J. C. Slater, *Int. J. Qu. Chem. Symp.* **9,** 7–21 (1975).
36. J. C. Slater, *The Self-Consistent Field for Molecules and Solids*, Vol. 4 (McGraw-Hill Co., New York, 1974).

37. J. C. Slater, *The Calculations of Molecular Orbitals*, Vol. 5 (McGraw-Hill Co., New York, 1979).
38. D. C. Langreth and J. P. Perdew, *Phys. Rev. B* **15**, 2884–2901 (1977).
39. O. Gunnarsson, J. Harris, and R. O. Jones, *J. Chem. Phys.* **67**, 3970–3979 (1977).
40. V. L. Morruzzi, J. F. Janak, and A. R. Williams, *Calculated Electronic Properties of Metals (Pergamon Press, New York, 1978).*
41. F. Stern, *CRC Crit. Rev. Solid State Sci.* **4**, 499–514 (1974).
42. T. Ando, *Phys. Rev. B* **13**, 3468–3477 (1976).
43. T. Ando, *Z. Phys. B***26**, 263–272 (1977).
44. P. Kneschaurek, A. Kamgar, and J. F. Koch, *Phys. Rev. B* **14**, 1610–1612 (1976).
45. R. Gordon and Y. S. Kim, *J. Chem. Phys.* **56**, 3122–3133 (1972).
46. J. R. Smith, *Phys. Rev.* **181**, 522–529 (1969).
47. P. Vashishta and W. Kohn, *Bull. Am. Phys. Soc.* **24**, 439 (1979).
48. M. M. Pant and A. K. Rajagopal, *Solid State Commun.* **10**, 1157–1160 (1972); (48a) A. K. Rajagopal and J. Calloway, *Phys. Rev. B* **7**, 1912–1919 (1973).
49. G., Pizzimenti, M. P. Tosi, and A. Villari, *Lett. Nuovo Cimento Serie 2* **2**, 81–84 (1971).
50. R. Dupree and D. J. W. Geldart, *Solid State Commun.* **9**, 145–149 (1971).
51. P. Vashishta and K. S. Singwi, *Solid State Commun.* **13**, 901–904 (1973).
52. O. K. Anderson and R. G. Wesley, *Mol. Phys.* **26**, 905–927 (1973).
53. S. H. Vosko and J. P. Perdew, *Can. J. Phys.* **53**, 1385–1397 (1975).
54. S. H. Vosko, J. P. Perdew, and A. H. MacDonald, *Phys. Rev. Lett.* **35**, 1725–1728 (1975).
55. A. H. MacDonald, J. P. Perdew, and S. H. Vosko, *Solid State Commun.* **18**, 85–91 (1976).
56. A. H. MacDonald and S. H. Vosko, *J. Low Temp. Phys.* **25**, 27–41 (1976).
57. J. F. Janak, *Phys. Rev. B* **16**, 255–262 (1977).
58. K. L. Liu and S. H. Vosko, *J. Phys. F* **8**, 1539–1556 (1978).
59. C. S. Wang and J. Callaway, *Solid State Commun.* **20**, 255–256 (1976).
60. L. M. Sander, H. B. Shore, and L. J. Sham, *Phys. Rev. Lett.* **31**, 533–536 (1973).
61. H. Bütner and E. Gerlach, *J. Phys. C* **6**, L433–L436 (1973).
62. T. M. Rice, *Phys. Rev. B* **9**, 1540–1546 (1974).
63. T. L. Reinecke and S. C. Ying, *Solid State Commun.* **14**, 381–385 (1974).
64. T. L. Reinecke, F. Crowne, and S. C. Ying, in *Proceedings of 12th International Conference on the Physics of Semiconductors*, Stuttgart, Germany, 1974, pp. 61–65, M. H. Pilkuhn, ed. (Teubner, Stuttgart, 1975).
65. P. Vashishta, R. K. Kalia, and K. S. Singwi, *Solid State Commun.* **19**, 935–938 (1976).
66. M. Morimoto, K. Shindo, and A. Morita, in *Proceedings of the Oji Seminar on the Physics of Highly Excited States in Solids*, Vol. 57, pp. 230–236, M. Ueta and Y. Nishina, eds. (Springer-Verlag, Berlin, 1976).
67. T. M. Rice, in *Solid State Physics*, H. Ehrenreich, F. Seitz, and D. Turnbull, eds., Vol. 32, pp. 1–86 (Academic Press, New York, 1977).
68. R. K. Kalia and P. Vashishta, *Solid State Commun.* **24**, 171–174 (1977).
69. M. Rasolt and D. J. W. Geldart, *Phys. Rev. B* **15**, 979–988 (1977); *Phys. Rev. B* **15**, 4804–4816 (1977).
70. J. H. Rose and H. B. Shore, *Phys. Rev. B* **17**, 1884–1892 (1978).
71. R. K. Kalia and P. Vashishta, *Phys. Rev. B* **17**, 2655–2672 (1978).
72. K. A. Brueckner, J. R. Buchler, R. C. Clark, and R. J. Lombard, *Phys. Rev.* **181**, 1543–1551 (1969).
73. G. Baym, H. A. Bethe, and C. J. Pethick, *Nucl. Phys. A* **175**, 225–271 (1971).

74. K. A. Brueckner, J. H. Chirico, and H. W. Meldner, *Phys. Rev. C* **4,** 732–740 (1971).
75. L. V. Keldysh, in *Proceedings of the 9th International Conference on Physics of Semiconductors*, Moscow 1968, pp. 1303–1312, S. M. Ryvkin and V. V. Shmastsev, eds. (Nauka, Leningrad, 1968).
76. Y. E. Pokrovsky, *Phys. Stat. Sol.* **A11,** 385–410 (1972).
77. V. S. Bagaev, *Springer Tracts Mod. Phys.* **73,** 72–90 (1951).
78. C. D. Jeffries, *Science* **189,** 955–964 (1975).
79. M. Voos and C. Benoit a la Guillaume, in *Optical Properties of Solids: New Developments*, B. O. Seraphin, ed., pp. 143–186 (American Elsevier, New York, 1976).
80. J. C. Hensel, T. G. Phillips, and G. A. Thomas, in *Solid State Physics*, H. Ehrenreich, F. Seitz, and D. Turnbull, eds., Vol. 32, pp. 88–314 (Academic Press, New York, 1977).
81. L. J. Sham and T. M. Rice, *Phys. Rev.* **144,** 708–714 (1966).
82. E. Hanamura, in *Proceedings of the 10th International Conference on the Physics of Semiconductors*, Cambridge, Massachusetts, 1970, S. P. Keller, J. C. Hensel, and F. Stern, eds., pp. 487–493, CONF-700801 (U. S. AEC Div. Tech. Information, Springfield, Virginia 1970).
83. M. Combescot and P. Nozieres, *J. Phys. C* **5,** 2369–2391 (1972).
84. W. F. Brinkman, T. M. Rice, P. W. Anderson, and S. T. Chui, *Phys. Rev. Lett.* **28,** 961–964 (1972).
85. W. F. Brinkman and T. M. Rice, *Phys. Rev. B* **7,** 1508–1523 (1973).
86. P. Vashishta, S. G. Das, and K. S. Singwi, *Phys. Rev. Lett.* **33,** 911–914 (1974).
87. P. Vashishta, P. Bhattacharyya, and K. S. Singwi, *Phys. Rev. B* **10,** 5108–5126 (1974).
88. B. Etienne, L. M. Sander, G. Benoit a la Guillaume, M. Voos, and J. Y. Prieur, *Phys. Rev. Lett.* **37,** 1299–1302 (1976).
89. Y. E. Pokrovsky and K. I. Svistunova, in *Proceedings of the Twelfth International Conference on the Physics of Semiconductors*, Stuttgart, Germany, 1974, pp. 71–75, M. H. Pilkuhn, ed. (Teubner, Stuttgart, 1975).
90. T. Ugumori, K. Morigaki, and C. Magashima, *J. Phys. Soc. Jpn* **46,** 536–541 (1979).
91. N. D. Mermin, *Phys. Rev.* **137A,** 1441–1443 (1965).
92. J. M. Luttinger, *Phys. Rev.* **119,** 1153–1163 (1960).
93. A. Ghazali and P. Leroux Hugon, *Phys. Rev. Lett.* **41,** 1569–1572 (1978).
94. A. A. Abrikosov, L. P. Gorkov, and I. E. Dzyaloshinski, *Methods of Quantum Field Theory in Statistical Physics* (Prentice-Hall, Englewood Cliffs, New Jersey, 1963).
95. R. Kubo, *J. Phys. Soc. Jpn Vol.* **12,** 570 (1957).
96. P. B. Roulet and P. Nozieres, *J. Phys. (Paris)* **29,** 167–180 (1968).

CHAPTER 3

DENSITY OSCILLATIONS IN NONUNIFORM SYSTEMS

S. LUNDQVIST

1. Introduction

The dynamical properties of an inhomogeneous electron gas is a subject with a history almost as long as quantum mechanics. Because the subject is not widely known it will serve as a suitable introduction to this chapter to remind the reader about a few of the major steps. An important early problem was the theoretical understanding of the stopping of a fast charged particle in matter. A charged particle excites the medium with excitation energies covering a wide spectrum from far ultraviolet to soft X-ray frequencies. The stopping power itself does not depend on the details of the spectrum but only on an average excitation energy. The idea came up that one might replace the full dynamical theory by a simplified picture in which the medium was considered as an inhomogeneous electron gas. The charged particle would excite oscillations in the electron gas around its ground-state density and the particle would lose energy by exciting the various modes of excitation in the nonuniform electron gas. These ideas were developed in a classical paper by Bloch[1] in 1933. He developed a dynamical extension of the Thomas–Fermi theory treated in Chapter 1, considering the hydrodynamical oscillations of the density around the Thomas–Fermi ground-state density. Applications by Jensen[2] to a simplified model treating the atom as a small metallic sphere agreed with stopping power data in its dependence on atomic number and supported the hydrodynamical model. The Bloch equations were actually not fully solved until after the Second World War. Many extensions to include,

S. LUNDQVIST • Chalmers University of Technology, S-41296 Göteborg, Sweden.

e.g., exchange and correlations have been made. After the development of the density functional method presented in Chapter 2, a hydrodynamical approach based on the density-functional scheme was proposed by Ying *et al.*[3]

The hydrodynamical theory suggested, as discussed already in the classical paper by Bloch, the existence of collective oscillations in atoms. However, the numerical solution of the Bloch equations by Wheeler *et al.*[4] showed that the spectrum was continuous for neutral atoms and the universal photoabsorption curve obtained by them showed no sign of a collective resonance. The further analysis of the hydrodynamical models indicated that the interplay between single-particle excitations and hydrodynamical motion plays an important role. In particular, any plasmalike motion tends to break up into single-particle excitations in the outer low-density region and a purely hydrodynamic approach to the problem is insufficient and incomplete.

An important step was the development of the dielectric theory, describing the dynamical properties in terms of a nonlocal response function such as the polarizability $\alpha(\mathbf{r},\mathbf{r}',\omega)$, dielectric function $\epsilon(\mathbf{r},\mathbf{r}',\omega)$, conductivity $\sigma(\mathbf{r},\mathbf{r}',\omega)$. It should be remembered that the microscopic response functions introduced by Lindhard[5] were introduced for the purpose of a better description of the interaction between matter and radiation or charged particles, e.g., for the energy loss properties. Several approaches have been made to construct approximate response function considering the medium locally as a uniform electron gas or an electron gas with a small density gradient. More important, however, is that the dielectric formulation has served as a starting point for a variety of many-body approaches.

The existence of strong collective effects in atomic systems has been observed in a variety of experiments during the last two decades. A variety of techniques have been employed, e.g., using synchrotron radiation, photoelectron spectroscopy, inelastic electron scattering, etc. Calculations based upon the random phase approximation with exchange (RPAE) and corrected for relaxation effects have given very good results in applications to atoms, and similar methods are being applied with equal degree of success also to molecules.

The applications show that the collective excitations are not simple general properties of atomic systems. There is a strong interplay between the single-particle excitations and the collective mean field interaction, and depending on this interplay one or the other of the two aspects may dominate. Most of the applications using many-body techniques do not even attempt to separate these two aspects. In this chapter we shall focus on the aspects which have to do with the electron gas. We shall therefore discuss a formulation in which the classical terms and the genuine single-particle aspects are kept separate. This gives some insight in each particular application of the relative importance of the collective and single-particle aspects, respectively.

We shall also discuss a recent extension of the density functional approach to the calculation of the photoabsorption cross section of atomic systems in the far ultraviolet region. The method uses the linear response theory of finite frequencies as a straightforward extension of the theory of linear response to a static external field. In the static case the procedure is formally justified on the basis of the Hohenberg–Kohn theorem dealt with in Chapter 2, but the straightforward extension to the high-frequency region with energies around 100 eV seems to be supported mainly by intuitive arguments. Applications to a number of atoms have been very successful and give results comparable with the most sophisticated many-body calculations published so far. The technique is relatively simple to apply and may become useful also in applications to molecular systems.

2. Hydrodynamical Equations for a Nonuniform Electron Gas

The hydrodynamical models express the collective motion of the electrons in terms of the deviations from the equilibrium density $n_0(\mathbf{r})$ assuming that all the relevant physical quantities, such as the potential energy, pressure, etc. can be expressed in terms of the electron density $n(\mathbf{r})$. The earliest density method to describe ground-state properties was the Thomas–Fermi theory discussed in Chapter 1 of this book. The Thomas–Fermi model forms the theoretical basis for the development of the hydrodynamical theory by Bloch.[1] We shall briefly outline the major ideas in his theory. We wish to point out that Bloch's theory serves with minor modification as a model for all other hydrodynamical models for nonuniform systems which shall be mentioned in this chapter.

2.1. *Bloch's Theory*

We assume that the ground state is described by the Thomas–Fermi theory and is characterized by the ground-state density $n_0(\mathbf{r})$ and the corresponding one-electron potential $V_0(\mathbf{r})$. The nonuniform gas is capable of oscillations about this steady state. We characterize the nonuniform gas by the following quantities: the density $n(\mathbf{r},t)$, the kinetic pressure in the electron gas $p = p(n)$, and the hydrodynamical velocity $\mathbf{v}(\mathbf{r},t)$. For the problems we shall consider we may assume irrotational flow and introduce a velocity potential $\phi(\mathbf{r},t)$ such that $\mathbf{v}(\mathbf{r},t) = -\nabla\phi(\mathbf{r},t)$.

The Hamiltonian function for the system can be written as

$$
\begin{aligned}
H = &\int d^3r \, \frac{1}{2} |\nabla\phi|^2 n(\mathbf{r},t) + \frac{3}{10}\left(\frac{3}{8\pi}\right)^{2/3} \int d^3r \, n^{5/3}(\mathbf{r},t) \\
&+ \frac{1}{2}\int d^3r \, d^3r' \, \frac{n(\mathbf{r},t)n(\mathbf{r}',t)}{|\mathbf{r}-\mathbf{r}'|} + \int d^3r \, V(\mathbf{r},t)n(\mathbf{r},t)
\end{aligned}
\tag{1}
$$

The first term is the hydrodynamic kinetic energy and the second is the internal kinetic energy of the Fermi gas, which can alternatively be expressed in terms of the kinetic pressure $p(n)$, using the well-known relation between the kinetic energy per particle and the kinetic pressure in a Fermi gas. The third term in equation (1) is the Coulomb self-interaction and the last term represents the interaction of the electrons with the nuclei and with external potentials.

The equations of motion are obtained from the variational principle

$$
\delta \int_{t_1}^{t_2} L \, dt = 0 \tag{2}
$$

where the Lagrangian function L is given by

$$
L = \int d^3r \, \frac{\partial\phi}{\partial t} n - H \tag{3}
$$

In this way Bloch obtained the Euler equations

$$
\begin{aligned}
&\frac{\partial n}{\partial t} - \nabla(n\nabla\phi) = 0 \\
&\frac{\partial\phi}{\partial t} = \frac{1}{2}|\nabla\phi|^2 + \frac{1}{2}\left(\frac{3}{8\pi}\right)^{2/3} n^{2/3}(\mathbf{r},t) + \int d^3r' \, \frac{\mathrm{n}(\mathbf{r}',t)}{|\mathbf{r}-\mathbf{r}'|} + V(\mathbf{r},t)
\end{aligned}
\tag{4}
$$

The first equation is the continuity equation and the second is the Bernoulli equation (which is obtained by integrating Newton's second law).

The equations are nonlinear and Bloch replaced them by a linear set of equations by expanding for small-amplitude motion as follows:

$$
\begin{aligned}
n(\mathbf{r},t) &= \mathrm{n}_0(\mathbf{r}) + n_1(\mathbf{r},t) \\
\mathrm{V}(\mathbf{r},t) &= \mathrm{V}_0(\mathbf{r}) + \int d^3r' \, \frac{n_1(\mathbf{r}',t)}{|\mathbf{r}-\mathbf{r}'|} + V_{\text{ext}}(\mathbf{r},t) = \mathrm{V}_0(\mathbf{r}) + V_1(\mathbf{r},t) \\
\phi(\mathbf{r},t) &= 0 + \phi_1(\mathbf{r},t)
\end{aligned}
\tag{5}
$$

The linearized equations have the form

$$\frac{\partial n_1}{\partial t} + \nabla(n_0 \nabla \phi_1) = 0$$

$$\frac{\partial \phi_1}{\partial t} = \frac{1}{3}\left(\frac{3}{8\pi}\right)^{2/3} \frac{n_1}{n_0^{1/3}} + \int d^3r' \frac{n_1(\mathbf{r}',t)}{|\mathbf{r}-\mathbf{r}'|} + V_{\text{ext}}(\mathbf{r},t) \tag{6}$$

It is implied that $V_{\text{ext}}(\mathbf{r},t)$ is a weak external field, of first order.

Equation (6) forms the basis of the calculation of the linear response to an external field. We shall be particularly concerned with the frequency-dependent polarizability and photoabsorption cross section. If we set $V_{\text{ext}}(\mathbf{r},t) = 0$, we have the case of free oscillations of the systems. The hydrodynamic theory gives density oscillations corresponding to plasmalike motion. The nature of the collective motion depends on the characteristics of the system: bulk plasmons in bulk metals, surface plasmons at metal surfaces, spherical plasmons in metallic spheres, special plasma type modes in atoms and molecules.

In order to proceed, we now consider the corresponding expansion of the Hamiltonian

$$H = H_0 + H_1 + H_2 + \cdots \tag{7}$$

H_0 is the energy of the static unperturbed system. For H_1 and H_2 we obtain

$$H_1 = \int d^3r\, n_0\, V_{\text{ext}} \tag{8}$$

and

$$H_2 = \int d^3r \left[\frac{n_0}{2} |\nabla\phi_1|^2 + n_1 V_{\text{ext}} + \frac{1}{6}\left(\frac{3}{8\pi}\right)^{2/3} \frac{n_1{}^2}{n_0{}^{1/3}}\right] + \frac{1}{2}\int d^3r\, d^3r' \frac{n_1(\mathbf{r},t)n_1(\mathbf{r}',t)}{|\mathbf{r}-\mathbf{r}'|} \tag{9}$$

We note the following properties:

(1) H_0 and H_1 do not contain the unknown functions n_1 and ϕ_1. Therefore, only H_2 is important for the linearized problem and we can write the variational problem as

$$\delta \int_{t_1}^{t_2} L_2 \, dt = 0$$

$$L_2 = \int d^3r \frac{\partial \phi_1}{\partial t} n_1 - H_2 \tag{10}$$

(2) Except for the term in Equation (9), which contains V_{ext}, the term H_2 is a quadratic form in the two unknown functions n_1 and ϕ_1. In the case of free oscillations, $V_{\text{ext}} = 0$, this property implies that the eigenmodes for free oscillations have some nice properties, and Bloch showed that this permitted a complete solution of the linear response to an external field in terms of the normal mode solutions.

Let us consider the case of free oscillations. The linearized equations putting $V_{\text{ext}} = 0$ are

$$\frac{\partial n_1}{\partial t} + \nabla(n_0 \nabla \phi_1) = 0$$

$$\frac{\partial \phi_1}{\partial t} = \frac{1}{3}\left(\frac{3}{8\pi}\right)^{2/3} \frac{n_1}{n_0^{1/3}} + \int d^3r' \frac{n_1(\mathbf{r}',t)}{|\mathbf{r}-\mathbf{r}'|} \tag{11}$$

The Hamiltonian is given by

$$H_2 = \int d^3r \left[\frac{n_0}{2} |\nabla \phi_1|^2 + \frac{1}{6}\left(\frac{3}{8\pi}\right)^{2/3} \frac{n_1^2}{n_0^{1/3}}\right] + \frac{1}{2} \int d^3r \, d^3r' \frac{n_1(\mathbf{r},t) n_1(\mathbf{r}',t)}{|\mathbf{r}-\mathbf{r}'|}$$

$$= \int d^3r \left(n_1 \frac{\partial \phi_1}{\partial t} - \phi_1 \frac{\partial n_1}{\partial t}\right) \tag{12}$$

The second form is obtained using the equations of motions after a partial integration.

We look for normal mode solutions of the form

$$n_1(\mathbf{r},t) = - \omega_i n_i(\mathbf{r}) \sin \omega_i t$$

$$\phi_1(\mathbf{r},t) = \phi_i(\mathbf{r}) \cos \omega_i t \tag{13}$$

Inserting these formulas in equation (12) we obtain

$$E_i = \tfrac{1}{2}\omega_i^2 \int d^3r\, n_i(\mathbf{r})\phi_i(\mathbf{r}) \tag{14}$$

Using the linearized equations of motion it is straightforward to show that the functions n_i and ϕ_j for two different normal modes are orthogonal. For $i=j$ we can of course normalize the product to unity. The functions n_i and ϕ_j therefore form a biorthogonal system, defined by the property

$$\int d^3r\, n_i(\mathbf{r})\phi_i(\mathbf{r}) = \delta_{ij} \tag{15}$$

A general solution for free oscillations can be expressed as a superposition of the normal mode functions as follows:

$$\begin{aligned} n_1(\mathbf{r},t) &= \sum_i a_i n_i(\mathbf{r}) \sin \omega_i t \\ \phi_1(\mathbf{r},t) &= \sum_i b_i\, \phi_i(\mathbf{r}) \cos \omega_i t \end{aligned} \tag{16}$$

The total excitation energy in a state of free oscillations can be written in the form

$$E = \tfrac{1}{2} \sum_i \omega_i^2 C_i^2 \tag{17}$$

the coefficients c_i being independent of time.

This concludes for the moment our discussion of the Bloch theory of hydrodynamical oscillations. In Section 2.3 we shall calculate the absorption of electromagnetic radiation by the hydrodynamical oscillations. Before discussing photoabsorption, we shall discuss briefly some extensions and simplifications of the hydrodynamical model.

2.2. Extensions and Simplifications of the Hydrodynamical Model

In the previous section we discussed the hydrodynamical theory for small oscillations around the Thomas–Fermi ground-state density, following essentially the classical paper by Bloch.[1] His theory serves as a model for all the different hydrodynamical models which have been introduced later and applied to the nonuniform electron gas. The use of the Thomas–Fermi theory has the particular advantage that the results for atoms scale with the atomic number Z so that the results have a universal character (this feature is discussed in the first chapter of this volume). On the negative side the Thomas–Fermi ground-state density is not a good approximation as it does not reflect shell structure, and the dynamical equations for the hydrodynamical oscillations are very approximate. Many steps have been taken to improve the original Thomas–Fermi method by including exchange, correlation, inhomogeneity corrections, etc. We refer to the discussion by March in the first chapter of this volume for these extensions of the method. The terms added to the Thomas–Fermi theory can easily be taken into account for the dynamical problems as well.

Each such term will correspond to an additional density-dependent term to be added to the Hamiltonian, and when deriving the Euler equations, it will give rise to an additional term in the Bernoulli equation. As an example consider the inclusion of exchange in the theory by adding to the Hamiltonian the total exchange energy of the system [of equation (6) in Chapter 1]

$$E_x = -\frac{3}{4}\left(\frac{3}{\pi}\right)^{1/3} \int d^3r\, n^{4/3}(\mathbf{r},t) \tag{18}$$

This adds a corresponding term to the hydrodynamical equations of motion. Including exchange we obtain the following linearized equations for free oscillations:

$$\begin{aligned} &\frac{\partial n_1}{\partial t} + \nabla(n_0 \nabla \phi_1) = 0 \\ &\frac{\partial \phi_1}{\partial t} = \frac{1}{3}\left(\frac{3}{8\pi}\right)^{2/3} \frac{n_1}{n_0^{1/3}} - \frac{1}{3}\left(\frac{3}{\pi}\right)^{1/3} \frac{n_1}{n_0^{2/3}} + \int d^3r' \frac{n_1(\mathbf{r}',t)}{|\mathbf{r}-\mathbf{r}'|} \end{aligned} \tag{19}$$

The second term on the right-hand side of the second equation represents the effect of exchange. It can be interpreted as the contribution to the pressure from the exchange hole around each electron. Exchange was introduced in the static Thomas–Fermi theory by Dirac[4] and gives the same local exchange potential as that introduced by Kohn and Sham[6] in the density functional method (see Chapter 2). The exchange modification of the hydrodynamical theory was introduced by Jensen.[2]

Most of the applications of the hydrodynamical theory have been based on a simplified version of the linearized equations. One of the basic assumptions of the Thomas–Fermi theory, discussed in Chapter 1, is that the density variation is slow enough that the change in the potential is small over a de Broglie wavelength of the electrons. In the dynamical equations one often goes one step further and neglects terms explicitly containing the gradient of $n_0(\mathbf{r})$. This implies that we consider the medium locally as a uniform electron gas. This results in the following linear equation for free oscillations:

$$\frac{\partial^2 n_1}{\partial t^2} + \left[\omega_{\mathrm{pl}}^2(\mathbf{r}) - \beta^2(\mathbf{r})\,\nabla^2\right] n_1 = 0 \tag{20}$$

where

$$\omega_{\mathrm{pl}}(\mathbf{r}) = \left[\frac{4\pi e^2 n_0(\mathbf{r})}{m}\right]^{1/2} \tag{21}$$

is the classical plasma frequency in an electron gas of density $n_0(\mathbf{r})$.

For normal modes of the form

$$n_1(\mathbf{r},t) = n(\mathbf{r})e^{i\omega t} \tag{22}$$

we obtain the equation

$$\left[\omega^2 - \omega_{\mathrm{pl}}(\mathbf{r}) + \beta^2(\mathbf{r})\nabla^2\right] n(\mathbf{r}) = 0 \tag{23}$$

The coefficent β is determined by the pressure term in the Bloch equations and is given in terms of the local Fermi velocity $v_F(\mathbf{r})$ as

$$\beta^2(\mathbf{r}) = \tfrac{1}{3} v_F^2(\mathbf{r}) \tag{24}$$

It should be remembered that the Bloch theory is an adiabatic theory, assuming that the electrons are always locally in equilibrium with a Fermi distribution corresponding to the instantaneous local density. However, when we discuss plasma-type oscillations we are often in the regime of high frequencies, where the motion is so fast that collisions cannot establish local equilibrium. A typical case is that of a plasmon in a degenerate electron gas. In the collisionless regime the distribution along the direction of the wave will change with the local change in density, but the two perpendicular directions will not be affected. In the random phase approximation (RPA) we obtain for plasmons in a uniform electron gas the long-wavelength dispersion relation

$$\omega^2 = \omega_{\rm pl}^2 + {}^3\!/_5 v_F^2 q^2 \tag{25}$$

where q is the wave number. This suggest that for high-frequency oscillations we may use

$$\beta^2(\mathbf{r}) = {}^3\!/_5 v_F^2(\mathbf{r}) \tag{26}$$

rather than the low-frequency form given by equation (23).

The general nature of the solutions to equation (23) can be studied by introducing the local wave number $q(\mathbf{r})$ defined by

$$q^2(\mathbf{r}) = \frac{\omega^2 - \omega_{\rm pl}^2(\mathbf{r})}{\beta^2(\mathbf{r})} \tag{27}$$

In regions of space where $\omega_{\rm pl}(\mathbf{r}) < \omega$, the local wave number is real and we have a wavelike solution. In regions where $\omega_{\rm pl}(\mathbf{r}) > \omega$, on the other hand, the local wave number is purely imaginary and we have essentially exponentially decaying amplitudes. This corresponds to the well-known fact that one cannot sustain a density oscillation at frequencies below the classical plasma frequency. Equation (7) also shows the peculiar behavior in systems where the density goes smoothly to zero. Since the denominator $\beta^2(\mathbf{r})$ goes to zero as $n_0^{2/3}(\mathbf{r})$, we see that the local wave number increases without limit when the density goes to zero. In a more realistic theory the oscillation would decay and damp out in the low-density region rather than oscillate with ever decreasing wavelength.

The density functional theory provides us with a technique to deal with hydrodynamical oscillations around the exact ground-state density $n_0(\mathbf{r})$. As discussed in Chapter 2 of this volume, the ground-state energy of an inhomogeneous electron system is a functional $E[n]$ of the density $n(\mathbf{r})$. The hydrodynamical theory can now be obtained by considering slow adiabatic perturbations of the ground state. The theory developed by Ying *et al.*[(3)] implies that we may replace the Thomas–Fermi formula for the total energy by the more general expression

$$E = \int d^3r \, \frac{1}{2} \, |\nabla\phi|^2 n(\mathbf{r},t) + \int d^3r \, d^3r' \, \frac{n(\mathbf{r},t) \, n(\mathbf{r}',t)}{|\mathbf{r}-\mathbf{r}'|}$$
$$+ \int d^3r \, V(\mathbf{r},t) n(\mathbf{r},t) + \int d^3r \, G\,[n(\mathbf{r},t)] \tag{28}$$

The functional $G[n]$ represents the sum of the kinetic, exchange, and correlation energies. The hydrodynamical equations can now be derived following the same procedure as in Section 2.1. The continuity equation is unchanged but the second of the Euler equations becomes

$$\frac{\partial \phi}{\partial t} = \frac{1}{2}|\nabla \phi|^2 + \frac{\delta G}{\delta n} + V(\mathbf{r},t) + \int d^3r' \frac{n(\mathbf{r}',t)}{|\mathbf{r}-\mathbf{r}'|} \tag{29}$$

The final form of the equation depends on the approximate form chosen for $G[n]$. The term $\delta G/\delta n$ represents the contribution from the pressure. In Bloch's theory only the kinetic pressure in the Fermi gas is taken into account. Here, we can also include the contribution from the interactions, represented by the exchange and correlation contributions to $G[n]$. The simplest choice is to use the local density approximation for $G[n]$. This gives the kinetic energy contribution as in Bloch's theory but also the contribution from exchange and correlation. Many recent applications have shown the importance of treating exchange and correlation, and the extension of the Bloch theory using the density functional approach represents a significant improvement.

2.3. *Application to Photoabsorption*

As mentioned in the Introduction, the formulation of the hydrodynamical model by Bloch[1] was motivated by the need to understand the stopping power by many-electron atoms. The first application was made by Jensen[2] using a very simplified atomic model, treating the atom as a small metallic sphere. The model gave the following results for the eigenfrequencies ω_n and the oscillator strengths fn:

$$\omega_n = k_n Z \text{ Ry}, \qquad f_n = q_n Z$$

where Z is the atomic number and k_n and q_n are numerical coefficients of order unity. The results were in good agreement with stopping power data. We note that the excitations of interest in the hydrodynamical approach fall in the spectral region between far ultraviolet and the characteristic X-ray spectra. In the prewar times this region was not accessible for detailed studies of, e.g., photoabsorption. Therefore interest in the photoabsorption spectrum did not come up until some time after the war, and has in later years become of great interest with the use of synchrotron radiation sources. An extensive analysis of the photoabsorption cross section of atoms in the far ultraviolet and soft X-ray region has been made by Wheeler[4] and co-workers using the hydrodynamical model by Bloch.

We shall first discuss the photoabsorption cross section in the hydrodynamical model and shall follow closely the discussion given in Ref. 4. We treated the normal modes of free oscillations in Section 2.1. We now turn to the question how the system absorbs energy from an external radiation field. We assume that the wavelength of the incident radiation is large in comparison with the size of the systems. The external field corresponding to a plane wave polarized in the z direction is given by the formula

$$V_{\text{ext}}(\mathbf{r},t) = zE_0 \sin \omega t \tag{30}$$

The potential will excite modes in the atom having the same symmetry, i.e., dipolar modes. It is clear that the perturbing field is irrotational, which justifies the use of a velocity potential in the hydrodynamical theory. It is convenient to expand the spatial part of $V_{\text{ext}}(\mathbf{r}_1 t)$ in the normal amplitudes of the velocity potential

$$V_{\text{ext}}(\mathbf{r},t) = \sin(\omega t) \sum_i V_i \phi_i(\mathbf{r}) \tag{31}$$

The expansion coefficients V_i are given by

$$V_i = \int d^3r \, V_{\text{ext}}(\mathbf{r}) n_i(\mathbf{r}) = E_0 \int d^3r \, ZN_i(\mathbf{r}) \tag{32}$$

i.e., it is determined by the dipole moment of the normal mode density oscillation $n_i(\mathbf{r})$.

The response of the system can be written as a superposition of normal modes with dipolar symmetry and time-dependent coefficients

$$\begin{aligned} n_1(\mathbf{r},t) &= \sum_i a_i(t) n_i(\mathbf{r}) \\ \phi_1(\mathbf{r},t) &= \sum_i b_i(t) \phi_i(\mathbf{r}) \end{aligned} \tag{33}$$

Substituting these expansions into the linearized equations of motion we obtain the following differential equations for the coefficents $a_i(t)$ and $b_i(t)$:

$$\begin{aligned} \dot{a}(t) &= -\omega_i^2 b_i(t) \\ \ddot{a}_i(t) + \omega_i^2 a_i(t) &= -\omega_i^2 V_i \sin \omega t \end{aligned} \tag{34}$$

The second equation shows that the amplitude of each normal mode behaves as a driven harmonic oscillator with frequency ω_i. The driving force is proportional to V_i, which according to equation (32) is proportional to the dipole moment of the oscillating charge in the mode i.

In order to calculate the absorption of energy from the external field, we assume that the external field is switched on at $t = 0$ and switched off after a time t which is much larger than ω_i^{-1}. The corresponding solution of equation (34) is

$$a_i(t) = V_i \frac{\omega\omega_i \sin \omega_i t - \omega_i^2 \sin \omega t}{\omega_i^2 - \omega^2} \tag{35}$$

After the field has been switched off, the state of system is a fixed superposition of free oscillations. The amplitude of the mode i at time t and thereafter is obtained by matching the solution obtained from Equation (35) with that for free oscillations and we obtain

$$\omega_i^2 C_i^2 = a_i^2(t) + \dot{a}_i^2(t)/\omega_i^2 \tag{36}$$

which gives the energy in the mode i. For $\omega \neq \omega_i$ this will be an oscillatory function of time according to equation (35). In order to obtain a real absorption, let us consider a band of frequencies around ω_i and integrate over a frequency internal enclosing ω_i. We then obtain an absorption which is proportional to time and the absorption at time t in the mode i is given by

$$E_i(t) = \frac{\pi}{4} V_i^2 \omega_i^2 t \tag{37}$$

The total absorption rate of the system can then be written as

$$\frac{dE}{dt} = \frac{\pi}{4} \sum_i V_i^2 \omega_i^2 \delta(\omega - \omega_i) \tag{38}$$

The photoabsorption cross section is defined as the rate of energy absorption divided by the flux of the incident radiation and we obtain

$$\sigma(\omega) = \frac{2\pi^2 e^2}{nc^2} g(\omega)$$

where

$$g(\omega) = \sum_i \omega_i^2 \left| \int d^3r\, z n_i(\mathbf{r}) \right|^2 \delta(\omega - \omega_i) \tag{39}$$

is the differential oscillator strength distribution.

These formulas show how the photoabsorption cross section can be calculated if the eigenfrequencies ω_i and eigenfunctions for the density oscillations are known. The procedure is very similar to the standard quantum mechanical approach.

From the theory by Bloch[1] one would expect a discrete spectrum of normal modes. In the model of Jensen,[2] in which the atom was treated as a uniformly charged sphere, the hydrodynamical equations were solved with the boundary condition that there is no radial current across the spherical boundary. The modes of this model correspond to the different plasmon modes of a spherical metallic particle and they form a discrete spectrum. Many approximate calculations for atoms using different versions of the hydrodynamical model, often with some kind of cutoff, have also given discrete modes for dipolar oscillations. The first complete analysis and solution of the Bloch equations was given by Wheeler *et al.*[4] The first report on the numerical solution of the Bloch equations and the photoabsorption of the Thomas–Fermi atom was given already in 1957, but the complete publication did not appear until 1973. Contrary to what might have been expected, their analysis showed that the absorption spectrum was continuous, rather than discrete. The origin of this behavior is to be traced to the long range of the Thomas–Fermi ground-state density. The solution starts at $r=0$ and rises to a maximum and then continues to oscillate more and more rapidly, with increasing r, just as was discussed in the previous section. As a consequence it is possible to fulfill the boundary conditions for any frequency. The major result of the work by Wheeler *et al.* is a universal photoabsorption curve, valid for all frequencies and all atomic numbers. The cross section is a smooth curve decreasing monotonously with increasing frequencies. Thus, there is no sign of any collective resonance behavior in the hydrodynamic description of the neutral Thomas–Fermi atom.

The collective oscillations about the Thomas–Fermi ground state have been further discussed by Walecka[8] for both neutral and ionized atoms. He confirms the conclusion by Wheeler *et al.* that there are no discrete resonances for the neutral Thomas–Fermi atom. However, in the ionized atom there is a discrete spectrum because of the finite radius of the Thomas–Fermi density (of Section 4 in Chapter 1). He carried out a variational calculation for the first excited states of several multipole modes for a 10% fractional ionization. Walecka pointed out that the low-lying density oscillations occur in the outermost region of the atom where the Thomas–Fermi model is least reliable, particularly in the regions of low electron density. For ionized atoms, the outermost electrons are tightly bound and the electron density is high out to a rather well-defined atomic radius. He suggests that one might consider the neutral atom as separated into a core part consisting of the tightly bound filled shells and a valence part consisting of the loosely bound outer electrons and then apply the hydrodynamical model to describe the collective oscillations of the core. Such a model has been studied with some success by Serr[9] who has presented results for photoabsorption, form factors and other atomic properties.

So far we have briefly mentioned some results for atoms obtained in the Thomas–Fermi approximation. The situation would to some extent improve by considering oscillations around the exact ground-state density and including, e.g., effects of exchange and correlations. This would introduce some structure due to the density variations associated with the shell structure of atoms. However, qualitatively the situation would not change in any essential way. In most nonuniform systems there are important contributions from density-fluctuations corresponding to single-particle excitations which occur in addition to the hydrodynamical oscillations and there will often be an appreciable interaction between these two types of contribution and in particular it leads to damping of the hydrodynamic oscillation. When this occurs the hydrodynamical theory will be inadequate to describe the physics of the system. Therefore, we have to go beyond the hydrodynamical approach and treat the dynamics of the systems in such a way that the interplay between the collective motion of electrons in the dynamical mean field and the single-particle excitations is properly taken into account. The following sections will describe the development of the theory along these lines.

3. Response of a Nonuniform System to an External Field

In order to develop the theory of density oscillations beyond the hydrodynamical approximation, we need a microscopic approach to account for important physical effects such as the coupling collective oscillation to the single-particle excitations. In developing such an approach it is not always useful to study the eigenmodes of free density oscillations, since they often have a rather short lifetime and a complicated line profile. It is usually more fruitful to study directly the response to a weak external field, which gives a direct approach to calculate the actual spectral profile. We shall summarize very briefly a few key formulas in this section. The linear response theory for nonuniform systems such as atoms have been treated by many authors. Out of the many references we choose to refer to a lucid paper by Roulet and Nozières[10] which also has a good discussion of collective motion in atoms.

In linear response theory the electron density $n_1(\mathbf{r},t)$ induced by a weak scalar field $V_{ext}(\mathbf{r},t)$ is given by

$$n_1(\mathbf{r},t) = \int d^3r' \, dt' \, R(\mathbf{r},\mathbf{r}', t-t') \, V_{\text{ext}}(\mathbf{r}', t') \tag{40}$$

The response function $R(\mathbf{r},\mathbf{r}',t-t')$ is defined as the average of the following retarded commutator:

$$R(\mathbf{r},\mathbf{r}',t-t') = -i\Theta(t-t') \langle [n(\mathbf{r},t),n(\mathbf{r}',t')]\rangle \tag{41}$$

The total electrostatic potential in the system is the sum of the external potential and the polarization potential arising from the induced charges, thus

$$V_{\text{tot}}(\mathbf{r},t) = V_{\text{ext}}(\mathbf{r},t) + V_{\text{pol}}(\mathbf{r},t) \tag{42}$$

with

$$V_{\text{pol}}(\mathbf{r},t) = \int d^3r' \, \frac{n_1(\mathbf{r}',t)}{|\mathbf{r}-\mathbf{r}'|} \tag{43}$$

It is often easier to calculate the response to the total electrostatic potential rather than the external potential and for that reason we introduce a new response relation

$$n_1(\mathbf{r},t) = \int d^3r' \, dt' \, P(\mathbf{r},\mathbf{r}',t-t') \, V_{\text{tot}}(\mathbf{r}',E') \tag{44}$$

The response function $P(\mathbf{r},\mathbf{r}',t-t')$ is often referred to as the irreducible polarization propagator.

We can now introduce the dielectric function $\epsilon(\mathbf{r},\mathbf{r}',t-t')$ and the inverse dielectric function $\epsilon^{-1}(\mathbf{r},\mathbf{r}',t-t')$ through the relations

$$\int d^3r' \, dt' \, \epsilon(\mathbf{r},\mathbf{r}',t-t') \, V_{\text{tot}}(\mathbf{r}',t') = V_{\text{ext}}(\mathbf{r},t) \tag{45}$$

and

$$\int d^3r' \, dt' \, \epsilon^{-1}(\mathbf{r},\mathbf{r}',t-t') \, V_{\text{ext}}(\mathbf{r}',t') = V_{\text{tot}}(\mathbf{r},t) \tag{46}$$

They are related to the response functions $P(\mathbf{r},\mathbf{r}',t)$ and $R(\mathbf{r},\mathbf{r}',t)$ through the formulas

$$\epsilon(\mathbf{r},\mathbf{r}',t) = \delta(\mathbf{r}-\mathbf{r}') - \int d^3r'' \, \frac{P(\mathbf{r}'',\mathbf{r}',t)}{|\mathbf{r}-\mathbf{r}''|} \tag{47}$$

and

$$\epsilon^{-1}(\mathbf{r},\mathbf{r}',t) = \delta(\mathbf{r}-\mathbf{r}') + \int d^3r'' \frac{R(\mathbf{r}'',\mathbf{r}',t)}{|\mathbf{r}-\mathbf{r}''|} \tag{48}$$

One usually works with the Fourier transform with respect to time and sometimes also with transforms with respect to one or both of the space vectors $\mathbf{r}$ and $\mathbf{r}'$. The Fourier transforms are defined as

$$F(\mathbf{q},\mathbf{q}',\omega) = \int d^3r\, d^3r'\, dt\, e^{i\omega t} e^{-i\mathbf{q}\cdot\mathbf{r}} e^{i\mathbf{q}\cdot\mathbf{r}'} F(\mathbf{r},\mathbf{r}',t) \tag{49}$$

The response function $R(\mathbf{q},\mathbf{q}',\omega)$ is directly related to the dynamical form factor $S(\mathbf{q},\omega)$, which is defined as

$$S(\mathbf{q},\omega) = \pi \sum_i |\langle i|n(\mathbf{q})|0\rangle|^2\, \delta(\omega-\omega_{oi}) \tag{50}$$

$|i\rangle$ is an exact eigenstate of the system, $n(\mathbf{q})$ is the Fourier component of the density $n(\mathbf{r})$ and $\omega_{0i} = E_i - E_0$ is the exact excitation energy. The relation between $S(\mathbf{q},\omega)$ and the response function R is simply

$$S(\mathbf{q},\omega) = -\operatorname{Im} R(\mathbf{q},\mathbf{q},\omega) \tag{51}$$

for $\omega > 0$. $S(\mathbf{q},\omega)$ describes the inelastic scattering of, e.g., electrons and photons in the Born approximation. The photoabsorption cross section is given by the formula

$$\sigma(\omega) = \frac{4\pi\omega e^2}{cq^2} S(\mathbf{q},\omega) \tag{52}$$

In the long-wavelength limit $|\mathbf{q}|\to 0$ we can alternatively express $\sigma(\omega)$ in terms of the oscillator strength function $g(\omega)$ through the formulas

$$\sigma(\omega) = \frac{2\pi^2 e^2}{mc} g(\omega) \tag{53}$$

where

$$g(\omega) = -\frac{\omega}{2\pi^2}\int d^3r\, d^3r'\, \operatorname{Im} \epsilon^{-1}(\mathbf{r},\mathbf{r}',\omega) \tag{54}$$

The dynamical structure factor $S(\mathbf{q},\omega)$ contains all the information about the density excitation spectrum of the system and the same is true for $g(\omega)$ in the limit $|\mathbf{q}|\rightarrow 0$. Much of the theoretical work in recent years have been focused on direct calculation of these quantities. However, we can also use these equations and look for normal modes of free oscillations. However, the free oscillations will in general be damped for a nonuniform system and we have then to look for solutions with a complex frequency ω. The free oscillations are characterized by the fact that we can have a field in the system even in the absence of a driving external potential $V_{\text{ext}}(\mathbf{r},\omega)$. According to equation (46) the self-oscillations are obtained from the equation

$$\int d^3r' \, \epsilon(\mathbf{r},\mathbf{r}',\omega) \, V_{\text{tot}}(\mathbf{r}',\omega) = 0 \tag{55}$$

which is a homogeneous integral equation to solve for the internal field $V_{\text{tot}}(\mathbf{r},\omega)$. This equation is the proper generalization of the condition $\epsilon(\mathbf{q},\omega) = 0$ for plasma resonances in a uniform electron liquid.

The dielectric function $\epsilon(\mathbf{r},\mathbf{r}',t)$ is given by equation (47) as soon as one has chosen an approximation for the irreducible polarization function $P(\mathbf{r},\mathbf{r}',t)$. From P one can then solve for the response function R which gives $\epsilon^{-1}(\mathbf{r},\mathbf{r}',t)$ according to equation (48) and the Fourier transform of R gives the dynamic form factor $S(\mathbf{q},\omega)$ using equation (51).

An approximation frequently used in many applications is the random phase approximation (RPA) which can be formulated in many different ways. In the dielectric formulation as presented here, the RPA is obtained by calculating the response to the total potential by using first-order perturbation theory starting from the noninteracting ground state and excitations. The elementary excitations are particle–hole excitations out of the ground-state configuration. The RPA result obtained for the irreducible polarization function is given by the formula

$$P_{\text{RPA}}(\mathbf{r},\mathbf{r}',\omega) = \sum_i \left[\frac{\phi(\mathbf{r})\phi_i^*(\mathbf{r}')}{\omega + i\delta - \omega_i} - \frac{\phi_i(\mathbf{r}')\phi_i^*(\mathbf{r})}{\omega + i\delta + \omega_i} \right] \tag{56}$$

Here ω_i is the particle–hole excitation frequency. The corresponding amplitudes for the density fluctuation is $\phi_i(\mathbf{r}) = u_p^*(\mathbf{r}) \, u_h(\mathbf{r})$, where $u_p(\mathbf{r})$ and $u_h(\mathbf{r})$ are the one-electron wave functions for the particle and hole states, respectively. The RPA theory can be extended by including exchange to obtain the random phase approximation with exchange (RPAE) which has been applied with great success to the photoabsorption spectrum of heavy atoms, particularly by Amusia[11] and Wendin.[12] These authors and others have also considered further refinements of the theory by including self-energy effects which account for the relaxation effects, which occur in the excitation process. We shall not discuss such effects in this chapter and refer the reader to the original papers.

These remarks conclude the presentation of the dielectric response theory. We would like to end this section by mentioning some approximate ways of calculating the response of a nonuniform system without actually attempting to solve the equations of motion. Many workers have introduced a local density approximation assuming that the nonuniform electron gas responds locally as a uniform electron liquid with the same density $n(\mathbf{r})$. The simplest of these approximations neglects completely the nonlocality of the response and describes the system locally by the classical Drude formula

$$\epsilon(\mathbf{r}) = 1 - \frac{\omega_{\text{pl}}^2(\mathbf{r})}{\omega^2} \tag{57}$$

Using equation (54) we obtain the result

$$g(\omega) = \int d^3r\, n_0(\mathbf{r})\delta(\omega - \omega_{\text{pe}}(\mathbf{r})) \tag{58}$$

This formula has a simple interpretation: the single-particle excitations are completely screened out and the atom absorbs at each frequency ω only in a shell where the frequency equals the local plasma frequency $\omega_{\text{pl}}(\mathbf{r})$. The f-sum rule is obviously satisfied by $g(\omega)$. Many improvements of this simple theory have been suggested in which one uses the nonlocal properties of a uniform electron liquid. Such calculations use the results of electron gas theory by using locally the results for the uniform electron with density $n_0(\mathbf{r})$ in the form $\epsilon(\mathbf{q},\omega, n_0(\mathbf{r}))$. The slightly surprising result is that the photoabsorption curves for a number of such approximations all show the same behavior and agree rather closely with one another. Indeed the calculations by Wheeler *et al.*,[4] the classical result given by Equation (58) and all calculations using different approximations for $\epsilon(\mathbf{q},\omega,n_0(\mathbf{r}))$ give equivalent results for a "universal" photoabsorption curve when applied to the Thomas–Fermi density. The results are equivalent in the sense that the difference between the various theoretical results is smaller in relation to the spread in experimental data for different atomic species. This is illustrated in Fig. 1, in which a typical "universal" photoabsorption curve is plotted together with some experimental cross sections.

It is clear from our discussion so far that the rather straightforward extensions of the uniform electron liquid theory do not describe in a satisfactory way the collective motion in nonuniform systems such as atoms. The approaches we have mentioned describe in an *average way* only the photoab-

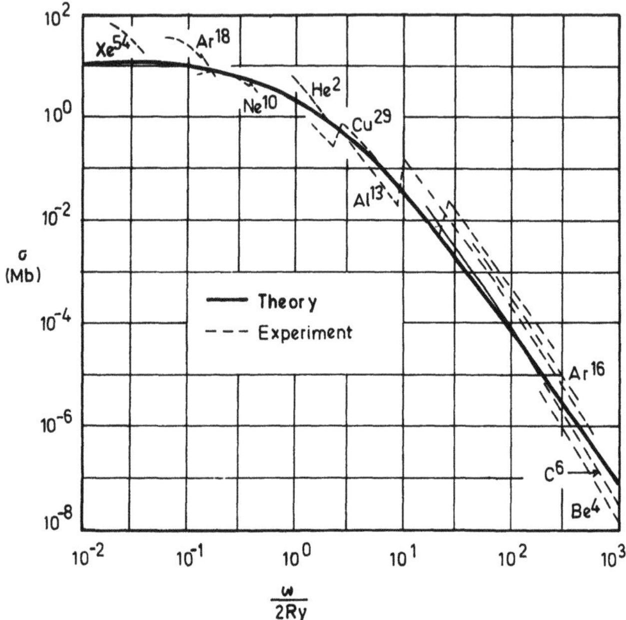

FIGURE 1. A typical "universal" photoabsorption curve compared with some experimental data. (From Ref. 28.)

sorption and stopping power. The results account reasonably well for, e.g., the variation with atomic number Z but they do not describe the actual dynamics of a given nonuniform system.

The statistical model including the description of collective motion has been discussed in an extensive series of papers by Kirzhnitz and co-workers[13] (the reference is to a major review paper containing references to the original papers). They have in particular discussed the limitations of the statistical model of matter. They conclude that the principles of the quasiuniform approach do not generally apply for nonuniform systems such as atoms. Instead they propose a different approach in the semiclassical limit by using the collisionless kinetic equation with a self-consistent field. They found that the answer can be expressed in terms of the classical trajectory of an electron in the self-consistent field (e.g., the Thomas–Fermi field). From the trajectories they construct the polarization function $P(\mathbf{r},\mathbf{r}',\omega)$ and the dielectric function itself. Independently, such an approach using the classical trajectories has been developed by Chakravarty *et al.*[14] to deal with the response of a nonuniform Fermi system. Their approach is described in Chapter 2 of this volume, Section 6.2.

Using the approach just described, Gadiyak, Kirzhnits, and Lozovik[15] calculated the dielectric function for trajectories in the Thomas–Fermi field and solved the eigenvalue equation for collective modes numerically for heavy atoms. They found two dipolar modes, one with the energy $\omega_1 = 13.74Z$ and the damping $\Gamma_1 = 3 \times 10^{-3}Z$ and the second with $\omega_2 = 36.04Z$ and $\Gamma_2 = 10^{-4}Z$. Energies and damping constants are given in electron volts. The corresponding collective oscillations are confined to the core region and are strongly damped inwards as well as outwards. The possible experimental verification of these results has not been discussed. We note that these resonances will occur at rather high energies for heavy atoms ($Z \geq 50$). The resonances of the d shell in Xe, Ba, La, which have been studied by Amusia,[11] Wendin,[12] and others using many-body techniques, occur at considerably lower energies, ~ 100 eV, in a region where accurate experimental data using synchrotron radiation are now available. The question about the existence of such universal collective modes at high energies for heavy atoms in the range above 500 eV does not yet seem to be settled. At lower energies both experimental data and theoretical work indicate that strong collective effects are not general properties of atomic systems but depend on the shell structure and are rather strongly coupled to single-particle excitations.

4. Linearized Quantum Equations for Density Oscillations

In the first part of this chapter we have discussed the hydrodynamic approach to collective motions in bounded systems. The hydrodynamical theory is based on the original Thomas–Fermi idea to treat the system locally as an electron gas characterized by its density. We concluded that an input of microscopic information is needed to account for important physical effects such as the decay of the collective motion by electron–hole excitations. In Section 3 we outlined the elements of the microscopic dielectric formulation which serves as a formal framework for further theoretical developments. We mentioned some applications using the quasiuniform approach to atomic systems, but also the alternative semiclassical approach by Kirzhnitz and coworkers,[13] who calculated the response of an atom using the classical trajectories in the self-consistent field. We may also mention a semiclassical limit of the dielectric theory in the RPA approximation discussed by Brandt and Lundqvist[16] valid in the limit of high quantum numbers. Their work presents a schematic discussion of the qualitative behavior in various frequency regimes but gives no numerical results.

To go beyond the previous discussion, we have to introduce a fully quantum mechanical description of the single-particle motion and the interaction between excitations. According to the scheme described in Section 4 one can introduce successively more and more accurate approximations by calculating the polarization function $P(\mathbf{r},\mathbf{r}',\boldsymbol{\omega})$ first in the RPA, then include exchange to get the results in RPAE, and finally going beyond these theories by including, e.g., self-energy and vertex corrections. Such calculations have been performed by Amusia,[11] Wendin,[12] and many other authors. The papers just referred to contain references to the wide literature in this field. We shall not discuss their line of approach further in this chapter since it would take us away from the central theme of this book which is the inhomogeneous electron gas. In the formulation using, e.g., diagrammatic techniques, the contact with the hydrodynamical approach is completely lost. Although one may see the effects of the dynamical self-consistency, the fact that everything is expressed in terms of single-particle energies, wave functions and matrix elements tend to obscure the possibility of establishing the relation with the classical description.

It is the purpose of this section to show that the linearized many-body approximations such as RPA, RPAE, and further improvements can all be cast in such a form that we can separate the purely classical contributions to the dynamics which only depend on the unperturbed local density $n_0(\mathbf{r})$ and the gradient $\nabla n_0(\mathbf{r})$ from the genuine quantum dynamics in the system, which depends on the single-particle energies and wave functions. We shall also show how the density functional scheme may be used in connection with the dynamical response theory. In this way we hope to bridge the gap between the classical and semiclassical theory discussed in the earlier section and the many-body approach and we hope to make it clear that the two approaches have some very important features in common.

4.1. *Separation of the Instantaneous Interactions from the Retarded Response*

We wish to derive a linearized quantum equation for density oscillations in such a way that one part of the equation has the appearance of a classical equation of motion for a nonuniform liquid and the other part contains the genuine quantum dynamics. The method is to derive directly an equation of motion for the second derivative of the perturbed density $n_1(\mathbf{r},t)$. Such an equation will have the classical terms appearing explicitly but will still include all the many-electron interactions. Such equations have been discussed by many authors. One particular form has been developed by March and Tosi[17] to discuss the propagation of plasmons in periodic lattices. We follow here the discussion in Ref. 18, which is based on the linear response theory in Section 3.

The major point is to rewrite the theory in a form such that the classical part of the response, which arises from the instantaneous Coulomb forces will appear explicitly and is separated to the retarded part, which depends on the dynamics of the electrons. In order to achieve this separation we differentiate the linear response formula for the induced charge density twice with respect to time and obtain on the right-hand side two contributions. One arises from the equal time contribution $t' = t$, which enters because of the unit step function in the density–density response function. The equal time commutator can be worked out from the commutation rules of the operators and does not depend on the dynamics. The other contribution comes from the retarded response and describes the effect of the actual dynamics of the electrons. In order to show the structure of the theory in its simplest form we put $V_{\text{ext}} = 0$ and look for free oscillations, which are described by the equation

$$\left\{\frac{\partial^2}{\partial t^2} + \omega_{\text{pl}}^2(\mathbf{r}) - \nabla n_0(\mathbf{r})\nabla v\right\} n_1 = - \int_{-\infty}^{t} dt' \, \ddot{P}T(t-t')vn_1(t') \tag{59}$$

where $v(\mathbf{r}-\mathbf{r}') = 1/|\mathbf{r}-\mathbf{r}'|$ and

$$\ddot{P}T(t-t') = -i\frac{d^2}{dt^2}\langle[n(\mathbf{r},t),n(\mathbf{r}',t')]\rangle_{\text{irr}} \tag{60}$$

We have used a shorthand matrix notation in equations (59) and (60). The subscript "irr" means that we have considered the response to the total acting field defined by the response function $P(\mathbf{r},\mathbf{r}',t)$.

The eigenvalue equation for free oscillations at a complex frequency takes the form

$$\left\{\omega^2 - \omega_{\text{pl}}^2(\mathbf{r}) + \nabla n_0 \nabla v\right\} n_1 = \ddot{P}T(\omega)vn_1 \tag{61}$$

The classical terms appear on the left-hand side. The classical restoring force is given by the local plasma frequency $\omega_{\text{pl}}(\mathbf{r}) = [4\pi n_0(\mathbf{r})]^{1/2}$. The term containing $\nabla n_0(\mathbf{r})$ is characteristic of a nonuniform system. In classical theory with discontinuous boundaries, this term is replaced by the usual boundary conditions.

It should be stressed that the equations (59) and (61) are *not* semiclassical approximations but in principle fully general linear quantum equations. The physical contents depend on the approximation chosen for $P(\mathbf{r},\mathbf{r},t)$ or $P(\mathbf{r},\mathbf{r}',\omega)$. In a hydrodynamical approach, the right-hand side would be of the form

$\alpha\,\nabla^2 n(\mathbf{r})$ where the density-dependent coefficient depends on whether we consider the low-frequency limit or oscillations of the order of the local frequency $\omega_{pe}(\mathbf{r})$. In the RPA we obtain the explicit formula

$$\tilde{\tilde{P}}T(\omega) = -\sum_i \omega_i^2 \left[\frac{\phi_i(\mathbf{r})\phi_i^*(\mathbf{r}')}{\omega + i\delta - \omega_i} - \frac{\phi_i(\mathbf{r}')\phi_i^*(\mathbf{r})}{\omega + i\delta + \omega_i} \right] \tag{62}$$

We wish to point some general properties of this formulation:

(1) Core contributions cancel out and give no restoring force for oscillations of the system. At any frequency ω we can define the core as those parts of the system for which the excitation frequences $\omega_{\text{core}} >> \omega$. The core contributions in the right-hand side of equation (61) cancel out against the core contributions to the classical terms on the left-hand side. This implies that at frequencies $\omega << \omega_{\text{core}}$ we can replace $n_0(\mathbf{r})$, $\nabla n_0(\mathbf{r})$, and $\omega_{pl}(\mathbf{r})$ with the corresponding contributions from the outer shells only. This feature shows the major response of the systems comes from parts which have excitation frequencies of the same order as the applied frequency ω.

(2) The response function is a complex function and the imaginary part describes the decay of the collective motion into particle–hole excitations.

(3) The detailed nature of the particle–hole spectrum is reflected in the response function and gives rise to structure in the spectrum, which may partly be a screened particle–hole spectrum with more or less pronounced collective resonances as was discussed in the preceding section.

The practical advantage of this formulation will be in applications where one expects that the classical theory should be a good first approximation. We note that the evaluation of the equal time contributions requires the use of closure properties of the one-electron states, exact relations involving the operators, etc. In an approximate application of many-body theory only part of these classical terms will be included.

We should mention that this formulation makes it easy to make contact with different versions of the hydrodynamic theory. In the low-frequency limit we would retrieve in essence the Bloch theory. As remarked earlier, however, we are rather interested in high frequencies. In the limit where ω is considerably higher than the typical electron–hole energies one can expand the response function in inverse powers of ω, with coefficients which can be calculated exactly (at least in principle). Such treatments have been published by several authors and we refer to Ref. 19 for a recent discussion. In application to surface plasmons and similar problems, the frequency does not always seem high enough to ensure that the high-frequency expansion will converge, which leads to some uncertainty about the applicability of the method.

4.2. Density Oscillations Based on the Density Functional Theory

The possible extension of the density functional theory to excited states and time-dependent phenomena has been discussed from different angles. The extension to a hydrodynamic approach for small adiabatic perturbations around the ground-state density by Ying *et al.*[3] has already been mentioned. It was suggested by Gunnarsson and Lundqvist[20] that the lowest excited state of a definite symmetry different from that of the ground state could be calculated using the standard density functional theory. In a recent paper Valone and Capitani[21] claim that functionals exist to calculate all bound excited states. This approach is discussed in Chapter 2 of this volume, Section 6.1, and in particular the rather severe limitations of the method. In Chapter 2 there is also a discussion of the transport properties. It is mentioned that the dynamical transport coefficients can be expressed as ground-state expectation values of the appropriate operators, and therefore are functionals of the unperturbed ground-state density. Unfortunately very little is known about the properties of these functionals except in particular limits such as high frequencies or very short wavelengths. Chapter 2 contains also a presentation of an extension of the Fermi liquid theory to systems where the density varies on a length scale large compared to the local p_F^{-1}. In this case a local density formulation is obtained and the density response to an external potential can be calculated from the classical trajectories in the self-consistent field, as was briefly mentioned in Section 3.

An attempt to formally construct a time-dependent generalization of density functional theory has been made by Peuckert.[22] He uses the action principle for the time-dependent Schrödinger equation for a many-electron system in the presence of a small time-dependent external field. He proceeds in analogy with the Hohenberg–Kohn theory and concludes that the action integral can be understood as a functional of the external potential $V_{ext}(\mathbf{r},t)$. However, in analogy with density functional theory, one would like to have the action functional expressed as a functional of the density $n(\mathbf{r},t)$. It is obvious that the external potential $V_{ext}(\mathbf{r},t)$ determines the local density $n(\mathbf{r},t)$ but the key question is whether this mapping can be inverted. Hohenberg and Kohn used the minimal property of the energy functional to prove that the mapping can be inverted in the static case, but the weaker stationary property of the action functional is not sufficient to prove a corresponding theorem in the neighborhood of $V_{ext}=0$. This is fulfilled *if* the inverse of the linear response matrix

$$R = \left.\frac{\delta n}{\delta V_{ext}}\right|_{V_{ext}=0} \tag{63}$$

exists. This is possible only if potentials which do not give any induced potentials to first order, i.e., fullfilling $RV_{ext}=0$, are excluded in the mapping. This brief summary of the arguments shows that there is no powerful theorem proved so far granting the extension of density functional theory to time-dependent problems. However, if the assumptions mentioned by Peuckert are satisfied, he shows that the time-dependent particle density $n(\mathbf{r},t)$ of the system can be calculated from one-electron wave functions of a time-dependent wave equation for noninteracting fermions in an effective one-particle potential.

Zangwill and Soven[23,24,25] have recently developed a time-dependent generalization of the density functional formalism and applied the method to the calculation of the frequency-dependent polarizability of closed-shell atoms. The method is a generalization of the linear response formalism in an intuitively reasonable way.

In the form of linear response theory used so far in this chapter we have considered the total field acting on an infinitesimal test charge. This field is the sum of the external field and the field due to the induced charges. The field on an electron differs from that on a test charge because of the local field due to the exchange-correlation hole around the electron. In the density functional scheme the effect of exchange and correlation is represented by the local exchange-correlation potential $V_{xc}(\mathbf{r})$. Zangwill and Soven used the local density approximation in their papers, but the theory is formally unchanged using a general form for $V_{xc}(\mathbf{r})$. It is a natural step to improve the linear response theory by adding the change in the exchange-correlation potential. Writing the density as $n(\mathbf{r},\omega) = n_0(\mathbf{r}) + n_1(\mathbf{r},\omega)$, we have to first order that

$$V_{xc}(\mathbf{r},\omega) = V_{xc}(\mathbf{r}) + \left.\frac{\partial V_{xc}}{\partial n(\mathbf{r})}\right|_{n=n_0} n_1(\mathbf{r},\omega) \tag{64}$$

We now approximate the effective perturbing field acting on an electron by

$$V_{eff}(\mathbf{r},\omega) = V_{ext}(\mathbf{r},\omega) + \int d^3r\, \frac{n_1(\mathbf{r}',\omega)}{|\mathbf{r}-\mathbf{r}'|} + \left.\frac{\partial V_{xc}}{\partial n}\right|_{n=n_0} n_1(\mathbf{r},\omega) \tag{65}$$

We now calculate the induced density due to $V_{eff}(\mathbf{r},\omega)$, using time-dependent perturbation theory based on the one-electron wave equations in the density functional theory generalized to include time-dependent fields. As mentioned earlier, the questions about the validity of such a generalization do not yet seem to be settled and we refer to the paper by Peuckert[22] for a detailed analysis and discussion. With this approach we obtain the relation

$$n_1(\mathbf{r},\omega) = \int d^3r' \, P_{\mathrm{TDLDA}}(\mathbf{r},\mathbf{r}',\omega) V_{\mathrm{eff}}(\mathbf{r}',\omega) \tag{66}$$

where

$$P_{\mathrm{TDLDA}}(\mathbf{r},\mathbf{r}',\omega) = \sum_i \left[\frac{\phi_i(\mathbf{r})\phi_i^*(\mathbf{r}')}{\omega + i\delta - \omega_i} - \frac{\phi_i(\mathbf{r}')\phi_i^*(\mathbf{r})}{\omega + i\delta + \omega_i} \right] \tag{67}$$

The response function $P_{\mathrm{TDLDA}}(\mathbf{r},\mathbf{r}',\omega)$ is identical in form with the RPA formula in equation (56). However, the wave functions for particles and holes as well as the corresponding excitation energies are now calculated from the one-body wave equation in the density functional theory with the exchange-correlation potential v_{xc} corresponding to the ground-state density $n_0(\mathbf{r})$. TDLDA stands for time-dependent local density approximation (the notation used by Zangwill and Soven), but the formula as such is independent of the particular choice of v_{xc}. To calculate the total dipole moment and the polarizability we have to express the induced density $n_1(\mathbf{r},\omega)$ in terms of the external potential

$$n_1(\mathbf{r},\omega) = \int d^3r' \, R(\mathbf{r},\mathbf{r}',\omega) V_{\mathrm{ext}}(\mathbf{r}',\omega) \tag{68}$$

For the response function R one obtains an integral equation involving P_{TDLDA}, similar to the corresponding integral equation in the RPA. We refer to Refs. 23–25 for a detailed discussion and the procedure how the equations are solved.

We mention that for $\omega = 0$ the theory of linear response to an external field is firmly based on the Hohenberg–Kohn theorem and the use of the density functional method in linear response is in principle an exact method to calculate the static polarizability $\alpha(0)$. The accuracy of the results depends only on the choice of exchange-correlation potential v_{xc}, and as already mentioned most calculations use the local density approximation.

The static case is discussed as the limit $\omega = 0$ by Zangwill and Soven. We also mention that Stott and Zaremba[26] independently discussed and solved these equations in the static case. The results obtained for $\alpha(0)$ agree well with each other as well as with experimental data. Recently Mahan[27] has used this approach to compute the polarizability corresponding to external fields of higher polarity. The applications in the dynamic case will be discussed in Section 5.

5. Applications to Nonuniform Systems

The purpose of this chapter has mainly been to review the classical work on density oscillations based on the Thomas–Fermi model and to give a brief summary of the extensions of the theory made later as well as some of the more recent ideas to deal with density oscillations in nonuniform systems. We have deliberately excluded the typical many-body approaches, such as the use of Green's function techniques, diagrammatic methods, etc., and focused on aspects where the concepts and methods in the theory of the inhomogeneous electron gas play an important role. In this last section, dealing with applications, we do not attempt to give a detailed report on the multitude of applications made. Most of the literature up to now has dealt with applications of the hydrodynamic model. As we have discussed in earlier sections, the limitations of this approach are quite severe and the results obtained in many applications are at most qualitatively correct. Many of these applications are treated in good reviews devoted especially to the particular area of physics where they belong. The field is now changing with regard to applications. The more powerful methods of self-consistent theories based on quantum theory are now being applied to these problems, and they are setting a new course and a new standard to treat time-dependent phenomena in nonuniform systems.

Instead of dealing with a large variety of applications we shall limit our discussion to a few remarks which do not have to do with the details of the treatment but are more often directly related to the physics of nonuniform systems. We have somewhat arbitrarily chosen three areas of application: (a) surface plasmons (b) small spherical particles and (c) atoms and molecules, which will be discussed in the following three subsections.

5.1. *Surface Plasmons in Jellium*

Surface plasmons are density oscillations localized to the surface region of a metal. They have infinite extension in the plane of the surface, which is here considered as an ideal planar surface. Perpendicular to the surface the density oscillation decays rapidly when we go into the interior as well as when we go out into the low-density region of the density profile.

We first recall the classical limit. Here we consider the case of a sharp interface between the metal and vacuum. We describe the metal by a classical dielectric function

$$\epsilon(\omega) = 1 - \omega_{\text{pl}}^2/\omega^2 \qquad (69)$$

where ω_{pl} is the average bulk plasma frequency of the metal. We consider the quasistatic limit, where retardation effects can be neglected. Classical theory gives the resonance condition $\epsilon(\omega) + 1 = 0$ and therefore the classical surface plasmon frequency is $\omega_s = \omega_{pl}/\sqrt{2}$.

The surface density oscillations have the following form, choosing $z = 0$ as the plane of the surface

$$n_1(\mathbf{r},t) = n_1(z,\omega)e^{i\mathbf{q}\cdot\mathbf{x}}e^{-i\omega t} \quad (70)$$

where $\mathbf{q}$ is the two-dimensional wave vector in the surface plane and $\mathbf{x} = (x,y)$ the two-dimensional position vector. In the classical model with a sharp surface, the space amplitude part is simply

$$n_1(z,\omega_s) = A\delta(z) \quad (71)$$

We can also consider a classical model with a continuous density profile by using the equations of motion (59) and (61) with the right-hand side equal to zero. This equation gives a solution for arbitrary $\mathbf{q}$, which reflects the shape of the density profile and permits an explicit solution at $|q| = 0$ of the form[18]

$$n_1(z,\omega_s) = A\frac{d}{dz}\frac{1}{\omega_s^2 - \omega_{pl}^2(z)} \quad (72)$$

where $\omega_{pl}(z)$ is the local plasma frequency at z. The amplitude has a singularity in the plane where $\omega_{pl}(z) = \omega_s$, which is a consequence of not having any damping mechanism in the equation.

We next turn to a discussion of collective surface modes using the hydrodynamical model. There is a recent extensive treatment by Barton[29] which includes many aspects of the hydrodynamical model that are often ignored or treated in a nonsatisfactory manner in the literature. The hydrodynamical theory gives a space amplitude for the surface plasmon which reflects the properties of the surface profile. The influence of the density profile on surface plasmons has been discussed by Equiluz *et al.*[30] They consider the hydrodynamical model based on the density functional scheme described in Section 2.2. For a sufficiently diffuse surface they not only obtain the usual surface plasmon, but also higher multipolar modes were obtained where the amplitude $n_1(z,\omega)$ has one or several nodes. The possible existence of such multipolar modes seem to depend very sensitively on the choice of surface profile.

Surface plasmons in the RPA have been studied in particular by Feibelman[31,32,33] and by Inglesfield and Wikborg.[34] Feibelman used the self-consistent potential and charge distribution by Lang and Kohn[35] and is therefore not at all a standard RPA calculation but is probably the earliest development and application of the theory discussed in Section 4.2. The work by Inglesfield and Wikborg, on the other hand, is a model RPA calculation based on the finite barrier model.

The plasmon amplitude $n_1(z,\omega)$ differs in a striking way from the amplitudes in classical theory or in the hydrodynamical model. Whereas the extension of the profile in the classical and hydrodynamical models roughly corresponds to the extension of the density profile, the RPA amplitude exhibits a long-range part which extends deep into the bulk, typically 25–50 Å for small $|\mathbf{q}|$. This behavior is of the same origin as the spatial oscillations in the longitudinal electromagnetic field which occur in the scattering of p-polarized light by a metal surface.[32,36,37,38] The form of the plasmon amplitude $n_1(z,\omega)$ is illustrated in Fig. 2. The deep tail in $n_1(z,\omega)$ is directly related to the strong nonlocality of the microscopic response functions in a quantum mechanical description, which is not included in a classical or hydrodynamical description. This strong nonlocality is related to the excitation spectrum of the Fermi gas and the existence of a Fermi surface. The oscillations in $n_1(z,\omega)$ can be considered as a frequency-dependent generalization of the Friedel oscillations in the static case. These features are discussed in detail in Ref. 38.

We now turn to the dispersion of surface plasmons. Irrespective of the shape of the surface profile, the long-wavelength limit gives the classical surface plasmon frequency $\omega_s = \omega_{pl}/\sqrt{2}$. We should mention that at very long wavelengths retardation effects become important. The surface density oscillation will couple strongly to the electromagnetic field and we have as a result a strongly dispersive mode usually called a surface polariton or surface

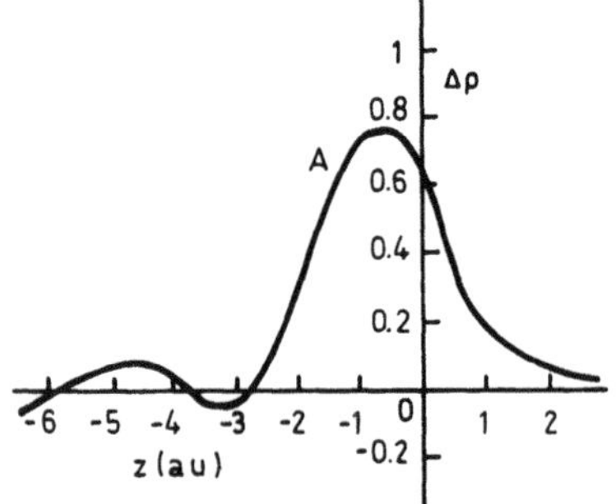

FIGURE 2. Typical surface plasmon amplitude (real part) for small q. (From Ref. 25.)

electromagnetic wave. The wavelength region for surface polaritons is of the order of the corresponding light wave. When discussing surface plasmons at long wavelengths we consider wavelengths shorter than the wavelength of the electromagnetic field but much longer compared with the lattice constant. In this region the coupling to the electromagnetic field is small and the quasistatic description can be used. The dispersion of surface plasmons is given by $\omega_s(q)$. There is an important general result for small q independent of any model, which was first found by Harris and Griffin[39] using the RPA and was soon thereafter derived in a general form by Garcia–Moliner and Flores.[40] The result is usually written in the form

$$\begin{aligned} \omega_s^2(q) &= \omega_s^2(0)\left[1 + q\frac{\int dz\, z n_1(z)}{\int dz n(z)}\right] \\ &= \omega_s^2(0)\,[1 + (a_1 + i\,a_2)q] \end{aligned} \tag{73}$$

Here $n_1(z)$ is the eigenfunction for the surface plasmon amplitude at $q=0$. The formula is valid for any model in the long-wavelength limit. It shows explicitly that the linear dispersion coefficient is given by the total dipole moment of the surface density oscillation. We note that $n_1(z)$ is generally complex since the surface plasmon is damped even at $q=0$, due to the decay into electron–hole pairs. The dipole moment is an important quantity for many surface properties as has recently been discussed by Feibelman[33] and by Apell.[41]

Experimental data seem to indicate that the dispersion coefficients a_1 and a_2 are *negative*. A negative coefficient means that the density fluctuation occurs essentially outside the edge of the positive background. (We have here chosen the coordinates so that the metal extends in the positive z direction.) Most model calculations, including the hydrodynamical model, give a *positive* dispersion coefficient. In Table 1 we have presented a few results for a_1 and a_2. We notice the remarkable spread in different models and that most of them even have a dipole of the opposite sign as the experimental values. The result obtained by Feibelman[31] using the Lang–Kohn results indicates that a self-consistent theoretical treatment of the density profile, and the dynamic response function is needed to obtain a correct result.

The formula (73) is only valid for small q compared to K_F or a^{-1}. In particular the negative dispersion at small q will change and give a positive dispersion at larger q values. Inglesfield and Wikborg[34] have calculated the plasmon dispersion curve in the RPA up to the cutoff wave number where the plasmon branch merges with the electron–hole continuum. Because of the broken translation invariance perpendicular to the surface there is a

TABLE 1. The Surface Plasmon Dispersion Coefficients a_1 and a_2 in Å[a]

Model	a_1	a_2
Hydrodynamic[42]	0.90	0.0
Semiclassical infinite barrier[43]	0.98	−0.06
Infinite barrier[44]	0.26	−0.05
Finite step[45]	0.32	−0.18
Lang–Kohn[31]	−0.42	−1.14
Experiment[46]	−0.14	−0.14
Experiment[47]	−0.36	−0.60

[a]Equation (73) (from Ref. 41).

coupling to electron–hole excitation at all values of q, which leads to a damping of the surface plasmons for all values of q.

5.2. *Small Spherical Particles and Spherical Shells*

The small metallic sphere was historically the first application of Bloch's hydrodynamical theory. Jensen[2] introduced a simplified model in which the atom was treated as a small metallic sphere with constant ground-state density. The hydrodynamical model was solved with the boundary condition that there should be no radial flow through the spherical boundary. Later work used more realistic models. However, the metallic sphere has continued to be of considerable interest as a model for a small metallic particle and for aggregates of such particles. We shall limit ourself to the study of metallic particles large enough to be treated as an approximately uniform electron gas with a spherical boundary. We assume that the density falls from the bulk value to zero over a distance small compared to the radius R of the sphere, so that we can, if we wish, replace the actual density profile by a step function.

The classical condition for resonant absorption of light by a small metal sphere is given by $\epsilon(\omega) + 2 = 0$. For a free-electron-like metal this gives $\omega_{sph} = \omega_{pl}/\sqrt{3}$, where ω_{pl} is the bulk plasma frequency. This frequency is just the first in a series of surface plasmon modes of the sphere. To study these modes we can use equation (59) with the right-hand side equal to zero, which immediately gives the following integral equation to solve for the plasma modes[18]:

$$n_1(\mathbf{r},\omega) = \frac{1}{\omega^2 - \omega_{pl}^2(\mathbf{r})} \frac{dn_0}{dr} \frac{d}{dr} \int d^3r' \frac{n_1(\mathbf{r}',\omega)}{|\mathbf{r}-\mathbf{r}'|} \tag{74}$$

For a spherical system we can expand in spherical harmonics:

$$n_1(\mathbf{r},\omega) = \sum_{lm} n_l(\mathbf{r},\omega) Y_{lm}(\theta\phi) \tag{75}$$

Assuming the density to be constant with a discontinuity on the surface, we can write

$$n_l(r,\omega) = A_l \delta(r-R) \tag{76}$$

showing that the density oscillation is localized to the surface of the sphere. We can consider two cases: (a) a metallic sphere or (b) a spherical hole in a uniform bulk metal. The solutions are

$$\omega_l^2 = \omega_{\text{pl}}^2 \frac{l}{2l+1} \qquad \text{(sphere)} \tag{77}$$

and

$$\omega_l^2 = \omega_{\text{pl}}^2 \frac{l+1}{2l+1} \qquad \text{(hole)} \tag{78}$$

where ω_{pl} is the bulk plasma frequency of the metal.

We note that these well-defined modes are a consequence of the discontinuity in the density at the surface. For a smooth density profile we would in general obtain a distribution of frequencies around the classical values. However, if R is very large compared with the region over which $n_0(\mathbf{r})$ varies, we can neglect the width of the spherical plasmons.

Closely related is the case of a spherical shell. Let us introduce d as the width of the shell and R as the average radius of the shell. Looking for modes confined to the surfaces we write

$$n_l(r,\omega) = A_l \delta(r-R+d/2) + B_l \delta(r-R-d/2) \tag{79}$$

The plasmon modes are found by solving equation (74) and we obtain

$$\omega_\pm^2 = \omega_{\text{sp}}^2 \left\{ 1 \pm \frac{1}{2l+1} \left[1 + 4l(l+1)x^{2l+1} \right]^{1/2} \right\} \tag{80}$$

where $\omega_{\text{sp}} = \omega_{\text{pl}}/2$ is the surface plasma frequency and

$$x = \frac{1-d/2R}{1+d/2R} \tag{81}$$

To see the meaning of these two modes we consider the case of a very thick shell which corresponds to the limit $d/2 \rightarrow R$ and $x \rightarrow 0$. By comparison with equations (77) and (78) we find that the ω_+ modes correspond to oscillations on the inner surface and the ω_- modes will correspond to oscillations localized on the outer surface. For a thick shell these modes are uncoupled, but for $x \neq 0$ the oscillations on the inner and outer surface will couple and the frequencies will shift as given by equation (80). The situation is analogous to the coupling of surface plasmons on the two sides of a thin film.

We now wish to consider the plasma oscillations of a small metal sphere using the hydrodynamic theory. We refer for details to a comprehensive study by Ruppin[48] which also gives references to earlier papers. Ruppin treats also the case of a sphere with a diffuse surface. For a spherical particle of homogeneous density one has to find the solutions of the equation

$$(\nabla^2 + k^2)n_1(\mathbf{r},\omega) = 0 \tag{82}$$

with

$$k^2 = \frac{\omega^2 - \omega_{\text{pl}}^2}{\beta^2} \tag{83}$$

The boundary conditions on the solution is that the radial component of the hydrodynamic velocity vanishes at the surface at all times.

We have two types of solutions for a spherical particle. For $\omega > \omega_{\text{pl}}$ the wave number k is real and we have modes in analogy with plasmons in bulk matter. The radial amplitudes are spherical Bessel functions, which show an oscillatory behavior with amplitudes extending throughout the volume of the sphere. For each value of l one obtains a discrete spectrum of levels, whose separations decrease with increasing radius R. Since the frequency is larger than the bulk plasma frequency the metal spheres are transparent in this frequency region.

The other category consists of the surface modes having $\omega < \omega_{\text{pl}}$. The interior of the sphere cannot sustain a plasma-type oscillation at $\omega < \omega_{\text{pl}}$ and the density oscillation is now confined essentially to a thin spherical shell around the surface of the sphere. The wave number is now imaginary and the equation for the plasma oscillations becomes

$$\begin{aligned} (\nabla^2 - \alpha^2)n_1(\mathbf{r},\omega) &= 0 \\ \alpha^2 &= \frac{\omega_{\text{pl}}^2 - \omega^2}{\beta^2} \end{aligned} \tag{84}$$

Instead of the spherical Bessel functions the radial solutions obtained will now be the modified spherical Bessel functions of the first kind. For a large sphere these solutions fall off rapidly (exponentially) inwards from the surface of the sphere. The solutions give a sequence of surface modes, one for each value of l (l = 1,2,3, . . .). In the limit of a large sphere, i.e., $\alpha R \rightarrow \infty$, we retrieve the classical solution given by equation (77), but for a finite R the frequencies shift upwards compared to the classical values.

We are not aware of any systematic study of plasmons in small metal spheres based on the RPA or similar methods. One can infer from the result for planar surfaces that the strong nonlocality of the response functions should give rise to characteristic differences compared with the results of hydrodynamic theory. The dispersion of the modes is expected to change and the spatial behavior of the plasmon amplitudes will be different. In particular the oscillations are expected to extend deeper into the bulk region.

5.3. *Atoms and Molecules*

The applications to atoms and molecules can be divided into two groups. In the first group one has applied either the hydrodynamic theory or some kind of quasiuniform assumption in an approximate dielectric theory or some other semiclassical approximation which does not take into account the shell structure and corresponding single-particle spectrum of the atom. We reviewed these theories in Sections 2 and 3 and concluded that these theories are in most cases only capable of giving an average description of the dynamical properties, such as the universal photoabsorption curve presented in Fig. 1. A theory which can describe in a quantitative way the properties of a particular atom must be based on a quantum description based on the one-electron excitation spectrum and corresponding wave functions. A number of such formulations and applications of many-body theory have been presented. The work in recent years has demonstrated the power of these methods and it seems that the many-body physics of atomic systems is now rather well developed and in its general features quite well understood. Since most of the recent work on atoms and molecules has used methods which have rather little connection with the theory of the inhomogeneous electron gas, we have not attempted to review these developments here. We refer to the proceedings of a recent Nobel symposium for an extensive presentation of methods and results.[(49)].

In Section 4 we presented two different approaches, which are directly related to the density methods, but which take into account the shell structure and one-electron spectrum at the same level as the standard many-body formulations.

The theory formulated in Section 4.1 separates the instantaneous (classical) Coulomb forces from the retarded response, which depends on the quantum properties of the electrons. No extensive application of this theory has yet been published. The only application has been to a simplified model of the dipolar oscillations of an atomic shell[50]. An outer (nl)-shell of a heavy atom has for the higher angular momenta a density distribution, which is rather well localized between an inner radius r_1 and an outer radius r_2. As a first approximation one may approximate the actual density distribution by a constant density between r_1 and r_2 and zero density outside. We then have the model of a spherical shell which was discussed in Section 5.2. Application of this model to the $4d^{10}$ shell in Xe results in two dipolar modes with the frequencies $\hbar\omega_- = 95$ eV and $\hbar\omega_+ = 160$ eV. About 80% of the oscillator strength resides in the lower frequency mode. It is interesting to note that a strong maximum in the experimental photoabsorption curve occurs around 100 eV. The close agreement between the numbers is not significant. However, the result shows clearly that the classical restoring force gives a frequency of the right order of magnitude. We also point out that the nature of the collective oscillation is qualitatively different for a spherical shell in comparison with bulklike plasmon modes in an almost uniform system. The oscillations occur near the boundary surfaces of the shell and the oscillations of the outer and inner surfaces will couple. The frequency ω_- would tend to the plasma frequency $\omega_{pl}/\sqrt{3}$ of a uniform sphere for a thick shell but is in this example smaller by about 25% because of the coupling to the inner surface. The frequency ω_- would go to zero if the thickness of the shell goes to zero.

The density functional approach described in Section 4.2 has been applied to the photoabsorption spectrum of a number of atoms: Ne, Ar, Kr, Xe, Ba, Cl, Ce, by Zangwill and Soven[23–25]. They also report some preliminary considerations with regard to applications to molecules such as N_2 and SF_6. Most of the atoms listed above have also been studied by Amusia[11] and by Wendin,[12] who used the RPAE and in a few cases approximations beyond the RPAE including relaxation effects. We show an example of their result in Fig. 3, which gives the absorption spectrum of the $4d^{10}$ shell in Xe. The dashed line gives the RPAE cross section,[11] the dotted line gives the experimental result, and the solid line shows the TDLDA result. We notice the very close agreement with experiments, considerably better than the RPAE results. Inclusion of relaxation effects improves the agreement with experiments[12] and give about as good agreement as the TDLDA result.

The $4d^{10}$ absorption spectrum in Ba is even more striking, as is shown in Fig. 4.[25] The TDLDA calculation agrees extremely well with the experimental curve except for the detailed structure near threshold and at the

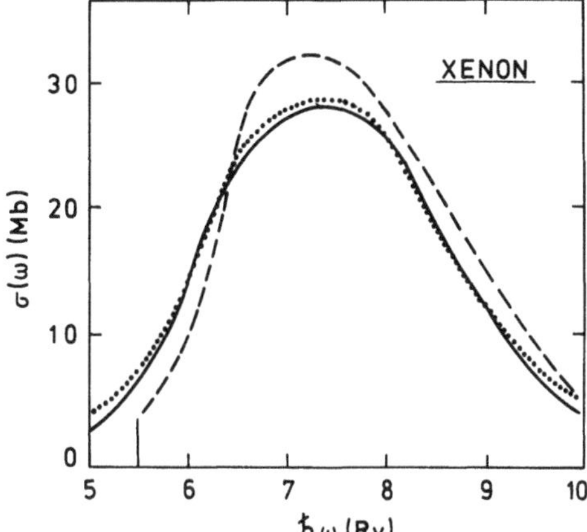

FIGURE 3. Photoabsorption cross section $\sigma(\omega)$ for the $4d^{10}$ shell in Xe. Dotted line, experiment; dashed line, RPAE; solid line, TDLDA. (From Ref. 25.)

absorption peak. The RPAE result[10] gives the peak position rather accurately but fails to account for the line shape. If relaxation effects are included[12] the agreement with the experimental curve will be about as good as the TDLDA result. The surprising feature is that the density functional scheme does not include relaxation effects and yet the TDLDA gives results which are comparable to those obtained if relaxation effects are added to the RPAE scheme, and it is definitely superior to the RPAE results.

The collective dynamics of the $4d$ shell in Xe is obtained by studying the

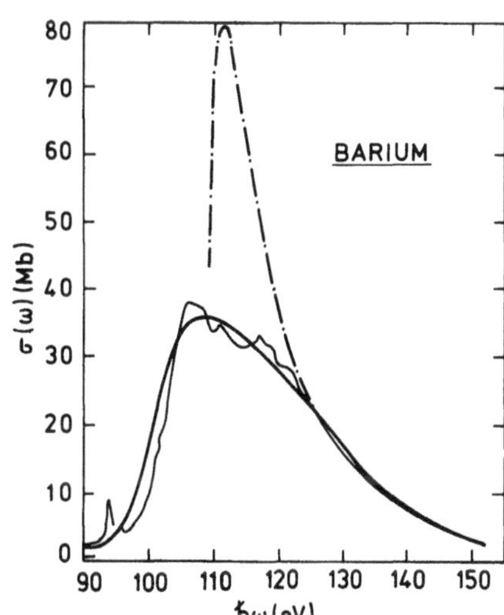

FIGURE 4. Photoabsorption cross section $\sigma(\omega)$ for the $4d^{10}$ shell in Ba. Solid line, TDLDA; dashed–dot line, RPAE. (From Ref. 25.)

properties of the effective electric field inside the atom. This is shown in Fig. 5 together with the $4d$ radial wave function. The field is shown at 5.5 Ry photon energy, which is below the resonance peak; at 7.5 Ry, which is at the resonance peak, and at 9.5 Ry, which is above the resonance. There is extremely strong screening at short distances within the $4d$ radius at 5.5 Ry. The field is about six times stronger than the external field and points in the opposite direction. Outside the $4d$ radius there is an antiscreening, i.e., an enhancement of the field. After passing through the resonance the situation is reversed at 9.5 Ry, above the resonance. There is now a strong antiscreening inside the $4d$ radius and a weak screening outside the $4d$ radius.

We notice that the system behaves qualitatively as a driven harmonic oscillator with regard to its frequency dependence. It changes from screening to antiscreening when we pass through the resonance frequency. The effective field plays an important role in calculating the cross section for photoemission. The matrix element for the process is simply obtained as the matrix element of the effective potential between unperturbed electron states. The strong spatial and frequency dependence of the field implies that the very strong induced fields will influence also the photoemission from other shells in the atom.[25]

The successful applications of the density functional method to collective motion in heavy atoms seem to suggest that the method could be successfully

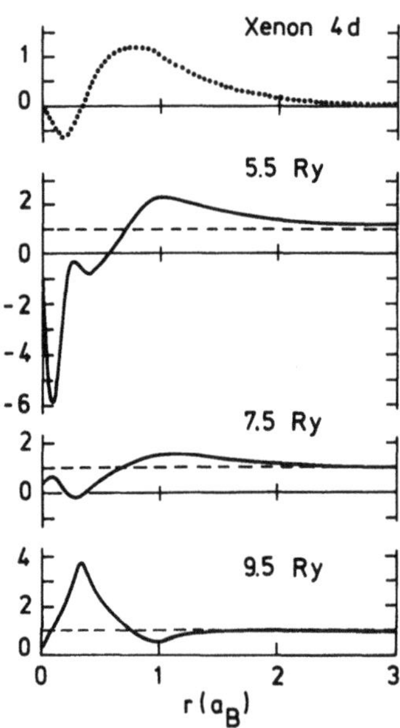

FIGURE 5. Radial $4d$ wave function for Xe. Effective electric field along the polar axis at three different frequencies. The dashed line is the constant external field.

applied also to the photoabsorption spectra of molecular systems, where rather little is known at present about the density oscillation spectrum.

References

1. F. Bloch, *Z. Phys.* **81,** 363 (1933).
2. H. Jensen, *Z. Phys.* **106,** 620 (1937).
3. S. C. Ying, J.R. Smith, and W. Kohn, *J. Vac. Sci. Technol.* **9,** 575 (1972).
4. J.A. Ball, J.A. Wheeler, and E. L. Fireman, *Rev. Mod. Phys.* **45,** 333 (1973).
5. J. Lindhard, *Dan. Mat. Fys. Medd.* **28,** 8 (1954).
6. P.A.M. Dirac, *Proc. Cam. Phil. Soc.* **26,** 376 (1930).
7. W. Kohn and L. Sham, *Phys. Rev.* **140,** A1133 (1965).
8. J.D. Walecka, *Phys. Lett.* **58A,** 81 (1976).
9. F.E. Serr, *Phys. Lett.* **62A,** 325 (1977).
10. P.B. Roulet and P. Noziéres, *J. Phys. (Paris)* **29,** 167 (1968).
11. M. Ya. Amusia and N. A. Cherepkov, *Case Studies in Atomic Physics* **5,** 47 (1976).
12. G. Wendin, in *Photoionization and Other Probes of Many-Electron Interactions*, F. Wuilleumier editor, NATO Advanced Study Institute series (Plenum Press, New York, 1976).
13. D.A. Kirzhnitz, Yu. E. Lozovik, and G.V. Shpatakovskaya, *Sov. Phys. Usp.* **18,** 649 (1976).
14. S. Chakravarty, M.B. Vogel, and W. Kohn, *Phys. Rev. Lett.* **43,** 775 (1979).
15. G.V. Gadiyak, D. A. Kirzhnits, and Yu. E. Lozovik, *Sov. Phys. JETP* **42,** 62 (1976).
16. W. Brandt and S. Lundqvist, *Ark. Fys.* **28,** 399 (1965).
17. N.H. March and M.P. Tosi, *Proc. R. Soc. (London)* **A330,** 373 (1972).
18. G. Mukhopadhyay and S. Lundqvist, *Nuovo Cimento* **27B,** 1 (1975).
19. G. Mukhopadhyay, *Physica Scripta* (in press).
20. O. Gunnarsson and B. .I. Lundqvist, *Phys. Rev. B* **13,** 4274 (1976).
21. S.M. Valone and J.F. Capitani, *Phys. Rev. A* **23,** 2127 (1981).
22. V. Peuckert, *J. Phys. C.: Solid State Phys.* **11,** 4945 (1978).
23. A. Zangwill and P. Soven, *Phys. Rev. Lett.* **45,** 204 (1980).
24. A. Zangwill and P. Soven, *Phys. Rev A* **21,** 1561 (1980).
25. A. Zangwill, Dissertation, University of Pennsylvania, 1981.
26. M.J. Stott and E. Zaremba, *Phys. Rev. A* **21,** 12 (1980).
27. G.D. Mahan, *Phys. Rev. A* **22,** 1780 (1980).
28. W. Brandt, L. Eder, and S. Lundqvist, *J. Quant. Spectrosc. Radiat. Transfer* **7,** 185 (1967).
29. G. Barton, *Rep. Progr. Phys.* **42,** 963 (1979).
30. P.J. Feibelman, *Phys. Rev. B* **9,** 5077 (1974).
32. P.J. Feibelman, *Phys. Rev. B* **12,** 1319 (1975).
33. P.J. Feibelman, *Phys. Rev. B* **14,** 762 (1976).
34. J.E. Inglesfield and E. Wikborg, *Solid State Commun.* **14,** 661 (1974).
35. N.D. Lang and W. Kohn, *Phys. Rev. B* **1,** 4555 (1970).
36. G. Mukhopadhyay and S. Lundqvist, *Phys. Scripta* **17,** 69 (1977).
37. P. Apell, *Phys. Scripta* **17,** 535 (1977).
38. P. Apell, *Phys. Scripta* **25,** 57 (1982).
39. J. Harris and A. Griffin, *Phys. Lett.* **34A,** 51 (1971).

40. F. Garcia-Moliner and F. Flores, *Introduction to the Theory of Solid Surfaces* (Cambridge University Press, 1979). The book contains references to the original papers.
41. P. Apell, *Phys. Scripta* **24,** 795 (1981).
42. R.H. Ritchie, *Phys. Rev.* **106,** 874 (1957).
43. R.H. Ritchie and A. L. Marusak, *Surf. Sci.* **4,** 234 (1966).
44. Ch. Heger and D. Wagner, *Z. Phys.* **244,** 449 (1971).
45. E. Wikborg and J. E. Inglesfield, *Phys. Scripta* **15,** 37 (1977).
46. A. Bagchi, C.B. Duke, P.J. Feibelman, and J.O. Porteus, *Phys. Rev. Lett.* **27,** 998 (1971).
47. C.B. Duke and V. Landman, *Phys. Rev. B* **8,** 505 (1973).
48. R. Ruppin, *J. Phys. Chem. Solids* **39,** 233 (1978).
49. I. Lindgren and S. Lundqvist, editors, "Many-Body Theory of Atomic Systems," *Phys. Scripta* **21,** No. 3/4 (1980).
50. G. Mukhopadhyay and S. Lundqvist, *J. Phys. B* **12,** 1297 (1979).

CHAPTER 4

APPLICATIONS OF DENSITY FUNCTIONAL THEORY TO ATOMS, MOLECULES, AND SOLIDS

A. R. WILLIAMS AND U. VON BARTH

1. Introduction

Following the discussion of the foundations of the theory of the inhomogeneous electron gas in Chapter 2, the focus of this chapter is the practical utility of density functional theory for the study of atoms, molecules, and solids. The emphasis on practicality means that the discussion will center on the local-density approximation and its generalization to spin-polarized systems (the local spin density approximation) discussed fully in Chapter 2.

Density gradient corrections to local density theory are, at first sight, a natural way to improve that theory. However, there are difficulties associated with such an approach, as discussed in Section 2.1.1 and it seems more basic to consider the representation of exchange plus correlation in terms of the pair correlation function (see also Section 2.1.2). This approach focuses directly on the size, shape, and position of the exchange-correlation hole with respect to the electron which it neutralizes.[1] Once the discussion of the local

A. R. WILLIAMS • IBM Thomas J. Watson Research Center, Yorktown Heights, New York 10598. U. VON BARTH • Institute for Theoretical Physics, University of Lund, Lund, Sweden.

density approximation is cast in these terms, it is readily seen that the strength of the local-density approximation are its precise maintenance of neutrality and the relative insensitivity with which the exchange-correlation energy depends on the shape of the exchange-correlation hole. The same analysis also reveals the most limiting aspects of the local-density approximation, namely, that it implicitly requires that the exchange-correlation hole remain centered on its electron—a situation which is not even approximately true in situations such as the outer surfaces of atoms and solids.

Another basic question addressed in this chapter is: How numerically accurate are calculations, based on the local-density approximation, for real systems? As might be imagined, this question can mean a spectrum of things. One indication of the breadth of this spectrum is the energy scale characterizing different problems to which the theory is expected to apply. It can be used to calculate the total energy of a free atom, a quantity which is often tens of thousands of electron volts. A second measurable quantity to which the theory can be applied is the energy required to remove the outermost electrons from free atoms and molecules. These energies are typically 10 eV in magnitude. A third class of problems to which the theory applies is the total-energy difference between free atoms and atoms in condensed or bonded systems, such as solids and molecules. Such bond energies are typically 1 eV. Continuing to smaller energies, we have total-energy differences between different condensed systems such as the heats of formation of solid compounds and the energies associated with different geometrical arrangements of the same atoms. The heats of formation of intermetallic compounds are often only 0.1 eV in magnitude and the energy scale associated with different crystal structures, phonons, and shear moduli is smaller yet ($\sim$ 0.01 eV). It might be noted that the sequence of physical problems we have just characterized in terms of energetic subtlety can also be characterized in terms of the portion of the atom involved in each physical effect. The decreasing energy scale can be roughly identified with increasing distance from the atomic nucleus. The relative ability of the local-density approximation to describe the interior and the exterior of atoms is one of the recurring themes of this chapter.

Also discussed in this chapter are the various ways in which calculations based on density-functional theory can be used to increase our understanding of physical systems. As with the question of numerical accuracy, the utility of density-functional theory can mean several things. The most direct use of electronic-structure calculations is the analysis of realistic models with the objective of calculating accurate values for measurable quantities. Much work of this type has been performed, and we shall discuss selected examples. The parameter-free character of calculations based on density-functional theory has led to a somewhat different use of electronic structure calculations, namely, the computation of quantities that are either difficult or impossible to measure. Such work can be divided into two types. In the first, the calculations are used as a microscope with which to study chemical bonding by "looking at" the electron density. Such calculations have provided the basic physical picture of the covalent bond in semiconductors,[2] of the density rearrangements associated with solid surfaces (see Chapter 5), of the different types of bonds by which atoms attach themselves to metal surfaces,[3] and of the density rearrangements associated with imperfections in an otherwise perfect crystalline solid.[4,5] Another use of electronic-structure calculations based on density-functional theory in which the principal focus is not measurable quantities is the study of the variation of quantities like heats of compound formation[6] caused by variation of constituent valence. Chemical trends of this kind are frequently very difficult to study experimentally. Perhaps the clearest example of this is the freedom such calculations provide of studying chemical sequences of molecules and compounds in which some of the members do not exist. In trying to determine the microscopic mechanisms which help and hinder compound formation, for example, it is just as important to understand why certain compounds do not form, as it is to understand those that do. Numerical experiments of this kind play a logical role in electronic-structure theory similar to that played by molecular-dynamics calculations in statistical mechanics. Such numerical experiments are interesting to the extent that the model studied is sufficiently realistic or in some sense "rich" enough to warrant the effort required to understand the results. While numerical experiments are never as realistic as physical experiments, they offer the advantage of far greater control over the physical parameters of the system.

Yet another way in which the density-functional theory framework has proved useful is in providing the conceptual basis for approximation schemes. Density-functional theory has been exploited in this way to obtain approximate theories of interatomic interactions in molecules,[7,8] intermetallic compounds,[9–11] and chemisorption systems.[12,13] These approaches attempt in various ways to characterize the environment of an atom in a condensed system in terms of the electron density surrounding the atom. Such approaches constitute an enormous simplification of the analysis of extended, low-symmetry systems, because they eliminate the need for a wave mechanical analysis of the system as a whole.

There are thus many aspects to the questions of how well density-functional theory works, why it works, and how it can be exploited. Our treatment of these questions will be far from exhaustive.

2. Local-Density Approximation: Interpretation and Extensions

2.1. *Pair-Correlation Function Description of Exchange-Correlation Energy*

2.1.1. Gradient Corrections. For the local-density (LD) approximation to the density-functional theory, the functional $E_{xc}[n_{\alpha\beta}]$ describing the effects of electronic exchange and correlation is given by[14] [see equations (105) and (100) in Chapter 2]

$$E_{xc}[n_{\alpha\beta}] = \int n(\mathbf{r})\epsilon_{xc}(n_\uparrow(\mathbf{r}); n_\downarrow(\mathbf{r}))\, d^3r \tag{1}$$

Here, $n_\uparrow(\mathbf{r})$ and $n_\downarrow(\mathbf{r})$ are the eigenvalues of the one-particle density matrix, $n_{\alpha\beta}(\mathbf{r})$, $n(\mathbf{r}) = n_\uparrow(\mathbf{r}) + n_\downarrow(\mathbf{r})$ is the electron density, and $\epsilon_{xc}(n_\uparrow, n_\downarrow)$ is the exchange-correlation energy per particle of a homogeneous spin-polarized electron gas with the spin densities $n_\uparrow$ and $n_\downarrow$. Thus, the LD approximation is exact in the limit of slowly varying spin densities[15] and one would expect it to work well only for systems in which the densities do not change appreciably over a distance corresponding to an inverse Fermi wave vector. Unfortunately, most real systems such as atoms, molecules, or solids show strong variations in the particle density over this distance. Still the LD approximation has been applied to many of these systems and we think it is fair to say that the results have been far more accurate than one had any reason to expect. Numerous examples of this success are given in other parts of this book. This section will be devoted to a discussion of the physics behind the success, but we will also point out some difficulties with the LD approximation and describe the attempts made to overcome these difficulties.

A very natural way to attempt to understand why a zeroth-order theory works is to estimate the first-order corrections in the hope that they will be small. Coming from the limit of a slowly varying density, these corrections are the so-called "density-gradient" terms. If only the lowest-order correction is included, the exchange-correlation functional becomes simply equation (60) of Chapter 2 in the case of no spin polarization,[15] the function $B_{xc}(n)$ in that equation being determined by the density–density response function of the homogeneous electron gas. The latter has been extracted from various approximate response functions,[16–22] but it has also been obtained by Geldart and Rasolt from a careful analysis of the diagrammatic expansion for the static dielectric function.[23,24]

As discussed in Chapter 2, a gradient expansion is expected to be valid if the inequalities (33) of Chapter 2 are satisfied. These inequalities express the requirement that the density variations be small over distances of the order of an inverse Fermi wave vector. This criterion is, however, seldom satisfied in real systems, as illustrated by the two examples in Fig. 1 taken from Ref. 1, which shows the relative density variations in bulk copper and in the surface region of semi-infinite jellium with a density corresponding to that of metallic sodium.

It is therefore not surprising that the results for the total energy of atoms turns out to be better in the local-density approximation than if also the lowest-order gradient correction is included. This is obvious from the work by Herman *et al.*[25] who tried to improve the total energies of atoms treating B_{xc} as an adjustable parameter. This optimal B_{xc} had a sign *opposite* that of the B_{xc} calculated from first pinciples.[23] As a next step, one can try to include higher-order gradient corrections of the form $\int C_{xc}(n)\,|\nabla^2 n|^2\,d^3r$ or $\int D_{xc}(n)\,\nabla^2 n\,|\nabla n|^2\,d^3r$.[15] For a system in which the particle density deviates little from its average value $\bar{n}$, we can use linear-response theory, which amounts to neglecting all gradient terms involving the density raised to powers greater than 2. The remaining subclass of gradient terms can be summed exactly, to infinite order giving[15]

$$E_{xc}[n] = \int n(\mathbf{r})\epsilon_{xc}(n(\mathbf{r}))\,d^3r - \tfrac{1}{4}\int\int K_{xc}(\mathbf{r}-\mathbf{r}';\,\bar{n})\,[n(\mathbf{r})-n(\mathbf{r}')]^2\,d^3r\,d^3r' \tag{2}$$

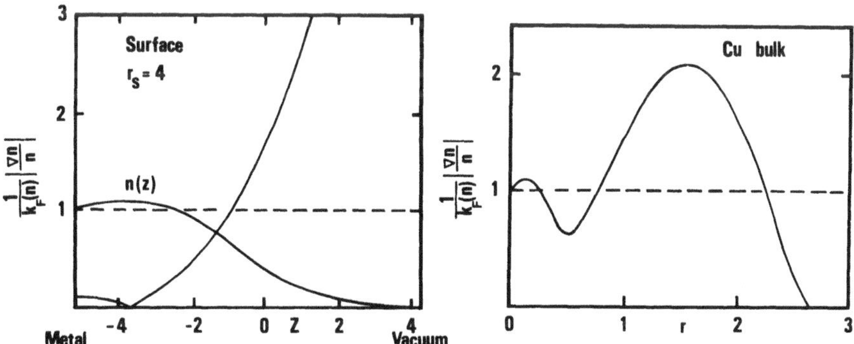

FIGURE 1. The dimensionless quantity $|\nabla n|/nk_F(n)$ giving the change in the density n over the local Fermi wavelength $k_F^{-1} \equiv (3\pi^2 n)^{-3}$ is here displayed as a function of the perpendicular distance z from the surface of semi-infinite jellium ($r_s = 4$) and as a function of the distance r from the nucleus in bulk copper. It is this quantity that should be much less than unity in order that the use of gradient corrections be reliable. The curve labeled $n(z)$ shows the surface density profile. Distances are measured in Bohr radii. The figure is taken from Ref. 1.

The kernel $K_{xc}(\mathbf{r},n)$ is given by $-(4\pi/q^2)\cdot G(q,n)$ where the local-field factor G, as defined by Singwi *et al.*,[26] can be obtained from the static density–density response function of the homogeneous electron gas of density n. If we choose the density argument in equation (2) to be the average density, then equation (2) is valid for arbitrarily rapid density variations, as long as $n(\mathbf{r})$ deviates little from n. It thus provides a possibility to test the quality of gradient corrections. These are proportional to the coefficients in the expansion of the Fourier transform $K_{xc}(q)$ of the kernel K_{xc} in powers of q. For instance, from Ref. 15, we have

$$K_{xc}(q) = K_{xc}(0) + 2B_{xc}q^2 + O(q^4) \tag{3}$$

In Fig. 2 (from Ref. 1), we show $K_{xc}(q)$ obtained from the dielectric function of Geldart and Taylor[27] together with the results for $K_{xc}(q)$ obtained by truncating its series expansion in q to order q^0, q^2, and q^4 corresponding to the LD approximation and the lowest- and next-higher-order gradient corrections, respectively. Looking at Fig. 2 and remembering that realistic density profiles have appreciable Fourier components with momenta larger than the Fermi momentum, it seems questionable whether the inclusion of the lowest-order gradient correction [equation (60) in Chapter 2] represents an im-

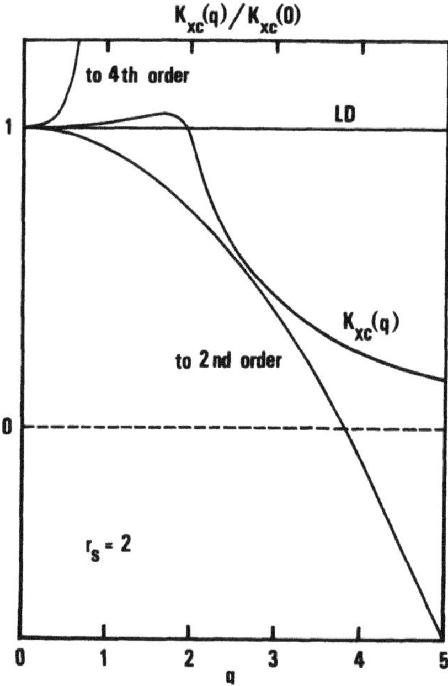

FIGURE 2. The local interaction K_{xc} describing the effects of exchange and correlation on the static density response function of the homogeneous electron gas [$r_s = 2$, see equations (2) and (3)] is here displayed as a function of wave vector q given in units of the Fermi wave vector. Also displayed is the same quantity obtained within the local-density approximation and obtained by also including the lowest (second) order and next higher (fourth) order gradient corrections. The figure demonstrates the futility of low-order gradient corrections. The figure is based on the dielectric function of Geldart and Taylor (Ref. 27), and appears in Ref. 1.

provement over the local-density result. If also the next-order gradient correction is included, the results are much inferior to those of the LD approximation. The argument given here against the relevance of gradient corrections is originally due to Geldart *et al.*[17] It is strictly valid only for systems with a nearly constant density. When the amplitude of the density variations becomes larger, gradient terms involving the density to powers higher than 2, e.g., $\int D_{xc}\nabla^2 n\, |\nabla n|^2\, d^3r$, will become important in the gradient expansion. Such terms originate from second- and higher-order response functions of the electron gas and they are not included in the infinite sum given by equation (2). There is, however, no reason to believe that gradient corrections would be more appropriate when the amplitude of the density variations increase.

Recently, Rose *et al.*[28] reported on a calculation of surface energies of simple metals in which they had included the first-order gradient correction in a self-consistent manner. They claimed improved agreement with experiment due to the gradient correction. The *opposite* conclusion had previously been reached by Lau and Kohn from a similar investigation.[29] These contradictory conclusions might be related to the fact that the gradient corrections to the surface energy in the two papers differ by less than the uncertainties in the experimental surface energies, and the relevance of these results regarding the usefulness of gradient corrections is unclear. Rasolt *et al.*[30,31] constructed a model system consisting of a surface with a finite barrier confining a semi-infinite jellium of Fermions having a Yukawa interaction. For this system they were able to compare the exact exchange contribution to the surface energy with the corresponding LD approximation both with and without the lowest-order gradient correction. Their results showed that the LD approximation was in error by approximately 15% and that the lowest-order gradient correction reduced this error by almost an order of magnitude. From this they concluded that gradient corrections are useful for the calculation of surface energies. This conclusion is, however, open to the criticism that it is based on the assumption of a short-range interaction, whereas the convergence of the gradient expansion in real systems is severely impaired by the long-range nature of the Coulomb interaction.[32]

The convergence of the gradient expansion has also been studied by Gunnarsson *et al.*[1] They applied equation (2) representing an infinite summation of a subclass of gradient terms, to the calculation of the total energies of atoms and surfaces. In contrast to the case of an almost constant density, discussed previously, the choice of density argument in the kernel K_{xc} of equation (2) is not unique for these strongly inhomogeneous systems. There are several possibilities which are physically reasonable and which reduce to the correct limit when the density variations are made small. Gunnarsson *et al.* discovered that the exchange-correlation energies of both atoms and surfaces are infinite for some of these choices, thus demonstrating the divergence of this particular, infinite subseries of gradient terms.

Perhaps the most conclusive argument against the relevance of gradient corrections has recently been presented by Langreth and Perdew.[32] In a previous paper,[33] they had analyzed both the exact exchange-correlation energy of a metallic surface and the LD approximation to it in terms of the wave vectors of the density fluctuations. They found that the LD approximation was quite accurate for large wave vectors but failed badly for small wave vectors. Using the surface-density profile of the infinite-barrier model, and treating correlation within the random-phase approximation, they also showed that a simple interpolation between the exact limiting behavior at small and large wave vectors gave an exchange-correlation energy very close to the exact result.[34] In Ref. 32 a similar analysis is carried out for the lowest-order gradient correction [equation (60) in Chapter 2]. It is shown that the wave-vector-dependent gradient correction has the wrong analytical behavior for small wave vectors. It is also shown that the largest contribution to the integrated wave-vector-dependent gradient correction comes from this region, resulting in a much too large correction to the LD approximation. This result is consistent with an earlier investigation by Perdew, Langreth, and Sahni[35] in which accurate results for the exchange-correlation part of the surface energy, obtained using the same wave-vector interpolation scheme mentioned above, were compared to the results obtained within the LD approximation and to those obtained by also including the lowest-order gradient correction. By means of an external potential, they were able to vary the steepness of the density profile of the surface, and they concluded that the gradient correction was appropriate only when the density gradients were much smaller than those of realistic surfaces.

It is clear from the discussion given above that a strong case can be made against gradient corrections as a way to improve the LD approximation for the total energies of systems in which the particle density varies as rapidly as in, e.g., atoms and metallic surfaces. The gradient expansion is, at best, asymptotically convergent; keeping only the lowest-order gradient term is a doubtful approximation. There are, however, other situations in which a gradient expansion might be more appropriate. This could be the case when there is hope for a cancellation of the errors due to strong inhomogeneities. The total-energy differences between different crystal structures or the energies of chemisorption might be examples. The discussion above was not primarily intended to demonstrate the inadequacy of gradient corrections, but rather to point out that the LD approximation gives quite reasonable answers even when gradient corrections are not appropriate. This means that the LD approximation cannot be justified as the zeroth-order approximation in a rapidly convergent expansion. It is rather an approximation in its own right and the reason for its success must be sought elsewhere.

2.1.2. Pair-Correlation Hole and Sum Rules. The most extensive investigation of the factors governing the success of the LD approximation is due to Gunnarsson *et al.*[1,36,37] Their discussion is based on an exact expression due to Almbladh[38] (cf. Chapter 2) relating the exchange-correlation energy $E_{xc}[n]$ to a quantity $\bar{g}(\mathbf{r},\mathbf{r}')$ which is the pair-correlation function of the system, integrated over the strength of the electron–electron Coulomb interaction [cf. equation (70) of Chapter 2]

$$E_{xc}[n] = \tfrac{1}{2} \int n(\mathbf{r})n(\mathbf{r}') \{\bar{g}(\mathbf{r},\mathbf{r}') - 1\}\, v(\mathbf{r}-\mathbf{r}')\, d^3r\, d^3r' \tag{4}$$

Here, $v(\mathbf{r}) = 1/r$ is the full-strength Coulomb interaction and $n(\mathbf{r})$ is the particle density. (For simplicity we have assumed the system to be non-spin-polarized.) The integral of the interaction-strength parameter λ has to be performed in the presence of an additional λ-dependent local external potential which vanishes at the physical value of λ but which is otherwise designed so as to make the particle density independent of λ. That it is possible to define such an external potential is postulated in the original derivation of density-functional theory.[15] A derivation of equation (4) can be found in Ref. 37.

Gunnarsson and Lundqvist[37] then defined the exchange-correlation hole to be

$$n_{xc}(\mathbf{r},\mathbf{r}') = n(\mathbf{r}') \{\bar{g}(\mathbf{r},\mathbf{r}') - 1\} \tag{5}$$

in terms of which the exchange-correlation energy is given by [cf. equation (73) of Chapter 2]

$$E_{xc}[n] = \tfrac{1}{2} \int n(\mathbf{r})v(\mathbf{r}-\mathbf{r}')n_{xc}(\mathbf{r},\mathbf{r}')\, d^3r\, d^3r' \tag{6}$$

According to this equation, the exchange-correlation energy arises from the Coulomb interaction of each electron (e.g., the one at $\mathbf{r}$) with a charge distribution $n_{xc}(\mathbf{r},\mathbf{r}')$, i.e., the exchange-correlation hole surrounding that electron. The hole is a consequence of the exchange and Coulomb interactions which cause a depletion of charge in the vicinity of each electron. The fact that the total amount of displaced charge corresponds to one unit of charge is expressed by the sum rule [equation (73) of Chapter 2] valid for all $\mathbf{r}$. The language of the present section can be used to give a definition of the LD approximation which is completely equivalent to equation (1). This equation results from a particular approximation for the exchange-correlation hole,

$$n_{xc}^{\mathrm{LD}}(\mathbf{r},\mathbf{r}') = n(\mathbf{r})\{\tilde{g}_h(\mathbf{r}-\mathbf{r}'; n(\mathbf{r})) - 1\} \tag{7}$$

Here, $\tilde{g}_h(r,n)$ is the pair-correlation function of the homogeneous electron gas of density n integrated with respect to the strength of the interaction. We note immediately that the exchange-correlation hole in the LD approximation fulfills the sum rule [equation (73) of Chapter 2] because this sum rule is valid also in the homogeneous case.

Because of the long range of the Coulomb interaction, we know that the interaction energy between two charge distributions only depends weakly on the shape of the distributions. The most important factor is the total amount of charge in each distribution. We therefore expect that a misrepresentation of the exchange-correlation hole will not result in a large error in the exchange-correlation energy as long as the sum rule [equation (73) of Chapter 2] is fulfilled. This is probably the most important single factor underlying the success of the LD approximation. We can carry the analysis one step further[37] and define a spherical average $\bar{n}_{xc}(\mathbf{r},R)$ of the hole $n_{xc}(\mathbf{r},\mathbf{r}')$ with respect to the point $\mathbf{r}$,

$$\bar{n}_{xc}(\mathbf{r},R) = \int n_{xc}(\mathbf{r},\mathbf{r}+\mathbf{R})\,\frac{d\Omega_{\mathbf{R}}}{4\pi} \tag{8}$$

It is then easy to see that

$$E_{xc}[n] = \frac{1}{2}\int n(\mathbf{r})\,d^3r \int v(R)\bar{n}_{xc}(\mathbf{r},R)\,d^3R \tag{9}$$

which means that the exchange-correlation energy only depends on the spherical average of the exchange-correlation hole.[37] This exact result is also very important for the quantitative success of the LD approximation, as can be appreciated from Figs. 3 and 4, both taken from Ref. 1. In Fig. 3, the exact exchange hole in the neon atom is compared to the corresponding quantity in the LD approximation for different positions of the electron considered (different $\mathbf{r}$). We have chosen to consider exchange only because in this case we have exact results to compare with. Exchange is also the dominant effect in an atom and the exchange hole has the same qualitative features as the full hole, e.g., it obeys the sum rule [equation (73) in Chapter 2]. The exchange-only version of the LD approximation is obtained by using the Hartree–Fock approximation for $\tilde{g}_h$ in equation (7). As seen in Fig. 3, the exact hole is always centered close to the nucleus whereas the LD-hole is always spherically symmetric and centered on the electron, as is evident from equa-

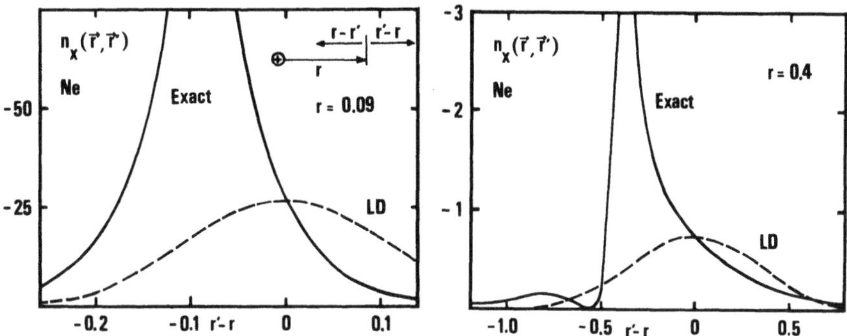

FIGURE 3. The exchange hole around an electron at r in the neon atom is here shown as a function of distance from the electron along a line connecting the electron and the nucleus. The full curves show the exact hole [equation (5)] and the dashed curves show the results obtained within the local-density approximation [equation (7)] which is seen to give a rather poor description of the hole. The two sets of curves correspond to the electron being at two different distances from the nucleus. Distances are measured in Bohr radii. The figure is taken from Ref. 1.

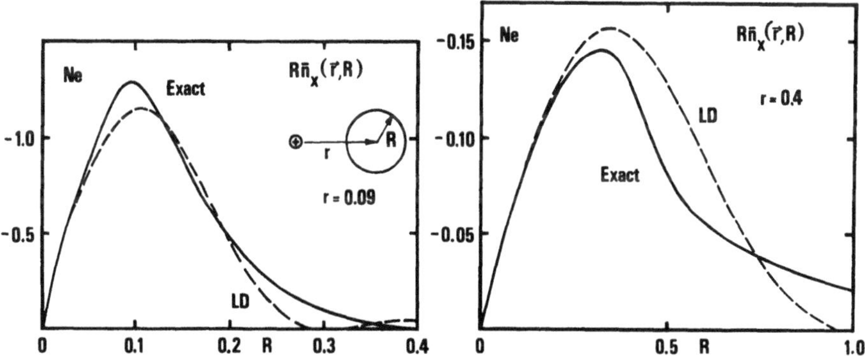

FIGURE 4. The spherical average [equation (8)] of the exchange hole surrounding an electron at r in the neon atom is here shown as a function of the distance from the electron. The full curves show exact results and the dashed curves were obtained using the local-density approximation. The hole is shown multiplied by R so that the area under each curve is directly proportional to the exchange-energy density [equation (10)]. When the curves shown here are compared with the corresponding curves in Fig. 3, it is seen that the local-density approximation gives a much more accurate description of the spherical average of the exchange hole than it gives of the hole itself. It is important to note in this context that the exchange energy depends only on the spherical average. Distances are measured in Bohr radii. (From Ref. 1.)

tion (7). Furthermore, the exact hole is rather localized whereas the LD-hole is much more extended. Thus, the LD approximation gives a rather poor representation of the exchange hole.

Looking instead at the more relevant spherically averaged hole in Fig. 4, we see that the picture has improved considerably. The LD approximation gives a rather accurate description of the spherical average of the hole and therefore also of the exchange energy according to equation (9). Since we have plotted $R\ \tilde{n}_{xc}(\mathbf{r},R)$, the area under each curve is directly proportional to the corresponding exchange-energy density $\epsilon_x(\mathbf{r})$, which is defined by

$$\epsilon_x(\mathbf{r}) = 2\pi \int_0^\infty \tilde{n}_x(\mathbf{r},R)R\ dR \tag{10}$$

and which determines the exchange energy E_x according to

$$\epsilon_x[n] = \int n(\mathbf{r})\epsilon_x(\mathbf{r})\ d^3r \tag{11}$$

We note the strong cancellation of errors that occurs for large and intermediate distances from the center of the hole. This cancellation is a consequence of the sum rule [equation (73) of Chapter 2] which implies that the area under the curve $-4\pi R^2\tilde{n}_x(\mathbf{r}, R)$ is 1. Thus, multiplying the curves in Fig. 4 by R would give two other curves of equal area which would intercept at the same place as the curves already shown. For these new curves we would have exact cancellation which is only partly destroyed by dividing by R.

The arguments given above will only undergo minor quantitative modifications if correlation effects are included. The basic conclusions remain the same. Correlations mainly reduce the spatial extent of the electron–gas pair-correlation function. As a consequence, the exchange-correlation hole of the local-density approximation will be more localized, thereby effectively improving this approximation for strongly inhomogeneous systems. This is the basic argument in favor of treating exchange and correlation effects together. At this point, we draw the reader's attention to the fact that obtaining correlation energies from a density-functional formalism is a much more difficult problem than that of obtaining exchange energies. As stressed by Tong,[39] correlation energies are connected also to the excitation spectrum and not just to wave functions, as is the case for exchange energies. Fortunately, the exchange energy is a factor of four larger than the correlation energy in an electron gas with a density corresponding to that of metallic sodium, and in atoms this ratio can easily be as large as thirty.

The arguments given by Gunnarsson and Lundqvist[37] to explain the quantitative success of the LD approximation for exchange-correlation energies of inhomogeneous systems can be summarized as follows: (i) the exchange-correlation energy is determined by the spherical average of the exchange-correlation hole; (ii) the LD approximation gives a relatively accurate description of this spherically averaged hole; and (iii) in addition, there is a systematic cancellation of errors due to the sum rule that expresses conservation of charge.

Before ending this section, we would like to stress that the arguments given above represent one possible way in which to explain the success of the LD approximation. There may be other ways that are equally valuable and informative. The discussion above is based on a description of electronic properties in real space. As mentioned in Section 2.1.1, Langreth and Perdew[32,33] have reformulated the problem of exchange and correlation in reciprocal space by decomposing the exchange-correlation energy into contributions from different wave vectors of the density fluctuations. In this language, the success of the LD approximation is due to its ability to accurately describe the density fluctuations with large wave vectors. There is now some indication[1,32,33] that the reciprocal-space approach may be a more useful starting point for improving on the local-density approximation, at least as far as surfaces are concerned.

2.1.3. Nonlocal Functionals. An implication of the previous section is that the many-body problem can be transformed into the problem of modeling the exchange-correlation hole of the system at hand. This is a rather difficult task for any specific system. The fact that we would like the model to be able to carry us continuously from the electron-gas limit to the atomic limit does not make the problem any easier. The knowledge of why the LD approximation works so well should, however, be useful in trying to improve it. Based on their earlier analysis, Gunnarsson and co-workers[1,40,41] have recently presented two "nonlocal" schemes to improve the LD approximation, both of which hold some promise. The schemes are "nonlocal" in the sense that the local exchange-correlation-energy density as defined by an equation analogous to equation (11), is made to depend on some average of the surrounding density, rather than just on the local density as in the LD approximation. Since exchange and correlation effects are inherently nonlocal, there is hope for an improved description of these effects in the new schemes. Note that they will still give rise to a local effective one-body potential as will any density-functional scheme.

The proposed methods have one feature in common with the LD approximation: the pair-correlation function of the homogeneous electron gas is used to model the exchange-correlation hole. The density argument and the density prefactor in equation (7) are, however, altered.

In the average-density (AD) approximation[40] both the argument and the prefactor are chosen to be an average of the density, $n(\mathbf{r})$, defined in terms of a density-dependent weight factor $W(\mathbf{r},n)$,

$$\bar{n}(\mathbf{r}) = \int W(\mathbf{r}-\mathbf{r}'; \bar{n}(\mathbf{r}))\bar{n}(\mathbf{r}')\, d^3r' \tag{12}$$

The AD-hole is given by

$$n_{xc}^{\mathrm{AD}}(\mathbf{r},\mathbf{r}') = \bar{n}(\mathbf{r})\, \{\tilde{g}_h(\mathbf{r}-\mathbf{r}'; \bar{n}(\mathbf{r})) - 1\} \tag{13}$$

and like the LD-hole and unlike the true hole, it is spherically symmetric and centered on the electron. The corresponding exchange-correlation energy is easily seen to be

$$E_{xc}^{\mathrm{AD}}[n] = \int n(\mathbf{r})\epsilon_{xc}(\bar{n}(\mathbf{r}))\, d^3r \tag{14}$$

from which the exchange-correlation potential $v_{xc}^{\mathrm{AD}}(\mathbf{r})$ is obtained by taking the functional derivative with respect to the density $n(\mathbf{r})$. This derivative will contain the functional derivative $\delta\bar{n}(\mathbf{r})/\delta n(\mathbf{r}')$ which is easily obtained from the definition of the average density [equation (12)]. By construction, the AD approximation obeys the sum rule [equation (73) of Chapter 2] irrespective of the choice of weight function. This follows from the corresponding sum rule for the homogeneous gas. The weight function $W(\mathbf{r},n)$ can thus be used to further improve the description of the exchange-correlation hole. Gunnarsson *et al.*[40] choose W such that the AD approximation becomes exact when the density is almost constant everywhere, i.e., such that the exchange-correlation energy of the AD approximation, in the linear-response limit, is given by equation (2). Consequently, in this limit, the AD approximation is exact, even for arbitrarily rapid density variations. It turns out that the chosen requirement uniquely determines the weight function W which is given by a first-order nonlinear differential equation involving coefficients determined by the wave-vector- and density-dependent static dielectric function of the homogeneous electron gas. This equation can be solved numerically and tabulated once and for all. A table can be found in Ref. 1.

The second scheme proposed by Gunnarsson *et al.*,[41] the weighted density (WD) approximation, has also been suggested by Alonso and Girifalco[42] for the exchange energy. In this scheme, the correct density prefactor $n(\mathbf{r}')$ from equation (5) is retained and the sum rule [equation (73) of Chapter 2] is satisfied for each $\mathbf{r}$ by a proper choice of density argument $\bar{n}(\mathbf{r})$ in the pair-correlation function of the homogeneous gas. The exchange-correlation hole becomes

$$n_{xc}^{\mathrm{WD}}(\mathbf{r},\mathbf{r}') = n(\mathbf{r}') \cdot \{\tilde{g}_h(\mathbf{r}-\mathbf{r}'; \bar{n}(\mathbf{r})) - 1\} \tag{15}$$

which, when inserted into equation (7) gives the corresponding exchange-correlation energy E_{xc}^{WD}. By taking the functional derivative of E_{xc}^{WD} with respect to the density $n(\mathbf{r})$, we obtain the exchange-correlation potential $v_{xc}^{WD}(\mathbf{r})$ of the WD approximation. The potential will contain the functional derivative $\delta\tilde{n}(\mathbf{r})/\delta n(\mathbf{r}')$ for which an explicit expression can be obtained from the sum rule [equation (73) of Chapter 2]. Thus, the exchange-correlation hole of the WD approximation is obtained by digging a hole in the true density. Consequently, this hole, like the true hole, is not spherically symmetric around the electron as is the case for the LD and AD approximations. Due to this feature the WD approximation is exact for any one-electron system. The only way in which the sum rule can be satisfied in a one-electron system is by choosing $\tilde{n}$ so small that $\tilde{g}_h$ vanishes almost everywhere and $n_{xc}^{WD}(\mathbf{r},\mathbf{r}')$ becomes $-n(\mathbf{r}')$, which, in this case, is the exact result. The WD approximation also gives the exact exchange energy for a two-electron system. The reason is that the Hartree–Fock approximation $\tilde{g}_h^{HF}$ to $\tilde{g}_h$ is always greater than ½ and, when there are only two electrons of opposite spin, one must choose $\tilde{n}$ very small in order to fulfill the sum rule. The exchange hole $n_x^{WD}(\mathbf{r},\mathbf{r}')$ then becomes $-\frac{1}{2}n(\mathbf{r}')$, which is the exact result for this case.

The AD approximation was designed to be exact for a gas of almost constant density, and we can assess the accuracy of the WD approximation in this limit by using it to calculate the static density–response function of the homogeneous electron gas. This has recently been done by von Barth,[43,44] who reports that one obtains a response function of a quality similar to that obtained by, e.g., the approach by Singwi *et al.*[26] Thus, to within our present knowledge of the density–response function of the electron gas, the AD and WD approximations are equivalent in the limit of an almost constant density. Note that they become identical in this limit, if the response function of the WD approximation is used as input to the AD approximation. Since the WD approximation is exact in the "opposite" limit of the hydrogen atom, we would expect the overall performance of the WD approximation to be superior to that of the AD approximation. It is therefore rather surprising that actual tests on atoms show the reverse to be true.[1] The error in the exchange-correlation energy in the usual LD approximation ranges from 8% to 5% for light to heavy atoms. The corresponding numbers for the AD approximation are 3% and 1%. The AD approximation thus represents a clear improvement over the LD approximation. The WD approximation, on the other hand, giving errors ranging from 15% to 6%, is inferior to the LD approximation. If only exchange energies are considered, the AD and WD approximations are of comparable accuracy and both are superior to the LD approximation. The former approximations give exchange energies which are too large (~ 5%), whereas the exchange energy is too small (~ 10%) in the LD approximation. The correlation energy of the WD approximation is larger than that of the LD approximation which is already about a factor of 2 too large. As a consequence, the error in the total energy of the WD approximation is increased when correlation is included. In the case of the AD approximation, the error is instead reduced when the correlation energy is added. This is because the correlation energy of the AD approximation is much too small and actually has the wrong sign for some atoms. This fact and the fact that the WD approximation should, by construction, be superior to the AD approximation suggest that the good performance of the latter might be fortuitous. In the LD, AD, and WD schemes, one tries to model the exchange-correlation hole by means of the pair-correlation function of the homogeneous electron gas. It is tempting to guess that this procedure is too simple-minded in a strongly inhomogeneous system like an atom. It could be that the errors inherent to this procedure are of the same order of magnitude as the energy differences between the three schemes.

Gunnarsson and Jones[45] have recently suggested a modification of the WD scheme in which the true pair-correlation function is modeled by an analytic function of the form

$$\tilde{g}_m(r,n) - 1 = A(n) \cdot \{1 - \exp[-B^5(n)/r^5]\} \tag{16}$$

The density-dependent parameters $A(n)$ and $B(n)$ are chosen so that $\tilde{g}_m$ obeys the sum rule imposed by charge conservation $[n\int d^3r(\tilde{g}_m - 1) = -1]$ and gives the correct exchange-correlation energy of the homogeneous electron gas. The exchange-correlation hole of the modified scheme is then given by equation (15), with $\tilde{g}_h$ replaced by $\tilde{g}_m$. The advantage of this scheme is that the long-range oscillatory tail of the electron-gas pair-correlation function, which certainly has no relevance for localized systems, such as atoms, e.g., is replaced by a rapidly decaying (r^{-5}) tail without destroying the basic correctness of the theory in the slowly varying limit. The scheme has been tested[45] in several atoms and the resulting exchange-correlation energies are in error by only a few percent. The new scheme thus represents a definite improvement over both the LD approximation and the original WD approximation. The choice of formula (16) is, however, somewhat ad hoc. It would therefore be valuable to investigate whether the results obtained using formula (16) are indeed due to the short-range nature of the parametrization and are not a fortuitous consequence of the particular analytical form employed in formula (16).

Gunnarsson, Jonson, and Lundqvist[1] have also tested the AD and WD schemes in a calculation of the exchange-correlation part of the surface energy of semi-infinite jellium. The results are quite disappointing. It is uncertain whether the AD approximation gives a convergent result for this energy, and in any case, the surface energy is much too large. The WD approximation gives a divergent exchange energy, but the total exchange-correlation energy is convergent although a factor of 2 smaller than the exact result within the

RPA approximation applied to the infinite-barrier model.[34] Consequently, the AD and WD approximations are much inferior to the LD approximation for this particular problem. The difficulty seems to stem from a slight misrepresentation of the exchange-correlation hole on the bulk side of the surface. Since the surface energy is an energy difference between an infinite and semi-infinite system, these small errors seem to have a tendency to add up rather than cancel. This problem ought to be less severe in the modified WD scheme of Gunnarsson and Jones,[45] because the exchange-correlation hole in their scheme is less extended.

One should keep in mind that an atom and a surface represent very severe test cases for any density-functional theory based on data from the homogeneous electron gas. We still can hope that the AD and WD schemes represent a substantial improvement on the LD approximation in systems where the density variations are less rapid. Thus, it would be interesting to apply these nonlocal schemes to calculations of molecular binding energies or cohesive energies of solids. These energies are essentially determined by the outer valence electrons[46] and one could hope that the new schemes would yield results for, e.g., the $3d$ metals of comparable accuracy as the results obtained by Moruzzi *et al.*[47] for the $4d$ metals. Another case for which the AD and WD schemes may prove useful is the calculation of binding energies and equilibrium distances for atoms and molecules adsorbed on surfaces. The difficulties encountered with these schemes in connection with the surface energy will not influence the result for binding energies since these are energy differences for the semi-infinite system, with and without the adsorbed species. The AD and WD schemes will, however, give a more realistic exchange-correlation potential in the surface region as compared to the LD approximation. Far outside a metallic surface a classical unit charge will experience an image potential of the form $-1/(4z)$, z being the distance from the surface. Since classicial arguments should apply also for an electron far enough outside a surface, we would expect the asymptotic form of the exchange-correlation potential to be

$$v_{xc}(\mathbf{r}) = -\frac{1}{4z} \tag{17}$$

a relation that can be rigorously proven. Due to its density prefactor in equation (7), the LD approximation, however, predicts an exponentially decaying potential outside the surface. Thus, the LD approximation cannot describe the van der Waal's interaction between, e.g., an atom and a surface. On the other hand, the AD and WD schemes, as well as the modified WD scheme,[45] all exhibit imagelike behavior, although only the latter gives the correct coefficient ($-\frac{1}{4}$) in equation (17). One can therefore hope that these schemes will produce more accurate potential surfaces than the LD approximation, at least for larger distances from the surface. A word of caution is appropriate here. The exact exchange potential decays exponentially away from the surface. Thus, the image potential [equation (17)] is entirely a correlation effect. In the AD and WD schemes, however, the imagelike exchange-correlation potentials are a consequence of the larger-r behavior of the pair-correlation function $\tilde{g}_h(r)$ of the homogeneous electron gas. Since this behavior is not very different from that of the Hartree–Fock approximation to $\tilde{g}_h(r)$[48] both the AD and WD schemes would produce image potentials even if only exchange effects were considered. Thus, one can say that the desirable imagelike behavior of these schemes is obtained for the wrong reason and their performance will have to be judged on the basis of numerical tests.

It is also of interest to discuss the asymptotic behavior of the exchange-correlation potential v_{xc} far away from an atom or an ion. Classical arguments and a rigorous microscopic derivation[43] lead to the asymptotic form

$$v_{xc}(\mathbf{r}) = -\frac{1}{r} - \frac{\alpha}{2r^4} \tag{18}$$

for a neutral atom. The quantity α is the static polarizability of the corresponding positive ion. The first term in equation (18) is a pure exchange effect. The second term is a pure correlation effect and should therefore be as difficult to reproduce as the image potential outside a surface. In the LD approximation v_{xc} decays exponentially away from the atom and thus neither of the two terms is given correctly by this approximation. The AD and WD approximations, on the other hand, both give a $1/r$ dependence in v_{xc} although the coefficient is not -1 as in equation (18) but rather -0.5.[1] In the case of the WD approximation, this partial failure can be traced[43,44] to the violation of the symmetry condition $g(\mathbf{r},\mathbf{r}') = g(\mathbf{r}',\mathbf{r})$ obeyed by the exact pair-correlation function [see equation (15)]. In the AD scheme, the origin of the erroneous coefficient seems to be a more subtle numerical issue. Since the charge density of an atom is very low in the asymptotic region, an erroneous coefficient cannot affect the total energy in any important way. In the case of negative ions (e.g., H^-, Na^-, Cl^-), however, this deficiency in the AD and WD schemes is of crucial importance. The potential from the nucleus and from all the electrons will have here a positive Coulomb tail ($+1/r$) which will be present, although with a reduced coefficient, in the total effective one-electron potential unless the exchange-correlation potential has a Coulomb tail with a coefficient less than or equal to -1. It seems likely that a positive Coulomb tail will produce a positive eigenvalue for the outermost one-electron orbital, which explains the instability of the negative ions in the LD approximation.[49,50] This argument suggests that the AD and WD approximations also suffer from the same deficiency. A slight modification of the WD scheme (not to be confused with the modification mentioned above) that retains the above-mentioned symmetry of the pair-correlation function has, however, recently been suggested.[43,44] The tail from the total Coulomb potential will then be exactly canceled by a term in the asymptotic exchange-correlation potential which ensures a total potential that decays faster than r^{-1}. Thus, there is some hope that this symmetrized scheme will predict stable negative ions.

Having discussed the virtues and drawbacks of the various nonlocal schemes, we now add a few comments on their computational feasibility. We have already described how to construct the local exchange-correlation potentials [$v_{xc}(\mathbf{r})$] of the AD and WD schemes, and in both cases it involves doing a small number of single three-dimensional integrals for each point in space where the potential v_{xc} is needed. In all integrals the particle density appears as a weight factor. In the case of spherical symmetry, often encountered in practice, these integrals become one-dimensional and both schemes are quite feasible, although the computational effort is larger than for the LD scheme. Any band-structure or atomic program based on the LD scheme is, however, easily converted to the new nonlocal schemes by merely replacing the subroutine that computes v_{xc}. The WD scheme requires a table of the electron-gas pair-correlation function for different densities and different interparticle separations. In the case of exchange only, there is a single analytical formula available[51]; if correlation effects are of interest, the parametrized pair-correlation function recently suggested by Rajagopal *et al.*[52] might be useful. In the modified WD scheme,[45] the necessary pair-correlation function is instead given by a simple analytical formula [equation (16)], which is a clear computational advantage. In the AD scheme, one needs to know the weight function which defines the average density discussed above and which is given in a table in Ref. 1. By making use of the following two facts, a substantial amount of computational work can, however, be avoided with only a minor loss of accuracy. (i) The charge density is relatively insensitive to different choices of the exchange-correlation functional. For instance, the charge density of an atom in the LD approximation including exchange differs by only a few per cent from the full Hartree–Fock result. (ii) Due to the variational principle, the total energy is insensitive to errors in the charge density. Thus, we recommend that the total energy of the nonlocal schemes be calculated by inserting the charge density of a self-consistent LD calculation into the nonlocal energy functionals.

We end this section with a brief discussion of an interesting local scheme due to Stoll *et al.*[53] which has proved to be much superior to the LD approximation for obtaining correlation energies of atoms and molecules.† For these rather localized systems, exchange is clearly the dominant effect, and it is thus reasonable to treat the exchange energy exactly (Hartree–Fock) and to use a local approximation only for the correlation part (E_c) of the total-energy functional. In analogy with equation (1) E_c is given by

$$E_c^{\mathrm{LD}}[n_{\alpha\beta}] = \int [n_\uparrow(\mathbf{r}) + n_\downarrow(\mathbf{r})] \cdot \epsilon_c(n_\uparrow(\mathbf{r}),n_\downarrow(\mathbf{r}))\, d^3r \tag{19}$$

in the LD approximation. [ϵ_c is the correlation part of ϵ_{xc}, and other quantities are defined beneath equation (1).] In a localized system, however, exchange is much more effective in keeping the electrons apart than in the electron gas. As a result, the correlations between parallel-spin electrons are much stronger in the electron gas than in the localized system, and the energy associated with the parallel-spin correlations is thus much overestimated by the LD approximation. As a matter of fact, it is not at all a bad approximation to neglect the parallel-spin-correlation energy altogether for a very localized system. Stoll *et al.*[53] therefore suggested that the LD approximation to the parallel-spin-correlation energy should be subtracted from equation (19). Thus,

$$\begin{aligned} E_c[n_{\alpha\beta}] &= \int [n_\uparrow(\mathbf{r}) + n_\downarrow(\mathbf{r})]\epsilon_c(n_\uparrow(\mathbf{r}),n_\downarrow(\mathbf{r}))\, d^3r \\ &\quad - \int n_\uparrow(\mathbf{r})\epsilon_c(n_\uparrow(\mathbf{r}),0)\, d^3r - \int n_\downarrow(\mathbf{r})\epsilon_c(0,n_\downarrow(\mathbf{r}))\, d^3r \end{aligned} \tag{20}$$

This simple approximation gives ~90% of the correlation energy in the atoms Be to Ar. Due to the complete neglect of the parallel-spin correlations the correlation energy comes out consistently too small (except for He, Li, and Be), but the error would have been larger if it had not been compensated by an overestimation of the anti-parallel-spin-correlation energy in the LD approximation. The correlation contributions to the binding energies of several diatomic molecules are also given accurately by the approximation (20). There are, however, cases (e.g., BeH and HF) for which the results are quite poor. This could be due to a lack of self-consistency {Stoll *et al.*[53] evaluated the correlation functional [equation (20)] using the Hartree–Fock spin densities}, but the difficulty could also be of a more profound nature. The first term in equation (20) gives the correct correlation energy in the limit of a slowly varying density, but the last two terms are finite and rather large. Thus, the procedure by Stoll *et al.* is not correct in the slowly varying limit.

†Two additional density-functional methods to improve on the LD approximation will be discussed at the end of Section 2.1.4. One of these, from Langreth and Perdew,[32] may very well prove to be the most useful method in the future.

2.1.4. Orbital- or Shell-Functional Methods. In this section we will discuss some methods of approximation which are not based on density-functional theory but which are closely related to it. In these methods the total energy is considered as a functional of some of the constituents of the total particle density like, e.g., the one-particle orbitals or the densities of the different atomic subshells. The minimization of the total energy with respect to these constituents then leads to one-electron Schrödinger-like equations, but with different effective local potentials for the different orbitals or subshells. This is in contrast to any density-functional scheme which always leads to the same effective potential for each orbital (or for each spinor in the spin-polarized case) and therefore to orthogonal orbitals. The problem with nonorthogonal orbitals in the orbital- or shell-based methods can be handled in different ways. The simplest procedure is to leave them nonorthogonal. One could also use a Schmidt process and force them to be orthogonal through the self-consisting procedure, although the computational effort would be considerable. It is, however, difficult to argue one way or the other and the ambiguity stems from the lack of proper theoretical foundation of the orbital- or shell-functional methods. A third possibility would be to evaluate the energy functionals of these methods using orbitals obtained from an ordinary density-functional calculation. This is probably the most appealing way to take advantage of the physically sound ideas which are the motivation for the orbital and shell methods.

In the first orbital-functional method due to Hartree[54] the quantity corresponding to the exchange energy E_x was written:

$$E_x = -\frac{1}{2}\sum_i \int n_i(\mathbf{r}) \cdot v(\mathbf{r}-\mathbf{r}') \cdot n_i(\mathbf{r}')\, d^3r\, d^3r' \tag{21}$$

where $n_i(\mathbf{r})$ is the density of the ith orbital. We have here dressed Hartree's method in modern density-functional language but the basic physical idea is the same. The minimization of the total energy, including E_x from equation (21), with respect to the orbitals leads to one-particle equations describing each electron moving in the field of the nuclei and all *other* electrons. Thus, an electron does not feel its own field; the self-interaction is absent. Hartree's method was then generalized by the Hartree–Fock method which takes due account of the Pauli exclusion principle. The static interelectronic Coulomb energy E_{Coul} and the exchange energy E_x can here be written as double sums over orbitals

$$E_{\text{Coul}} = \frac{1}{2}\sum_{ij} \langle ij \mid v \mid ij\rangle \tag{22}$$

$$E_x = -\frac{1}{2}\sum_{ij} \langle ij \mid v \mid ji\rangle \tag{23}$$

and the self-interaction $\frac{1}{2}\langle ii\,|v|\,ii\rangle$ in E_{Coul} is again canceled by a term in the exchange energy. Unfortunately, this cancellation is partly destroyed when the LD approximation is applied to the exchange energy,

$$E_x \cong \int n(\mathbf{r})\epsilon_x(n_\uparrow(\mathbf{r}), n_\downarrow(\mathbf{r}))\, d^3r \tag{24}$$

The reason is that we here use $\epsilon_x\,(n_\uparrow, n_\downarrow)$, the exchange energy per particle of a homogeneous electron gas with spin-up and spin-down densities $n_\uparrow$ and $n_\downarrow$. For this infinite system the self-interaction is negligible. Lindgren and Rosen[55,56] investigated possible corrections to equation (24) when the number of electrons is made finite and small. They arrived at the following physically rather appealing approximation to the exchange energy:

$$\begin{aligned} E_x = & -\frac{1}{2}\sum_i \int n_i(\mathbf{r})v(\mathbf{r}-\mathbf{r}')n_i(\mathbf{r}')\, d^3r\, d^3r' \\ & -\sum_i \int n_i(\mathbf{r}) \cdot \epsilon_x(n_i(\mathbf{r});0)\, d^3r \\ & + \int n(\mathbf{r})\epsilon_x(n_\uparrow(\mathbf{r}); n_\downarrow(\mathbf{r}))\, d^3r \end{aligned} \tag{25}$$

The first term exactly cancels the self-interactions in the static interelectronic Coulomb energy as in the Hartree approximation. The third term is the ordinary LD approximation and the second term is the LD approximation to self-exchange. It is this part of the LD exchange energy which is supposed to cancel the self-interaction and it is therefore subtracted to avoid double counting. Lindgren and Rosen have tested their so-called Hartree–Slater method on a number of atoms and the method gives atomic exchange energies with an accuracy of a few per cent. This is almost an order of magnitude better than the accuracy of the LD approximation which is typically 10%–15%.[57] They also found that the one-electron eigenvalues of the Hartree–Slater method are closer to those of Hartree–Fock theory and therefore to experimental excitation energies at least for shallow valence states in atoms.

The Hartree–Slater method has recently been generalized in order to include correlation effects.[58] In this new scheme the exchange-correlation energy is approximated by an expression analogous to that in equation (25) but with ϵ_x replaced by ϵ_{xc}, the exchange-correlation energy per particle of the homogeneous electron gas. This is equivalent to using the following rather ad hoc approximation for the correlation energy E_c of the inhomogeneous system:

$$E_c = -\sum_i \int n_i(\mathbf{r})\epsilon_c(n_i(\mathbf{r});\, 0)\, d^3r + \int n(\mathbf{r})\epsilon_c(n_\uparrow(\mathbf{r});\, n_\downarrow(\mathbf{r}))\, d^3r \tag{26}$$

in an obvious notation. Note that the cancellation of the self-interaction is a pure exchange effect. Consequently, arguments based on this requirement cannot be used as a guide to find accurate correlation-energy functionals. Surprisingly accurate atomic correlation energies are, however, obtained from equation (26)—at least for the light atoms He, Li, and Be, where one obtains 1.36, 1.71, and 2.80 eV compared to 1.14, 1.22, and 2.56 eV obtained from experiment (see Ref. 58.). Part of the explanation for this remarkable success is to be found in the fact that the new scheme is, by construction, exact for any one-electron system, e.g., for the hydrogen atom. With only one orbital the two terms in equation (26) cancel and the correlation energy E_c vanishes. Also the last two terms in equation (25) cancel leaving the first term to cancel the static Coulomb interaction energy. In the limit of a slowly varying density, the orbital-dependent terms in the present scheme become negligible and the scheme becomes equivalent to the LD approximation which is exact in this limit. That these arguments do not give the full picture can be understood as follows. The WD scheme discussed earlier is also exact for the hydrogen atom and it is actually superior to the present orbital scheme in the slowly varying limit. Still, the WD scheme gives atomic correlation energies which are typically a factor or 2 too large. Equation (26) represents a possible interpolation for the correlation energy between the hydrogenic and slowly varying limits. It remains an interesting project for future research to find out why this particular choice is so successful.

Gunnarsson *et al.*[1] have also suggested improved versions of the AD and WD schemes discussed in the previous subsection. These shell-partitioned AD and WD schemes (ADS and WDS) are based on the observation that, in atoms, the original AD and WD schemes overestimate intershell exchange and correlation effects. This is due to the density sampling procedure in both schemes. Gunnarsson *et al.* were able to reduce this unphysical effect by splitting the total exchange-correlation energy into a sum of contributions from each atomic subshell. The AD and WD techniques were then applied only within each subshell. When tested on atoms, the shell-partitioned methods give much better exchange energies than the unpartitioned methods, the error being typically of the order of 1%. For the total exchange-correlation energy there is, however, almost no improvement. Only the WDS scheme represents a slight improvement over the WD scheme but the results are still much worse than those of the LD scheme. As discussed earlier, atoms represent unnecessarily difficult test cases and one would like to see the new schemes applied to solids. On physical grounds, it would then seem most appropriate to treat the core electrons as one shell and the valence electrons as the other. The effective one-electron potential would then be the same for all valence electrons and it would be possible to treat these separately in a self-consistent fashion. The problem of nonorthogonability between core and valence electrons would probably not be more severe here than in the rather successful frozen-core approximation[46] or in the pseudopotential approach to density-functional theory.[59] Strictly speaking, the ADS and WDS methods are theoretically ill-defined, but they may provide an empirical and accurate way in which the core and valence electrons can be treated separately. This feature may be of importance as previously suggested by Hedin and Lundqvist[60] By considering the core electrons as a nonlocal external potential acting on the valence electrons, an attempt has been made to justify procedures based on a separate treatment of core and valence electrons.[61] In this theory the one-particle density matrix rather than its diagonal, the particle density, enters as the fundamental variable, but the theory does not result in a scheme for practical calculations. In Ref. 46 a more pragmatic attitude was taken when it was rigorously shown that the total-energy functional in the LD approximation could be generalized to a functional with the core and valence charge densities as two independent variables. The proof can easily be modified to the case of the WDS scheme or the ADS scheme, thus providing a theoretical foundation for these schemes.

We end this section by briefly mentioning two other somewhat related methods to improve on the LD approximation. The first, due to Langreth and Perdew,[33] has been touched upon briefly in Section 2.1.1. The method is discussed in more detail in Chapter 5 by Lang on surface applications, and we merely mention that the method is based on a decomposition of the exchange-correlation energy into contributions from different three-dimensional wave vectors of the density fluctuations. It is shown that the LD approximation gives an adequate description of the contribution from large wave vectors and an exact result is derived for the small wave vector contributions. A scheme for interpolating between these two limits is suggested which proves quite successful for the calculation of surface energies. The interpolation scheme has recently been challenged by Rasolt *et al.*[62] However, this work, because it is based on a wave-vector decomposition of only the exchange contribution to the surface energy, is of questionable relevance to the work of Langreth and Perdew. The results of their somewhat arbitrary interpolation scheme have now essentially been reproduced by the average-gradient approximation, also due to Langreth and Perdew.[32] This approximation is obtained from rather appealing physical arguments associated with the wave-vector decomposition of the first-order gradient correction. Essentially the same decomposition has also been carried out by Rasolt and Geldart.[63] From this work, the accuracy of the wave-vector interpolation scheme appears to be somewhat fortuitous, but the work does not cast doubt on the average-gradient approximation. The schemes by Langreth and Perdew do not yet allow one to construct an effective local one-electron potential and can therefore not be carried to self-consistency. Their total-energy functional can still be minimized with respect to the density by using a parmetrized charge density and varying the parameters as suggested in Ref. 35. The Langreth–Perdew scheme is probably the most useful improvement on the local-density approximation thus far introduced.

The second method, due to Kohn and Hanke,[64] is based on the decomposition of the Coulomb interaction into a strong but short-ranged part and a weak but long-ranged part. The long-ranged part is then treated by hydrodynamic theory and for the short-ranged part a gradient expansion is suggested. No calculations based on this theory have yet been carried out, and it is therefore difficult to judge both the accuracy and the computational feasibility of the theory, although the basic idea is physically appealing.

2.2 Extension to Spin-Polarized Systems

In the original density-functional theory introduced by Hohenberg, Kohn, and Sham,[15,65] physical quantities were considered as functionals of a single quantity, the electron density as discussed fully in Chapter 2. As a result, the original theory could not be used to describe spin-dependent phenomena, such as magnetism. The possible extension of the work to spin-polarized systems was, however, suggested in the original work of Kohn and Sham.[65] The formal justification of this extension was given by von Barth and Hedin,[14] and also by Rajagopal and Callaway.[66] Von Barth and Hedin considered an electronic system subject to an arbitrary, spin-dependent, but local external potential. We denote such a potential by $w_{\sigma\sigma'}(\mathbf{r})$. It was shown that, for such a system, the fundamental variable (i.e., the analog of the electron density in the non-spin-polarized theory) is the one-particle density matrix $n_{\sigma\sigma'}(\mathbf{r})$ defined by

$$n_{\sigma\sigma'}(\mathbf{r}) = \langle 0 \mid \psi^{+}_{\sigma'}(\mathbf{r})\psi_{\sigma}(\mathbf{r}) \mid 0\rangle \tag{27}$$

Here, the quantities $\psi_{\sigma}(\mathbf{r})$ and $\psi_{\sigma}(\mathbf{r})$ and $\psi^{+}_{\sigma}(\mathbf{r})$ are the usual field operators corresponding to the annihilation and creation of an electron with spin σ at $\mathbf{r}$ and $|0\rangle$ is the ground state of the system. A basic distinction between the spin-polarized and original theories is that the spin-polarized case lacks the analog of the one-to-one correspondence between electron densities and external potentials, on which the original theory was built. It is nonetheless true, however, that the ground-state energy in the spin-polarized case is a functional of the density matrix, and it is easily shown[14] that this functional is minimized by the correct density matrix. (Proof of this statement requires that the class of density matrices considered be restricted to those which can be constructed from an N-electron many-body wave function.) In analogy with the development of the non-spin-polarized theory, this variational principle leads to an equivalent single-particle self-consistent-field scheme for the evaluation of the ground-state energy and the density matrix. The scheme consists of a single-particle Schrödinger equation

$$\sum_{\sigma'} \left\{ -\frac{1}{2}\nabla^2 \delta_{\sigma\sigma'} + v_{\sigma\sigma'}(\mathbf{r}) \right\} \phi_{i\sigma'}(\mathbf{r}) = \epsilon_i \phi_{i\sigma} \tag{28}$$

in which the effective potential $v_{\sigma\sigma'}(\mathbf{r})$ is given by

$$v_{\sigma\sigma'}(\mathbf{r}) = w_{\sigma\sigma'}(\mathbf{r}) + \delta_{\sigma\sigma'} \sum_{\sigma''} \int d^3r' \, n_{\sigma''\sigma''}(\mathbf{r}') \cdot v(\mathbf{r} - \mathbf{r}') + v^{xc}_{\sigma\sigma'}(\mathbf{r}) \tag{29}$$

where $v(r) = 1/r$ is the Coulomb interaction and where the exchange-correlation contribution $v^{xc}_{\sigma\sigma'}(\mathbf{r})$ and the density matrix are defined in analogy with the non-spin-polarized theory,

$$v^{xc}_{\sigma\sigma'}(\mathbf{r}) \equiv \frac{\delta E_{xc}}{\delta n_{\sigma\sigma'}(\mathbf{r})} \tag{30}$$

and

$$n_{\sigma\sigma'}(\mathbf{r}) = \sum_{\epsilon_i < \epsilon_F} \phi_{i\sigma}(\mathbf{r}) \phi^*_{i\sigma'}(\mathbf{r}) \tag{31}$$

The expression for the total ground-state energy is also completely analogous to that of the non-spin-polarized theory,

$$\begin{aligned} E = \sum_{\epsilon_i < \epsilon_F} \epsilon_i - \frac{1}{2} \sum_{\sigma\sigma'} \int d^3r \, n_{\sigma\sigma}(\mathbf{r}) \int d^3r' \, n_{\sigma'\sigma'}(\mathbf{r}') \cdot v(\mathbf{r} - \mathbf{r})' \\ + E_{xc} - \sum_{\sigma\sigma'} \int d^3r \, n_{\sigma\sigma'}(\mathbf{r}) v^{xc}_{\sigma'\sigma}(\mathbf{r}) \end{aligned} \tag{32}$$

As in the non-spin-polarized theory, practical calculations require an explicit form for E_{xc}, which is inevitably approximate. Once again, in analogy with the non-spin-polarized theory, a local-density approximation [equation (1)] to E_{xc} is intuitively attractive. This approximation requires only a knowledge of the contribution of exchange and correlation to the total energy of an interacting, but homogeneous, electron gas $\epsilon_{xc}(n,\zeta)$ as a function, not only of density n, but of spin density, $\zeta \equiv (n_\uparrow - n_\downarrow)/n$, as well. In order to facilitate computations, von Barth and Hedin[14] proposed the following approximate interpolation formula for the electron-gas data required by the formalism

$$\epsilon_{xc}(n,\zeta) = \epsilon_{xc}^{P}(n) + [\epsilon_{xc}^{F}(n) - \epsilon_{xc}^{P}(n)] \cdot f(\zeta) \tag{33}$$

where the function $f(\zeta)$ gives the exact spin dependence of the exchange energy,

$$f(\zeta) = \tfrac{1}{2}(2^{1/3} - 1)^{-1} \{(1 + \zeta)^{4/3} + (1 - \zeta)^{4/3} - 2\} \tag{34}$$

The exchange-correlation energies $\epsilon_{xc}^{P}(n)$ and $\epsilon_{xc}^{F}(n)$ appearing in equation (33) are those of the unpolarized (paramagnetic) and completely polarized (ferromagnetic) homogeneous electron gas. This simple formula [equation (34)] accurately approximates the spin dependence of the exchange-correlation energy given by both the random-phase approximation and by more sophisticated electron-gas calculations.[67] The density dependence of both $\epsilon_{xc}^{P}(n)$ and $\epsilon_{xc}^{F}(n)$ is conveniently parameterized in terms of the function

$$F(x) \equiv (1 + x^3) \ln\left(1 + \frac{1}{x}\right) + \frac{x}{2} - x^2 - \frac{1}{3} \tag{35}$$

introduced by Hedin and Lundqvist.[60] In their work, the correlation energy was written simply as $-0.045F(x)$ Ry, where $x \equiv r_s/21.0$ (r_s is the usual electron-gas density parameter). This description of the electron-gas data cannot, however, reproduce the known density dependence in the high- and low-density limits. This is an unfortunate shortcoming, because the electron densities encountered in an atom, for example, span the entire high-density/low-density range. Although it would certainly be desirable not to have to worry about the parametrization of the electron-gas data, when studying atoms, molecules and solids, e.g., it is noteworthy that such errors are fortunately systematic (as opposed to random) and tend therefore to subtract out of many calculated properties of interest. Arbman and von Barth[68] have shown that a much improved description of the electron-gas data results if the function $F(x)$ defined equation (35) is used twice in the description of $\epsilon_{xc}^{P}(n)$ and $\epsilon_{xc}^{F}(n)$. That is,

$$\epsilon_{xc}^{P}(n) = \epsilon_{x}^{P}(n) - c_1^P F(r_s/r_1^P) - c_2^P F(r_s/r_2^P) \tag{36}$$

and similarly for $\epsilon_{xc}^{F}(n)$. The four parameters introduced in this way (c_1^P, r_1^P, c_2^P, r_2^P) can be set so that equation (36) reproduces the high-density result obtained by Gell-Mann and Bruckner,[69] the low-density result of Wigner[70] while also providing a good fit to the results of numerical calculations (see, e.g., Refs. 26, 71, and 72) in the metallic-density range ($2 \lesssim r_s \lesssim 5$).* The exchange energy $\epsilon_x^P(n)$ appearing in equation (36) is given by $\epsilon_x^P(n) = -0.9163/r_s$ Ry; the analogous result for polarized systems is $\epsilon_x^F(n) = -1.1545/r_s$ Ry. Note that, while the errors introduced by the use of the local-density approximation are usually larger than those due to the imperfect description of the electron-gas data, there are important exceptions[73] that make it desirable to improve the description, particularly as improved electron-gas results become available. Such improved descriptions have been proposed by Vosko *et al.*[76]

Once the density dependence of the electron-gas data is represented analytically by a form, such as that in equation (36), the density derivative required by the exchange-correlation potential {recall that $v_{xc}(n) \equiv (d/dn)\,[n\epsilon_{xc}(n)]$} can be taken easily. Note that, quite independent of the accuracy of representations such as equation (36), properties such as the virial theorem (see Section 3.2.1) depend crucially on the consistency which results when $\epsilon_{xc}(n)$ and $v_{xc}(n)$ are obtained from the *same* representation of the electron-gas data.

The spin-polarized version of density-functional theory outlined above is, in principle, able to describe complicated magnetic structures such as antiferromagnets[74,75] and even incommensurate spiral structures, but only to the extent that velocity-dependent forces ignored in equations (28) and (29) are in reality negligible. The coupling between a magnetic field and the orbital angular momentum is ignored in these equations; similarly, the spin-orbit interaction is neglected. Diamagnetism, which is caused by a term in the many-body Hamiltonian that is quadratic in the externally applied vector potential, is therefore also beyond the scope of the theory presented so far. These limitations are serious in some systems. Effects due to the spin-orbit interaction, for example, are important in systems containing heavy atoms. Similarly, in the rare-earth metals exhibiting spiral structures, the orbital magnetic moment is almost certainly not quenched.

Density-functional theory can be generalized to account for such velocity-dependent forces, as demonstrated by Rajagopal and Callaway.[66] This work is based on a relativistic formulation in which the four-current density appears as the fundamental variable. Since this approach is based on the Dirac treat-

*A particular set of numercial values for the parameters which yields the correct limiting behavior, a good fit to the correlation energies of Singwi *et al.* and which obeys approximate RPA scaling[14] between $\epsilon_{xc}^P(n)$ and $\epsilon_{xc}^F(n)$ is $r_1^P, r_2^P, r_1^F, r_2^F = 32.5, 0.8, 30.6, 0.1$ Bohr radii and $c_1^P, c_2^P, c_1^F, c_2^F = 0.0352, 0.0270, 0.0278, 0.0034$ Ry.

ment of relativistic electrons, the spin degrees of freedom enter in a natural way. When the nonrelativistic limit is taken in this formulation, the effective-single-particle Schrödinger equation which results contains the appropriate velocity-dependent terms. This formulation has been applied by Rajagopal[(77)] and by McDonald and Vosko[(78)] to obtain explicit exchange potentials appropriate for use in the effective-one-particle Dirac equation arising in the local-density description of the ground-state properties of systems such as molecules and solids containing heavy atoms.

Before going on to discuss applications of the spin-polarized theory, we briefly mention three other extensions of density-functional theory. Mermin[(79)] has generalized the theory to finite temperatures, and Peuckert[(80)] has constructed a density-functional theory which is applicable to time-dependent external fields and therefore to excited states. Theophilou[(81)] has recently shown that the sum of the energies and of the densities of a finite number of the lowest-lying states of a many-body Hamiltonian can be obtained from a density-functional theory. By considering in succession just the ground state, then the ground state together with the first excited state, and so on, the properties of the excited states can be obtained by subtraction. Practical calculations based on these three extensions of the theory have not yet been carried out. Another generalization, that to the ground states of each of a set of different symmetries[(37)] and to states of a specified mixture of symmetries,[(82,83)] is discussed in Section 2.3.

2.2.1 Applications of Spin-Polarized Theory. Applications of the spin-polarized local-density approximation (called the local-spin-density approximation) have produced numerous informative results, many of which are summarizeed in an excellent review by Pettifor.[(84)] The local-spin-density approximation provides a parameter-free microscopic theory of itinerant magnetism.[(85–87)] Calculations based on this approximation account very well for the ground-state magnetic properties of the ferromagnets Fe, Co, and Ni.[(87–94)] The theory can also be reduced, using appropriate approximations,[(85–87)] to the Stoner model of itinerant magnetism. Calculations of the paramagnetic susceptibility of metals[(87)] correctly identify Fe, Co, and Ni as the only systems possessing divergent susceptibilities, the hallmark of ferromagnetism. Calculations that allow for more complicated types of magnetic order (e.g., antiferromagnetic) have begun to appear,[(95–97,75)] with encouraging results. Spin-polarized calculations for compounds such as NiAl[98] exhibit the effect of alloying on magnetic properties. The spin-polarized theory has been particularly successful in revealing a common aspect of a large number of magnetic systems, that of the mechanical implications of magnetic order.[(88,92)] Recent calculations have applied the local-spin-density approximation to isolated magnetic impurities in metals[(99,100)] and to thin transition-metal films.[(101,102)]

This section consists of a brief review of these applications of the spin-polarized theory.

2.2.2. Spin-Polarization Effects in Free Atoms. The inclusion of spin-polarization effects in calculations for free atoms is important in several ways. A particularly important manifestation of these effects is their role in the calculation of bond energies, e.g., molecular-dissociation and cohesive energies. In atomic sodium, for example, the inclusion of spin-polarization effects lowers the calculated total energy by ~ 0.3 eV. Neglect of this effect would destroy the close agreement between the calculated[103] (1.10 eV) and the measured (1.13 eV) cohesive energies for metallic sodium. The importance of spin-polarization effects in this context was first pointed out by Gunnarsson, Lundqvist, and Wilkins.[36] The example of sodium illustrates both the power and the limitations of the local-density approximation. The cohesive energy of Na illustrates the accuracy with which chemical-bonding properties can be calculated using the approximation, but the ionization potential of Na illustrates the extent to which the accurate description of chemical-bonding properties depends on the systematic cancellation of errors introduced by the local-density approximation in the treatment of the atomic core. The energy required to remove the outermost electron of an alkali atom, such as Na, might seem to be a far less subtle effect than the condensation of atoms to form a metal. Nonetheless, the error in the calculated cohesive energy is only .03 eV, whereas the error in the calculated ionization potential is 0.42 eV. The point is that the atomic core changes less on going from the free atom to the metal than it does in going from the atom to the ion.

A second way in which the inclusion of spin-polarization effects is important to calculations for free atoms concerns difficulties introduced by use of the local-density approximation. Consider the intuitive notion that an electronic system is in its lowest energy state when all the occupied levels lie lower in energy than all the unoccupied levels. Density-functional theory confirms this intuition, which is succinctly stated by the relation $\partial E/\partial n_i = \epsilon_i$, where E is the total energy, ϵ_i is the eigenenergy of the ith level, and n_i is the occupation of the ith level. That is, if level i is occupied and level j is not and if the orbital eigenenergy ϵ_i lies above ϵ_j, then the total energy of the system will be lowered by transferring charge from level i to level j. In *density-functional theory*, this type of energy reduction leads unambiguously to a uniquely defined ground state, because the orbital eigenenergy ϵ_i rises as charge is removed from the level and conversely ϵ_j falls. Unfortunately, when the *local-density approximation* is used, the variation of the orbital eigenenergies with occupation has the wrong sign. That is, the following situation can, and frequently does, occur. For a given configuration (occupation of levels), an occupied level will lie above an unoccupied level, but, when charge is shifted to the lower-lying level, the newly occupied level rises and the emptied level falls, thereby maintaining the unwanted condition in which an unoccupied level lies below an occupied level. (In this situation, the total energy is minimized by fractionally occupying the two levels in the particular way that causes them to be degenerate.[103] The origin of this unphysical behavior in the local-density approximation is discussed in Section 4.3; what is relevant here is the fact that the inclusion of spin-polarization effects reduces to a considerable extent this unfortunate aspect of the local-density approximation. The simplest system exhibiting this problem is the free atom B, which possesses a single p electron. If the outermost electron is placed in the p_x orbital, for example, then the corresponding eigenenergy lies above those of the p_y and p_z states, if spin polarization is ignored, but below, if spin-polarization is included.[104] It should be noted that despite these effects, calculations based on the local-density approximation provide a description of free atoms that is adequate for many purposes. For example, interconfigurational total-energy differences for transition-metal atoms are given much more accurately by calculations based on the local-density approximation than by calculations based on the Hartree–Fock approximation.[105]

2.2.3. Applications of Spin-Polarized Theory to Solids. The spin-polarized version of the local-density approximation has been particularly successful in describing the ground state of the transition-metal ferromagnets.[89,90] The calculations correctly indicate that Fe, Co, and Ni are the only transition metals which order ferromagnetically.[87] Beyond that, the calculations provide an accurate quantitative description of both the magnetic properties of these systems (the magnetic moment and the hyperfine field as a function of pressure) and nonmagnetic properties such as the lattice constant and the bulk modulus. Values for the magnetic properties calculated by Janak[87,92–94] are compared with measurements in Fig. 5. The agreement between theory and experiment for the hyperfine field is particularly encouraging, because the hyperfine field is a measure of the magnetization density $\rho_\uparrow(\mathbf{r}) - \rho_\downarrow(\mathbf{r})$ at the nucleus. Therefore, the hyperfine field measures the magnetization of the $3d$ shell only through the coupling of the latter to the s electrons. One might expect this coupling (particularly the coupling between the $3d$ and $1s$ states) to depend in an important way on the nonlocal character of the exchange interaction. It is therefore somewhat surprising that a theory based on a local approximation to the exchange interaction does as well as Fig. 5 indicates.

The ground-state magnetic susceptibility has also been calculated for many metals. The calculations employ a variational formulation of the susceptibility due to Vosko and Perdew[85] which has been discussed in Chapter 2. The comparison of the calculated susceptibilities with measurements takes

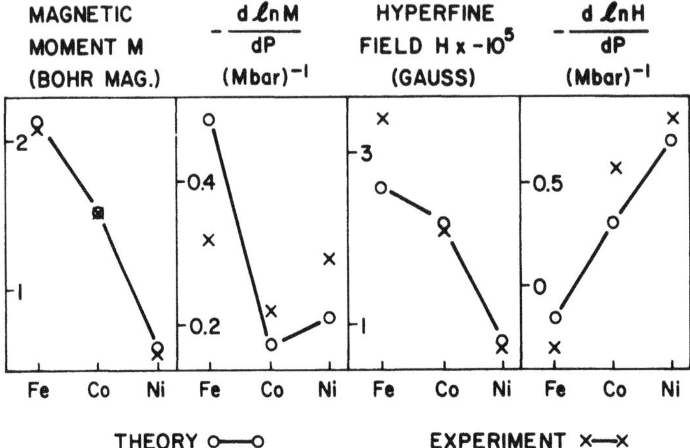

FIGURE 5. Comparison of measured and calculated ground-state magnetic properties for Fe, Co, and Ni. The calculated properties are due to Janak (see Refs. 87, 92, 93, and 94). Since the calculations involve no empirical adjustment the correct accounting for both the magnitude and the atomic-number dependence of these measurements provides strong support of their interpretation in terms of itinerant magnetism as described by the local-spin-density formalism.

two forms in this context. In simple metals, calculated values can be compared directly with the results of electron spin resonance experiments. Such a comparison has been made by Vosko *et al.*[106]; the results have been summarized in Table 1. In other systems, the relative magnitudes of the orbital and spin contributions to the measured susceptibility are unknown, making the comparison of the calculated spin susceptibilities with measurements difficult. The fact that the paramagnetic spin susceptibility should be singular for ferromagnets constitutes a check of the theory.[87] When the spin susceptibility is expressed as follows,

$$\chi = \frac{\chi_0}{1 - IN(\epsilon_F)} \tag{37}$$

then the condition for spontaneous ferromagnetic order becomes the magnitude of the quantity $IN(\epsilon_F)$ compared to unity. Figure 6, which displays this quantity for 31 metals, shows that the theory correctly identifies Fe, Co, and Ni as the only ferromagnets. Furthermore, the calculations from which Fig.

TABLE 1. Spin-Susceptibility χ/χ_0 for Alkali Metals[a]

	Li	Na	K	Rb	Cs
Theory	2.66	1.62	1.79	1.78	2.20
Experiment	2.50	1.65	1.701	1.724	2.24

[a]Theoretical values (Ref. 106) obtained using the local-spin-density approximation in conjunction with a variational expression for χ. References to the experimental work can be found in Ref. 106.

6 was created contain the answer to the question: *Why* are none of the 4*d* transition metals ferromagnetic? Is it because the Fermi-level state densities $N(\epsilon_F)$ are smaller in the 4*d* metals or is it because the intra-atomic exchange integral I is smaller? The calculations[87] indicate that the product $IN(\epsilon_F)$ in the 4*d* series remains smaller than one, even if *either* I or $N(\epsilon_F)$ is replaced by

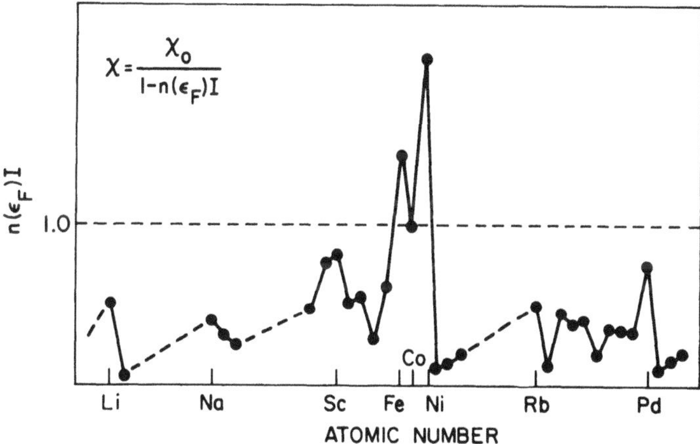

FIGURE 6. Condition for spontaneous ferromagnetic order in the local-spin-density theory. Shown is the calculated variation (Ref. 87) with the atomic number of the product of the Fermi-level state density $N(\epsilon_F)$ and, the intra-atomic correlation-corrected exchange integral I, usually called the Stoner parameter. The calculations involve no adjustable parameters. The local-spin-density theory relates the Stoner parameter I to the density and magnetization dependence of the exchange-correlation energy of the uniform, but interacting electron gas. The calculations correctly indicate that Fe, Co, and Ni are the only elements of the 31 shown which spontaneously order ferromagnetically.

its $3d$ counterpart. In other words, it is the magnitude of *both* I and $N(\epsilon_F)$ that is responsible for ferromagnetism in the $3d$ series.

The representation of the susceptibility given by equation (37) indicates that the local-spin-density theory of itinerant magnetism can be cast[86] in the form of the semiempirical theory due to Stoner.[107] In the Stoner theory the exchange energy is taken to be proportional to the square of the magnetization—the constant of proportionality being called the Stoner parameter I. This model is the basis of many discussions of ferromagnetism, in particular, Slater's prediction[108] in 1936 that Ni is ferromagnetic. The Stoner model has been discussed in terms of Hartree–Fock theory by Wohlfarth,[109] who also calculated the contribution of correlation to the Stoner parameter. Gunnarsson[86] has shown that the local-spin-density approximation can be used to derive the Stoner model and to evaluate the Stoner parameter I, treating exchange and correlation on an equal, albeit approximate, footing. The derivation indicates the approximations which are otherwise implicit in the use of the model.

A particularly nice result obtained using the local-spin-density approximation concerns the mechanical implications of magnetic order. The presence or absence of magnetic order results from a competition between kinetic and exchange energies. A system can exploit the exchange interaction by preferentially occupying the states of one spin. Because, however, the configuration in which the spin-up and spin-down levels are equally populated minimizes the kinetic energy, the system must pay a kinetic-energy price in order to magnetize. The system can reduce this kinetic-energy price by expanding, because the kinetic energy decreases with increasing volume; this is the origin of the lattice dilatation associated with magnetic order. This simple physical picture has been confirmed and quantified by calculations employing the local-spin-density approximation.[88,92] Janak[92] showed that the lattice dilatation caused by magnetic order in the $3d$ transition-metal elements is responsible for the anomalously small bulk moduli of these systems. (See Section 3.2.) The results of calculations by Andersen and co-workers[110] shown in Fig. 7 indicate that the same physical effect manifests itself in a variety of magnetic systems. What is nice about this application of the theory is that it exhibits several of the different ways in which the theory can be used. That is, first, a comparison of calculations[90] for existent systems (nonmagnetic $4d$ transition metals) with those for nonexistent systems (nonmagnetic $3d$ transition metals) indicated anomalous behavior. [See the lowest panels (bulk modulus) of Fig. 17.] Then, spin-polarized calculations for the $3d$ metals[88,92] identified the microscopic origin of the anomaly. Finally, spin-polarized calculations for related, but more complicated, systems[110–112] indicated the broader relevance of the effect.

Another type of analysis made possible by the local-spin-density approximation is that of the effects of alloying on magnetic systems. In some cases, the magnetic moment of a ferromagnetic element is reduced and finally eliminated by the admixture of a nonmagnetic material. This process has been traced in calculations by Hackenbracht and Kübler[98] for the case of NiAl. In other cases, nonmagnetic constituents can combine to form a ferromagnet. Calculations by de Groot *et al.*[113] have described this process in the case of $ZrZn_2$.

An interesting physical phenomenon which has received greater attention as the result of local-spin-density calculations is the possible existence of more than a single ferromagnetic phase in the same material. The existence of two magnetic states in the case of fcc iron was postulated by Tauer and Weiss[114] in an effort to understand the thermodynamics of the bcc $\rightarrow$ fcc phase transition. Calculations by Roy and Pettifor[115] suggest the existence of two ferromagnetic fcc phases in iron. More detailed calculations by Janak[93] indicate that multiple magnetic phases are possible in both Fe and Co. The first calculations to exhibit such multiple phases were those of Hattox *et al.*[116] for V. Although V was correctly found to be paramagnetic at its equilibrium volume, two ferromagnetic phases were found to exist for volumes 1% and 3% greater than this volume.

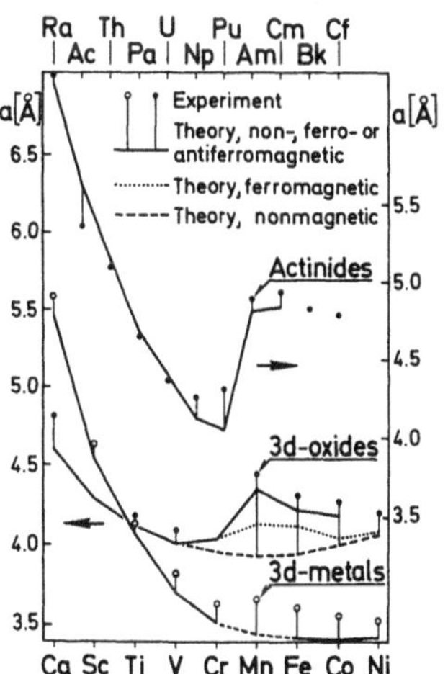

FIGURE 7. Comparison of measured and calculated lattice constants (interatomic separations). In each of the three chemical sequences, the onset of magnetic order is seen to produce a lattice dilatation. The calculations employ the self-consistent LMTO method of Andersen (Ref. 156) for the solution of the single-particle equations and the local-spin-density approximation for the treatment of exchange and correlation. (From Ref. 110.)

The local-spin-density approximation has recently been applied to systems that order antiferromagnetically. Calculations now exist for δ Mn,[96] for Cr[74,97,75] and for several transition-metal oxides.[110] Such calculations are only possible at present for relatively simple crystal structures, such as the CsCl structure in which the cube-corner atoms of a body-centered-cubic lattice are permitted to have a magnetization antialigned with respect to that of the cube-center atoms. In the case of Cr, for example, even though this simple crystal structure ignores the sinusoidal modulation of the magnitude of the magnetic moment, calculations based on this crystal structure show that antiferromagnetic order is energetically stable with respect to both ferromagnetic and paramagnetic order. The calculations[74] also exhibit the "nesting wave vector" separating the portions of the Fermi surface responsible for the antiferromagnetic character of the order.

The reduced symmetry of antiferromagnetic systems compared to ferromagnetic systems complicates the single-particle calculations required to implement the local-spin-density theory. Systems in which the translational symmetry of bulk crystals is broken by either a surface or a point imperfection are yet more difficult. Nonetheless, a few calculations of this type have been performed. Two groups have studied ferromagnetic slabs of Ni,[101,102] that is, films which are infinite and periodic in two dimensions and several atomic layers thick in the third dimension. Part of the interest in this type of system stems from the reduced coordination (the number of nearest neighbors) at a metal surface. Simple (tight-binding) models of d bands in transition metals indicate that the reduced coordination of the surface should cause the d band to narrow and the local state density to rise in the vicinity of the crystal surface. Since the kinetic-energy price associated with magnetic order is inversely proportional to the state density (see below), we might therefore expect an enhancement of the magnetic moment in the atomic layers nearest the film surface. The calculations performed thus far for Ni indicate that this is a relatively small ($\sim 10\%$) effect and that its precise magnitude is sensitive to the thickness of the slab.

The problem of magnetic impurities in nonmagnetic metallic hosts has been the focus of many experimental studies. The local-spin-density approximation has recently been applied to this problem by Zeller *et al.*[99,100] The local-spin-density calculations present an internally consistent picture of the width of the impurity d resonance, the energy splitting of the spin-up and spin-down resonances, the number of electrons each resonance can accommodate and the local magnetic moment. Where the comparison of the calculated properties and measurements is feasible, the two are in good agreement. The spin-resolved state densities for Cr in Cu, Fe in Cu, Mn in Cu, and Mn in Ag are shown in Fig. 8. Note, in particular, the progressive filling of the minority-spin resonance with increasing valence as we go from Cr to Mn to Fe. Note also the extent to which the majority-spin states are excluded from the energy region of the Cu *d* band and the width of the minority-spin resonance due to hybridization of the impurity *d* level with the *s-p* band of Cu. Such a realistic view of these systems would be impossible without both the local-spin-density formalism and the sophisticated Green's-function treatment[99] of the single-particle equations describing the isolated transition-metal impurity in a *d*-band host.

2.3. Symmetry Dependence of Total Energy Functional

For some atomic properties it is important to consider the total energy with a particular symmetry that is determined by orbital and spin angular momentum properties. Some examples are treated below.

2.3.1. Multiplet Structure and Valence Exchange-Correlation Energies. As stated several times in this chapter, the local-density approximation was designed for systems with a slowly varying electronic density. We would therefore expect the approximation to work much better for the more spread-out valence electrons than for the tightly bound core electrons. Fortunately, many interesting physical properties such as the cohesive energy of a solid and the binding energy and the equilibrium distance of a molecule are mainly a consequence of rearrangements among the valence electrons.[46] Still the density of these electrons can have a substantial spatial variation, and it would be nice if we, in some way, could monitor the accuracy of the approximation as we go from localized to more extended electron orbitals. Since density-functional theory is a theory of the total energy and the total charge density, this would perhaps seem to be a difficult task, but the multiplet splittings of atoms and molecules offer an interesting possibility in this regard. These splittings are almost entirely due to exchange and correlation effects among the outer valence electrons. By comparing the splittings obtained from the local-density approximation with the experimental splittings for cases in which the valence electrons are more or less localized, we can estimate the accuracy of the approximation as a function of the degree of localization. The original density-functional theory is a theory of the ground state, and some of the multiplet states of, e.g., an atom, are certainly excited states. However, these excited states usually have a symmetry different from that of the ground state, and we can then obtain their energies from the symmetry-dependent density-functional theory proposed by Gunnarsson and Lundqvist.[37] Their idea was to split Hilbert space into orthogonal subspaces with different sym-

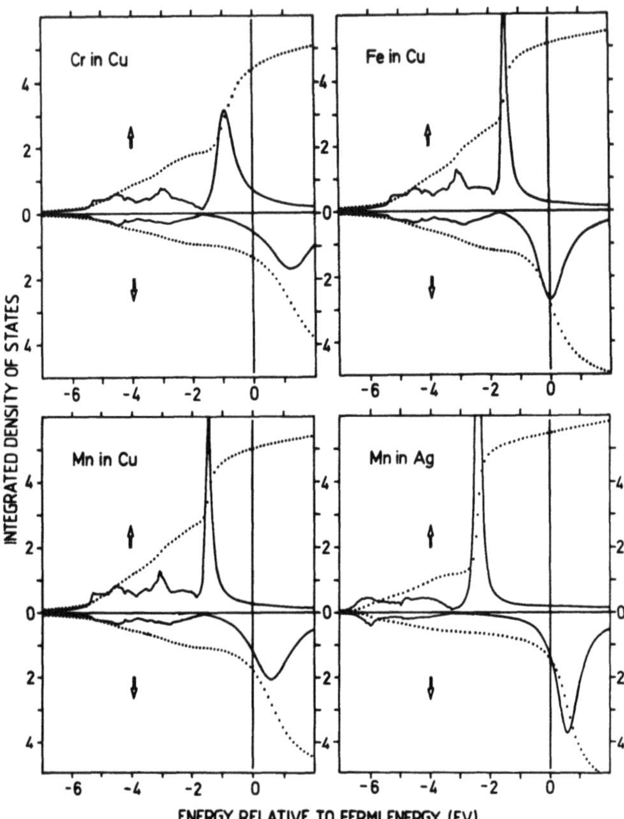

FIGURE 8. Densities of majority- and minority-spin states on transition-metal impurities in noble-metal hosts. Solid lines are state densities; dotted lines are integrated state densities. The impurity atom substitutes for a noble-metal atom in each case. The calculations which produced these results are based on the local-spin-density treatment of exchange and correlation; they assume the electron density and the effective one-electron potential to be spherically symmetric within nonoverlapping spheres and constant otherwise. The single-particle equations are solved using the self-consistent Green's-function technique of Zeller *et al.* (Ref. 99).

metries and to define a density-functional theory for each such subspace. The total energy of the system with a particular symmetry S that, e.g., could designate the angular momentum and spin quantum numbers of a light atom, is then a functional of those spin-density matrices that can be realized by a many-body wave function of the same symmetry:

$$E^S[n^S_{\alpha\beta}] = T_0[n^S_{\alpha\beta}] + \int n(\mathbf{r})w(\mathbf{r})\,d^3r + \tfrac{1}{2}\int n(\mathbf{r})v(\mathbf{r}-\mathbf{r}')n(\mathbf{r}')\,d^3r\,d^3r' + E^S_{xc}[n^S_{\alpha\beta}] \tag{38}$$

Here, as usual, $T_0[n^S_{\alpha\beta}]$ is the kinetic energy of noninteracting electrons having a spin-density matrix $n^S_{\alpha\beta}(\mathbf{r})$, $n(r)$ is the charge density $[n(\mathbf{r}) = \Sigma_\alpha n^S_{\alpha\alpha}(\mathbf{r})]$, $v(\mathbf{r}) = 1/r$ is the Coulomb interaction, and $E^S_{xc}[n^S_{\alpha\beta}]$ is the exchange-correlation functional into which the many-body aspects of the problem have been transferred. The external potential $w(\mathbf{r})$, which would be $-Z/r$ in the case of an atom with nuclear charge Z, must of course commute with the symmetry operator $\hat{S}$. The functional for the total energy implicitly depends on the symmetry via the symmetry dependence of the density matrix, but there is also, in principle, an explicit dependence on symmetry in the exchange-correlation part of the functional. The lowest energy in a particular symmetry channel S is obtained by minimizing the functional E^S with respect to all density matrices obtainable from many-body wave functions of symmetry S. It is hoped that this minimization can be achieved, as in the original theory,[65] by solving an equivalent one-particle problem with an effective potential obtained by functionally differentiating the last three terms in equation (38) with respect to the density-matrix $n^S_{\alpha\beta}$. The resulting one-particle orbitals would then be used to generate Slater determinants which can be combined to form a many-body wave function with the proper symmetry. From this we would obtain the one-particle density matrix which allows us to compute the effective one-particle potential, and we would have a self-consistency procedure for determining the density matrix and the total energy. Such a procedure would permit, for example, the description of an alkali atom in which the valence s electron has been excited to a state of p symmetry. The difficulty rests, as usual, in our inadequate knowledge of the exchange-correlation functional E^S_{xc} and, as we shall see, this problem can become much more severe when the lowest state of symmetry S is not the absolute ground state.

For the lack of a better alternative we may consider the consequences of using the local-density approximation for E^S_{xc} [equation (1)]. Two specific examples will illustrate the inadequacy of this approximation in connection with the procedure outlined above. Consider first the excited $1s2s$ configuration of atomic helium which gives rise to four states, one ^1S-state denoted $|0,0\rangle$ and three ^3S-states denoted $|1,M_S\rangle$, $M_S = -1,0,1$. It can easily be seen that the density matrices corresponding to these states are diagonal and spherically symmetric. They have the form

$$\begin{pmatrix} \tfrac{1}{2}n(r) & 0 \\ 0 & \tfrac{1}{2}n(r) \end{pmatrix}$$

for the states $|0,0\rangle$ and $|1,0\rangle$ and

$$\begin{pmatrix} n(r) & 0 \\ 0 & 0 \end{pmatrix}$$

for the state $|1,1\rangle$. Because there is no explicit symmetry dependence in the local-density approximation to E_{xc}^S, only an implicit dependence through the density matrix, states with the same form of the density matrix will acquire the same energy and the same density matrix upon minimization. Consequently, the scheme predicts the same energy for the states $|0,0\rangle$ and $|1,0\rangle$ and, as a matter of fact,[83] a different energy for the state $|1,1\rangle$ which should be degenerate with the state $|1,0\rangle$. This clearly does not make sense. The difficulty stems from the fact that the local-density approximation is not invariant under rotations in spin-space thus producing different energies for different M_S quantum numbers. Another example is furnished by the p^2 configuration of the carbon atom which gives rise to the three terms ^3P, ^1D, and ^1S. The states belonging to the ^1S and the ^1D terms are singlets and therefore have no net spin density. Furthermore, the ^1S state has a spherical charge density, and it is possible to form a linear combination of the five degenerate ^1D states that according to the prescription above, also will have a spherical (and therefore ultimately the same) charge density. Consequently, the present scheme has the undesirable feature of giving the same energy for the ^1S and the ^1D. It will also predict energy differences between the degenerate ^3P states which are of the same order of magnitude as the whole ^3P–^1D splitting. These errors reflect the fact that the local-density approximation is not invariant under rotations in real space.

The failure of the straightforward generalization of the density-functional theory outlined above suggests that the symmetry dependence of the exchange-correlation functional must be accounted for. There is, however, presently no available prescription for a symmetry-dependent E_{xc}^S. Fortunately, there exists an alternate procedure based on the notion of mixed symmetry states[83] that allows us to keep the simple local-density approximation for E_{xc}. As observed by Ziegler *et al.*,[82] one obtains rather accurate total energies for those particular symmetry states that would reduce to single Slater determinants if there were no Coulomb interaction between the electrons. The energies of the states consisting of several determinants, on the other hand, often turn out to be quite poor. The reason for this can be understood from the exact expression for the exchange-energy functional obtained by using the Hartree–Fock pair-correlation function in equation (4). Knowing that the pair-correlation function has the four spin components

$g_{\sigma\sigma'}$ where $\sigma,\sigma' = \uparrow,\downarrow$, this expression allows us to consider the exchange energy as a sum of four spin components $E^X_{\sigma\sigma'}$, where $\sigma,\sigma' = \uparrow,\downarrow$. Now, the crucial point to realize is that any single-determinantal state will always have $g_{\uparrow\downarrow} = g_{\downarrow\uparrow} = 1$, and consequently the unequal-spin exchange energies $E^X_{\uparrow\downarrow}$ and $E^X_{\downarrow\uparrow}$ vanish for such a state. In the local-density approximation one replaces the true pair correlation function with that of the homogeneous electron gas which, in the Hartree–Fock approximation, is obtained from a single Slater determinant. Therefore, the unequal-spin exchange energies $E^{LDX}_{\uparrow\downarrow}$ and $E^{LDX}_{\downarrow\uparrow}$ are always zero in the local-density approximation. As a result, states which reduce to single Slater determinants in the absence of interactions will have this part of their exchange energies given exactly by the local-density approximation. A state consisting of several determinants can, however, have large unequal-spin exchange energies, and this contribution to the total energy is completely lost in the local-density approximation.

To overcome this difficulty von Barth[83] constructed a rigorous density-functional theory of mixed-symmetry states. These states are defined as linear combinations of those states of pure symmetry which are lowest in energy

$$|D_i\rangle = \sum_j \alpha_{ji} \, | S_j,0\rangle \tag{39}$$

Here, $|S_i,0\rangle$ is the "ground state" of the S_i symmetry channel, and the significance of the method lies in choosing the coefficients α_{ij} such that the mixed-symmetry state $|D_i\rangle$ reduces to a single Slater determinant in the non-interacting case. The energy expectation value $E(D_i)$ of this state is given by

$$E(D_i) = \sum_j |\alpha_{ji}|^2 E_0(S_j) \tag{40}$$

where $E_0(S_i)$ is the lowest energy in the S_i symmetry channel. The quantity $E(D_i)$ can, however, also be estimated from the local-density approximation by minimizing the functional (38) with respect to density matrices of a form appropriate to $|D_i\rangle$. If we can find as many mixed-symmetry states as there are pure-state energies, we can invert the system of linear equations [equation (40)] to get the pure-state energies $E_0(S_i)$. In cases of high symmetry, such as an atom, we generally get an overdetermined set of equations which can be solved by a min–max procedure. There are, however, low symmetry cases in molecular physics for which the number of single-determinantal states are too few to enable us to determine all pure-state energies.[82]

Several people [83,118,73,119] have carried out calculations of multiplet splittings for many different atoms using the method of mixed-symmetry states. They find that the local-density approximation including only exchange reproduces the full Hartree–Fock results with a remarkable accuracy, the error being typically of the order of 0.1 eV. These results permit us to conclude that the charge density of the outer valence electrons in atoms and molecules (and in solids?) is sufficiently slowly varying to allow for a quantitative description of their exchange energies within the local-density approximation.

More interesting, however, is the fact that when correlations are included in the local-density approximation, one also obtains reasonable correlation-energy contributions to the multiplet splittings. As mentioned in Section 2.1.4, the total correlation energy of an atom is typically a factor of 2 too large in the local-density approximation.[57] In the case of the multiplet splittings, the corresponding error is only of the order of 20%–50% depending on the diffuseness of the orbital; furthermore, the correlation energy is usually *too small.* There are two things that indicate that this encouraging result is significant and not fortuitous: (i) The larger correlation-energy error in carbon relative to silicon is consistent with the relative localization of the valence p orbitals in the two atoms ($2p$ in C, $3p$ in Si). This is the behavior one would expect from a beginning convergence of the local-density approximation with respect to the rapidness of the density variations; (ii) Within perturbation theory one can show[83] that the local-density results for the multiplet splittings can be described in terms of reduced Slater integrals (usually denoted F_k). The formula for the reduction term only involves quantities associated with the correlation part of the local-density energy functional. This is in accord with what is usually expected from a theory of correlations. It is also consistent with an observation by Wood[118] that the correlation energy cannot be obtained by choosing a particular α in the $X\alpha$ energy-functional. On the other hand, the perturbation treatment also points to a limitation of the local-density theory of multiplet splittings. The theory no longer applies when the correlations become so strong that the splittings cannot be described in terms of reduced Slater integrals. Another deficiency has been pointed out by Gunnarsson and Jones,[119] who show that there is a peculiar asymmetry between particles and holes in the theory. For example, the results for several sp configurations are consistently better than those for the sp^5 configurations. They also found a recipe for restoring the symmetry which considerably improved the results for the sp^5 configurations, but there is not, as yet, formal justification for this procedure.

In summary of this section, we stress that the high accuracy with which both exchange and correlation energies of the outer valence electrons of atoms can be obtained from the local-density approximation, as demonstrated by the accuracy of the atomic multiplet splittings, strongly supports the use of this approximation for obtaining those physical properties of molecules and solids which primarily depend on rearrangements of the outermost electrons.

2.4. Almost Homogeneous Systems: Linear Response

The density-functional approach to the calculation of linear-response functions of inhomogeneous many-electron systems is an interesting and useful area of research. The principal reason for this interest is the fact that density-functional theory presently seems to offer the only practical way in which to incorporate the effects of exchange and correlation into the response functions of inhomogeneous systems with, e.g., band structure effects. Another important aspect of this line of work is that it provides an independent check on the accuracy of different approximations within density-functional theory. In the homogeneous limit the density–density response function of the electron gas obtained from elaborate diagrammatic methods can, for instance, be compared with the same quantity obtained from density-functional theory. A density-functional theory is usually designed to be exact in the limit of a slowly varying density and will therefore give a response function with the correct long-wavelength limit.[17] Deviations at short wavelengths are, therefore, a measure of the quality of the approximations.

By means of density-functional theory we can in principle calculate the spin-density matrix $n_{\sigma\sigma'}(\mathbf{r})$ [defined by equation (27)] of the ground state of any electronic system subject to a local spin-dependent external potential $w_{\sigma\sigma'}(\mathbf{r})$. The latter describes the coupling of the charge and the spin of the electrons to external electric and magnetic fields.[65,14,66] In practice, we must of course construct some approximation for the functional dependence of the total energy on the density matrix $n_{\sigma\sigma'}(\mathbf{r})$, and, as demonstrated in several chapters of this book, we can obtain the ground-state energy and density matrix from a relatively simple and accurate method based on an equivalent one-electron formulation. In principle, density-functional theory can also be used to obtain different time-dependent correlation functions such as the density–density response function needed for optical absorption, the spin–spin response function giving information on the spin-wave excitations, or the electron self-energy describing the Fermi surface and other quasiparticle properties. This is due to the fact that the Hamiltonian and therefore all expectation values with respect to its eigenvectors are functionals of the density matrix.[65,14,66] With a few exceptions—such as the electronic self-energy,[120] to which we shall return in a later section on excitation energies, and the density response function, which has recently been discussed by Chakravarty *et al.*[121]—no practical schemes for obtaining these correlation functions directly from density-functional theory have yet been developed. Fortunately, we can obtain the *static* response functions from the ground-state scheme because the latter allows us to calculate the ground-state density matrices in the presence of two external potentials differing by an infinitesimal quantity $\delta w_{\sigma\sigma'}(\mathbf{r})$. By definition, the static linear spin-density response function $\chi_{\pi\pi',\sigma\sigma'}(\mathbf{r},\mathbf{r}')$ measures the proportionality between the perturbation δw and the resulting change δn in the density matrix:

$$\delta n_{\sigma\sigma'}(\mathbf{r}) \equiv \sum_{\pi\pi'} \int d^3r' \; \chi_{\sigma\sigma',\pi\pi'}(\mathbf{r},\mathbf{r}')\delta w_{\pi\pi'}(\mathbf{r}') \tag{41}$$

In the case of external potentials with a very slow variation in time such that we can consider the system to remain in the instantaneous ground state of the time-dependent Hamiltonian, we can use the ground-state scheme to obtain information on the low-frequency response of the system. There are, however, very few cases in which the adiabatic approximation is strictly valid and, as we shall see, extreme care is necessary in order to avoid spurious results.

Within density-functional theory the most general derivation of the spin-density response function χ, defined in equation (41), has been presented by Rajagopal.[122] This derivation also shows in a transparent way how the frequency dependence enters into the response function when the perturbation varies with time. Several other authors have worked through the derivation in various restricted cases.[14,17,60,123,85,124–126,87,127–129] In order to demonstrate the basic ideas, we will here give a derivation for the case of a uniform direction of magnetization, i.e., the density matrix is assumed to be diagonal ($n_{\sigma\sigma'} = n_\sigma\delta_{\sigma\sigma'}$). The spin-up and spin-down densities n_σ, for $\sigma = \uparrow, \downarrow$, are given by the self-consistent equations[14]

$$[-\tfrac{1}{2}\nabla^2 + V^\sigma_{\text{eff}}(\mathbf{r})] \cdot \phi_{i\sigma}(\mathbf{r}) = \epsilon_{i\sigma}(\mathbf{r})\phi_{i\sigma}(\mathbf{r})$$

$$V^\sigma_{\text{eff}}(\mathbf{r}) = w_\sigma(\mathbf{r}) + \sum_{\sigma'} \int d^3r' \; v(\mathbf{r} - \mathbf{r}')n_{\sigma'}(\mathbf{r}') + \frac{\delta E_{xc}}{\delta n_\sigma(\mathbf{r})} \tag{42}$$

$$n_\sigma(\mathbf{r}) = \sum_{\epsilon_{i\sigma}<\epsilon_F} |\phi_{i\sigma}(\mathbf{r})|^2$$

A small external perturbation δw changes the effective one-electron potential from V_{eff} to $V_{\text{eff}} + \delta V_{\text{eff}}$ thereby giving rise to new one-electron orbitals $\{\phi_i\}$. The resulting change δn_σ in the density of electrons with spin σ is easily obtained from ordinary first-order perturbation theory as

$$\delta n_\sigma(\mathbf{r}) = \sum_{\sigma'} \int d^3r' \; \chi^0_{\sigma\sigma'}(\mathbf{r}, \mathbf{r}')\delta V^{\sigma'}_{\text{eff}}(\mathbf{r}') \tag{43}$$

where

$$\chi^0_{\sigma\sigma'}(\mathbf{r}, \mathbf{r}') = \delta_{\sigma\sigma'} \cdot \sum_{\epsilon_{i\sigma}<\epsilon_F} \sum_{\epsilon_{j\sigma}<\epsilon_F} \frac{\phi_{i\sigma}(\mathbf{r})\phi^*_{i\sigma}(\mathbf{r}')\phi^*_{j\sigma}(\mathbf{r})\phi_{j\sigma}(\mathbf{r}')}{\epsilon_{i\sigma} - \epsilon_{j\sigma}} + \text{c.c.} \tag{44}$$

i.e., the response function for noninteracting electrons. The expression given here pertains to a finite system and also to an infinite system provided the integral over the continuous variables i and j are performed before the $\mathbf{r}$ and $\mathbf{r}'$ dependence of $\chi^0_{\sigma\sigma'}$ is considered. According to equation (42), the change in the effective potential is

$$\delta V^{\sigma}_{\text{eff}}(\mathbf{r}) = \delta w_\sigma(\mathbf{r}) + \sum_{\sigma'} \int d^3r' \left\{ v(\mathbf{r}-\mathbf{r}') + \frac{\delta^2 E_{xc}}{\delta n_\sigma(\mathbf{r})\delta n_{\sigma'}(\mathbf{r})'} \right\} \delta n_{\sigma'}(\mathbf{r}') \tag{45}$$

and since by definition [equation (41)]

$$\delta n_\sigma(\mathbf{r}) = \sum_{\sigma'} \int \chi_{\sigma\sigma'}(\mathbf{r},\mathbf{r}')\delta w_{\sigma'}(\mathbf{r}') \, d^3r' \tag{46}$$

we arrive at the following integral equation for the spin-density response function $\chi_{\sigma\sigma'}(\mathbf{r},\mathbf{r}')$

$$\begin{aligned} \chi_{\sigma\sigma'}(\mathbf{r},\mathbf{r}') = \chi^0_{\sigma\sigma'}(\mathbf{r},\mathbf{r}') + \sum_{\sigma1\sigma2} \int d^3r_1 \, d^3r_2 \, \chi^0_{\sigma\sigma1}(\mathbf{r},\mathbf{r}_1) \\ \times \{v(\mathbf{r}_1-\mathbf{r}_2) + K^{xc}_{\sigma2\sigma2}(\mathbf{r}_1,\mathbf{r}_2)\}\chi_{\sigma2\sigma'}(\mathbf{r}_2,\mathbf{r}) \end{aligned} \tag{47}$$

where

$$K^{xc}_{\sigma\sigma'}(\mathbf{r},\mathbf{r}') = \frac{\delta^2 E_{xc}}{\delta n_\sigma(\mathbf{r})\delta n_{\sigma'}(\mathbf{r}')} \tag{48}$$

The derivation given here clearly demonstrates how the full response of the system is obtained by letting the electrons respond as free particles to an effective field which consists of the external field, the induced Coulomb field and an additional field due to the electrons' desire to avoid each other. This exchange-correlation interaction is rather short ranged[26]—it is a local field from the induced distortions of the exchange-correlation hole. Equation (48) shows how this interaction is related to the functional for the total exchange-correlation energy.

We now specialize the discussion to a non-spin-polarized system in which case there are only two independent response functions. These are χ_{nn} and χ_{mm}, describing, respectively, the charge-density response to an external potential and the spin-density response to an externally applied magnetic field. They are given by

$$\begin{aligned} \chi_{nn} &= \chi_{\uparrow\uparrow} + \chi_{\downarrow\downarrow} + \chi_{\uparrow\downarrow} + \chi_{\downarrow\uparrow} \\ \chi_{mm} &= \chi_{\uparrow\uparrow} + \chi_{\downarrow\downarrow} - \chi_{\uparrow\downarrow} - \chi_{\downarrow\uparrow} \end{aligned} \tag{49}$$

Note that in the unpolarized case there can be no charge response to a magnetic field and no spin-density response to an external potential ($\chi_{nm} = \chi_{mn} = 0$). The four equations (47) now reduce to

$$\begin{aligned} \chi_{nn}(\mathbf{r},\mathbf{r}') = \chi^0(\mathbf{r},\mathbf{r}') &+ \int d^3r_1\, d^3r_2\, \chi^0(\mathbf{r},\mathbf{r}_1) \\ &\times \{v(\mathbf{r}_1-\mathbf{r}_2) + K_{xc}(\mathbf{r}_1,\mathbf{r}_2)\}\, \chi_{nn}(\mathbf{r}_2,\mathbf{r}) \end{aligned} \tag{50}$$

and

$$\chi_{mm}(\mathbf{r},\mathbf{r}') = \chi^0(\mathbf{r},\mathbf{r}') + \int d^3r_1\, d^3r_2\, \chi^0(\mathbf{r},\mathbf{r}_1)\, I_{xc}(\mathbf{r}_1,\mathbf{r}_2)\, \chi_{mm}(\mathbf{r}_2,\mathbf{r}') \tag{51}$$

where

$$\begin{aligned} K_{xc}(\mathbf{r},\mathbf{r}') &= \frac{\delta^2 E_{xc}}{\delta n(\mathbf{r})\delta n(\mathbf{r}')} \\ I_{xc}(\mathbf{r},\mathbf{r}') &= \frac{\delta^2 E_{xc}}{\delta m(\mathbf{r})\delta m(\mathbf{r}')} \end{aligned} \tag{52}$$

The quantity $m(\mathbf{r})$ is the spin-density given by

$$m(\mathbf{r}) = n_\uparrow(\mathbf{r}) - n_\downarrow(\mathbf{r}) \tag{53}$$

and we note that the same noninteracting response function χ^0 ($\chi^0 = \chi^0_\uparrow + \chi^0_\downarrow = 2\chi^0_\uparrow$) appears in both integral equations defining the electric (χ_{nn}) and magnetic (χ_{mm}) response functions. This is a reflection of the fact that these two response functions are identical for noninteracting electrons. The frequently considered random-phase approximation (RPA) corresponds to neglecting K_{xc}, i.e., the distortions of the exchange-correlation hole in equation (50). Thus, the essence of this approximation is to allow the electrons to respond as free particles to the induced Hartree field. "The time-dependent Hartree approximation" is therefore a more illuminating name for the RPA in the case of time-dependent external fields. In the random-phase approximation with exchange (RPAE), only distortions of the exchange hole are considered [$K_{xc} \rightarrow K_x$ in equation (50)], and this approximation is therefore preferably referred to as the time-dependent Hartree–Fock theory. Since a magnetic field cannot induce a net charge, there is no induced Coulomb field in equation (51) for χ_{mm}. Thus, the RPA for the magnetic response function corresponds to the theory for noninteracting particles. In the full theory, however, there is an exchange-correlation interaction due to the spin-polarization of the exchange-correlation hole.

In the local-density approximation given by equation (1) the local-field interactions become δ-function interactions:

$$
\begin{aligned}
K_{xc}^{LD}(\mathbf{r},\mathbf{r}') &= \mu'_{xc}(n(\mathbf{r})) \cdot \delta(\mathbf{r}-\mathbf{r}') \\
I_{xc}^{LD}(\mathbf{r},\mathbf{r}') &= \nu_{xc}(n(\mathbf{r})) \cdot \delta(\mathbf{r}-\mathbf{r}')
\end{aligned}
\tag{54}
$$

Here, $\mu'_{xc} = \{\partial^2[n\epsilon_{xc}(n,m)]/\partial n^2\}$, for $m = 0$ and $\nu_{xc} = \{\partial^2[n\epsilon_{xc}(n,m)]/\partial m^2\}$ for $m = 0$, and we have considered the exchange-correlation energy per particle of the homogeneous electron gas, ϵ_{xc}, as a function of density n and spin-density m. In view of the short range of the exchange-correlation interaction, this does not appear to be unreasonable, but the accuracy of this approximation as well as of more sophisticated versions of density-functional theory can now easily be tested in the limit of small, but possibly rapid, density variations. One just has to compare the different approximate local fields obtained from density-functional theory to those of more elaborate theories of the electric and magnetic responses of the homogeneous gas. In the homogeneous case all quantities appearing in equations (50) and (51) become functions of the relative distances only, and the integral equations are trivially solved in reciprocal space giving

$$
\begin{aligned}
\chi_{nn}(q) &= \chi^0(q) \cdot \{1-[v(q) + K_{xc}(q)]\chi^0(q)\}^{-1} \\
\chi_{mm}(q) &= \chi^0(q) \cdot \{1-I_{xc}(q)\,\chi^0(q)\}^{-1}
\end{aligned}
\tag{55}
$$

It follows from the compressibility sum-rule[130] and from the corresponding sum-rule[130] for the magnetic response function that K_{xc} $(q = 0) = \mu'_{xc}$ and $I_{xc}(q = 0) = \nu_{xc}$. The local-density approximation [equation (54)] is therefore exact in the long-wavelength (slowly varying) limit as one would expect from its construction. Now, according to available electron-gas theories,[26,27] both $K_{xc}(q)$ and $I_{xc}(q)$ have a rather weak wave-vector dependence up to the Fermi wave vector beyond which they have some structure. At large wave vectors they tend to zero as $1/q^2$. Figure 2 shows a typical K_{xc}, which in this case was obtained from the static density response function by Geldart and Taylor.[27] In the local-density approximation both K_{xc}^{LD} and I_{xc}^{LD} [equation (54)] remain at their zero-wave-vector value for all q. We can thus conclude that the local-density approximation is reasonable for static disturbances having appreciable Fourier components only below the Fermi wave vector.

More sophisticated approximations within density-functional theory will presumably produce superior linear-response functions, and it would be interesting to know the performance of, e.g., the AD and WD approximations by Gunnarsson *et al.*[1] in the limit of small, but possibly rapid, density variations. The AD approximation, being based on the electron-gas linear-response function, will of course produce the same response function as was used to construct the approximation. The WD approximation, on the other hand, is completely defined by the electron-gas pair-correlation function and the sum rule [equation (73) in Chapter 2] and the corresponding exchange-correlation interaction $K_{xc}^{WD}(q)$ has to be worked out numerically. This has recently been done by von Barth[43] for a modification of the WD approximation in which the pair-correlation hole was approximated by

$$n_{xc}^{s}(\mathbf{r},\mathbf{r}') = n(\mathbf{r}')\,\{\tilde{g}_h(\mathbf{r}-\mathbf{r}';\ \tfrac{1}{2}[\tilde{n}(\mathbf{r}) + \tilde{n}(\mathbf{r}')]) - 1\} \tag{56}$$

Here, as with the WD procedure discussed in Section 2.1.3 $\tilde{n}(\mathbf{r})$ has to be chosen such that the sum rule [equation (73) in Chapter 2] is obeyed for each $\mathbf{r}$. The virtue of this modification is that it retains the symmetry $\tilde{g}(\mathbf{r},\mathbf{r}') = \tilde{g}(\mathbf{r}',\mathbf{r})$ of the exact pair-correlation function, which is essential for obtaining the correct $-1/r$ dependence of the exchange-correlation potential far away from a neutral atom. The exchange-correlation interaction $K_{xc}^{s}(q)$ obtained in the symmetrized scheme depends somewhat on the approximation employed for the electron-gas pair-correlation function, but the results are reasonable for all wave vectors. In particular, the large q-limit of K_{xc}^{s} is $4\pi\{\tilde{g}(0) - 1\}/q^2$ which is to be compared with the exact result[131] $8\pi\{g(0)-1\}/(3q^2)$. Thus, K_{xc}^{s} has the correct $1/q^2$ decay for large q although the coefficient is slightly off. The symmetrized scheme has also been generalized by Rajagopal and von Barth[44] in order to obtain the magnetic exchange-correlation interaction

$I_{xc}(q)$, but no numerical results have been obtained so far. Since we do not expect the symmetrization to have a major effect on the wave-vector dependence of the exchange-correlation interaction, we expect the WD scheme to give results of similar accuracy. We thus think it is fair to say that both the AD and the WD schemes and probably also the recent modifications of the WD scheme by Gunnarsson and Jones[45] are essentially correct in the limit of small density variations, at least to within our presently rather imprecise knowledge of the interacting electron gas. The testing ground for the new density-functional schemes should therefore be molecules and solids.

Solutions to the integral equations (50) and (51) for real solids would enable us to study the combined effect of exchange and correlation and of band structure on the response of these systems to static electric and magnetic fields. The equations are, however, very difficult to solve for inhomogeneous systems. In a free-electron-like material, a representation in terms of plane waves would be appropriate, and the integral equations become equations for matrices whose rows and columns are labeled by the reciprocal-lattice vectors. These matrix equations can then be solved by matrix inversion, and the feasibility of such calculations has been demonstrated by Singhal and Callaway[126] in the case of aluminum. As expected from the free-electron-like behavior of this metal, neither the exchange-correlation interaction nor the so-called local-field corrections seem to be very important. Note that we have here followed common jargon and used the term local-field correction for the difference between the result for χ obtained by full matrix inversion and that obtained by neglecting the off-diagonal elements of the matrices. This effect should not be confused with the effect of the local field arising from exchange and correlation. Hanke and Sham[132] have demonstrated that both effects are important in a material like diamond. It would therefore be of interest to have density-functional results for diamond and similar materials, within, e.g., the local-density approximation. Recently, Zangwill and Soven[133] and also Stott and Zaremba[134] have managed to solve the integral equation (50) for the electric response of different atoms and have obtained good static polarizabilities using the local-density approximation. These calculations are now being extended also to molecules. It should be noted that the work presented by Hanke and Sham and by Zangwill and Soven was primarily aimed at the frequency-dependent response which is formally outside the scope of the ground-state theory; we return to this important point below.

From a solution to the integral equation (51) we can, for instance, obtain the spin-susceptibility enhancement factor and the effect of exchange and correlation on the Knight shift of real inhomogeneous systems. Again, the equation is quite difficult to solve exactly even when the local-density approximation is used for I_{xc}. In the special case of the spin-susceptibility enhancement factor, however, Vosko and Perdew[85] have shown that there exists a variational principle which allows us to obtain a lower bound to this quantity from an approximate solution to equation (51). In this case we are concerned with the spin–density $m(\mathbf{r})$ caused by an external homogeneous magnetic field B. The spin–density $m(\mathbf{r})$ is then given by

$$m(\mathbf{r}) = \int \chi_{mm}(\mathbf{r},\mathbf{r}')d^3r' \cdot B \tag{57}$$

In the local-density approximation, equation (51) becomes

$$m(\mathbf{r}) = \chi^0(\mathbf{r})B + \int \chi^0(\mathbf{r},\mathbf{r}') \cdot v_{xc}(n(\mathbf{r}')) \cdot m(\mathbf{r}')\, d^3r' \tag{58}$$

where

$$\chi^0(\mathbf{r}) = \int \chi^0(\mathbf{r},\mathbf{r}')\, d^3r' = 2\sum_i |\,\phi_i(\mathbf{r})\,|^2\delta(\epsilon_F - \epsilon_i) \tag{59}$$

will be proportional to the ordinary Pauli susceptibility upon integration over $\mathbf{r}$. [The last equality follows from the definition of χ^0, equation (44), and the factor of 2 reflects the spin degeneracy.] We see again the same picture of noninteracting particles responding to an exchange-correlation enhanced effective field given by $B + v_{xc}m$, and we note that this effective field has a spatial dependence although the external field does not. By assuming that the full spin density $m(\mathbf{r})$ will have the same spatial dependence as it would have in the absence of exchange and correlation, only enhanced by a uniform factor β, we obtain the following approximate solution to the integral equation:

$$m(\mathbf{r}) = \beta \cdot \chi^0(\mathbf{r}) \cdot B \tag{60}$$

The approximate enhancement factor β which, according to Vosko and Perdew, represents a lower bound to the true spin-susceptibility enhancement factor, is obviously given by

$$\frac{1}{\beta} = 1 - \frac{1}{N(\epsilon_F)}\int [\chi^0(\mathbf{r})]^2 v_{xc}[n(\mathbf{r})]\, d^3r \tag{61}$$

where $N(\epsilon_F) = \int \chi^0(\mathbf{r})d^3r$ is the density of states at the Fermi level ϵ_F. Janak[87] evaluated β from equation (61) for a large number of metals and found a negative β for only cobalt, iron, and nickel indicating the ferromagnetism of these materials. Janak also tested the accuracy of the approximate procedure outlined above by comparing β for chromium to the "exact" spin-susceptibility enhancement factor obtained from a self-consistent solution to the local-density single-particle equations in the presence of an external magnetic field. The two numbers differed by only 10%, demonstrating the usefulness of the variational procedure for the enhancement factor. The procedure is, however, not adequate for obtaining the Knight shifts of metals which are proportional to the spin density at the nucleus, $m(0)$.[135] As shown by Zaremba and Zobin,[136] the exchange-correlation field can have a substantial variation in the core region thereby causing a spin polarization of the core electrons which could not occur [equation (59)] in the presence of a strictly homogeneous field. As a result, the spatial dependence of the total spin-density can be quite different from what it would be in the absence of the exchange-correlation interaction, thereby invalidating the assumption underlying the approximate theory above. Note that the core region contributes very little to the average magnetization $\int m(\mathbf{r})d^3r$, which is accurately given by the approximate theory. There is a variational principle for $\int m(\mathbf{r})d^3r$ but not for the magnetization at the nucleus [$m(0)$] on which the Knight shifts depend.

We end this section on linear response theory with a short discussion of the possibility of obtaining a description of collective excitations within the ground-state theory. We have in mind those excitations whose energies are given by the position of the poles of the frequency-dependent electric or magnetic linear-response function, such as, e.g., plasmons or spin waves. As discussed in the beginning of this section, these dynamical-response functions are strictly outside the scope of the ground-state theory. They can, however, be obtained formally, e.g., from the integral equations (50) and (51), by considering the dynamic response of noninteracting electrons to the external, time-dependent field and the instantaneously induced Hartree and exchange-correlation fields. The key word here is instantaneous, which indicates that the exchange-correlation field at a particular time is considered to be the same as it would have been if the strength of the external field at this particular time had remained the same at all times. Physically, this implies that the exchange-correlation hole would always be able to keep up with the electron, which clearly cannot be true for rapid disturbances. Formally, it means that $K_{xc}(\mathbf{r},\mathbf{r}')$ in equation (50) will be replaced by $K_{xc}(\mathbf{r},\mathbf{r}') \cdot \delta(t - t')$, whereas one would have $K_{xc}(\mathbf{r},t,\mathbf{r}',t')$ in a full theory. As a consequence, the only frequency dependence in this extension of the ground-state density-functional theory enters through the frequency dependence of the noninteracting response function $\chi^0(\mathbf{r},\mathbf{r}',\omega)$. Goodman and Sjölander[137] have, however, demonstrated that introducing the frequency dependence in this simple way can lead to spurious results for the dynamic response functions, at least for higher frequencies. As our dynamic density-functional response functions by construction become exact when the frequency tends to zero, we can nevertheless hope for a reasonable description of the response to disturbances which vary slowly in time. In particular, we would expect to obtain accurate dispersion relations for collective excitations with small excitation energies such as zero sound ($\omega_q = s \cdot q$) or spin waves ($\omega_q = D \cdot q^2$). Since the energies of these excitations vanish in the long-wavelength limit, it is tempting to argue that the zero-sound velocity s and the spin-wave stiffness coefficient D are given exactly by the ground-state theory. This argument is, however, incorrect as illustrated by the case of zero sound in ^{3}He. The energy ω_q of a zero-sound wave of wave-vector q is given by the pole in the complex frequency plane of the density-response function which is obtained from the dynamic generalization of equation (55). Since we in this case consider a neutral Fermi liquid (^{3}He), the long-range Coulomb interaction in the denominator is absent and we have the following condition for a pole in the density response function $\chi_{nn}(q,\omega)$:

$$K_{xc}(q)\chi^0(q,\omega_q) = 1 \tag{62}$$

From the analytic structure of $\chi^0(q,\omega)$ (the free-particle response function or the Lindhardt function; see, for instance, Ref. 138) when ω and q simultaneously tend to zero with a fixed ratio s, the long-wavelength limit of equation (62) becomes

$$\frac{\lambda}{2} \ln \frac{\lambda + 1}{\lambda - 1} = \frac{1}{1 - \kappa/\kappa_0} \tag{63}$$

We have also used here the fact that $\lim_{q\to 0} K_{xc}(q) = \mu'_{xc} = (\kappa_0 - \kappa)/(n^2 \cdot \kappa \cdot \kappa_0)$ [See the discussion following equation (5).] The quantities κ and κ_0 are the interacting and noninteracting compressibilities of the helium liquid, and the symbol λ is short for $s \cdot m/k_F$, where m is the bare mass of the helium particles and k_F is the Fermi momentum. By solving the transcendental equation (63) for λ we obtain the velocity of zero sound, s, within the density-functional formalism. We can then compare this velocity to the exact value of s obtained from Fermi-liquid theory.[130] It so happens that this theory leads to a transcendental equation very similar to equation (63). There are, however, two important differences. In the exact theory κ/κ_0 in equation (63) is replaced by $(m\kappa/m^*\kappa_0)$ and λ is replaced by $m^*\lambda/m$, where m^* is the effective mass due to the interaction between the quasiparticles. The zero-sound velocity of density-functional theory therefore differs from the true velocity to the extent that the effective mass differs from the bare mass. This failure of density-functional theory has its origin in the assumption of quasiparticles responding as free particles to an effective field. Real quasiparticles have an energy and momentum-dependent self-energy giving rise to an effective mass which is different from the bare mass. The situation here is quite analogous to that of the specific heat of the electron gas for which density-functional theory fails to provide the correct answer, as discussed in the section on excitation energies (Section 4.2).

We now turn to a discussion of the spin-wave stiffness coefficient D which is given by the pole of the transverse magnetic response function, $\chi_{+-}(q,\omega)$. This response function gives the magnetization induced by a small magnetic field applied perpendicular to the prevailing direction of magnetization, and it can therefore not be obtained from equation (47), which was derived under the assumption of all spins being either parallel or antiparallel to a given direction. The derivation of the transverse magnetic response function within density-functional theory is, however, very similar to the derivation that led to equation (47), and details can be found in, e.g., Refs. 122, 125, 127, and 128. The derivation leads to the following analytical structure of χ_{+-} for sufficiently small q and ω:

$$\chi_{+-}(q,\omega) = \frac{A(q,\omega)}{\omega - Dq^2} + B(q,\omega) \tag{64}$$

The functions A and B are bounded and, in addition, the static long-wavelength limit of A is given by $A(0,0) = (N_\downarrow - N_\uparrow)/(N_\downarrow + N_\uparrow)$, where $N_\uparrow$ and $N_\downarrow$ are the total number of spin-up and spin-down electrons. In the event that the exact transverse susceptibility has the same analtical structure as was suggested by Edwards and Fisher,[139] we can obtain the exact static transverse susceptibility $\chi_{+-}(q)$ from the zero-frequency limit of equation (64). Thus,

$$\chi_{+-}(q) = -\frac{A(q,0)}{Dq^2} + B(q,0) \tag{65}$$

or

$$\frac{1}{D} = \frac{N_\uparrow + N_\downarrow}{N_\uparrow - N_\downarrow} \lim_{q\to 0} q^2 \cdot \chi_{+-}(q) \tag{66}$$

which clearly shows that the spin-wave stiffness coefficient D is a ground-state (i.e., static) property of the system. This was also used by Liu and Vosko[128] to construct a variational expression for D

We have given above two examples of collective excitations with dispersive energies tending to zero in the long-wavelength limit. In the case of spin waves we argued that density-functional theory provides the exact spin-wave stiffness coefficient whereas the frequency-dependent extension of the theory can give only an approximate answer for the velocity of zero sound in ^{3}He. This difference was connected to the different analytical behavior of the corresponding response functions.

As a final example of collective excitations, we will show how density-functional theory can be used to obtain the dispersion of plasmons in the electron gas. Since the energy of a state with an excited plasmon is at least the classical plasma frequency ω_p above the ground-state energy, we do not expect to get the exact answer, but density-functional theory does indeed yield a very accurate plasmon dispersion relation. The energy ω_q of a plasmon with wave-vector q is given by the pole of the electric response function $\chi_{nn}(q,\omega)$. From the frequency-dependent extension of equation (55), we find

$$[v(q) + K_{xc}(q)]\cdot\chi^0(q,\omega_q) = 1 \tag{67}$$

which, together with the explicit expression[138] for the Lindhard function $\chi^0(q,\omega)$, gives

$$\omega_q^2 = \omega_p^2 + q^2\,[\tfrac{6}{5}\epsilon_F + (n\kappa)^{-1} - (n\kappa_0)^{-1}] \tag{68}$$

Here, ϵ_F, κ, and κ_0 are the free-electron Fermi energy and the interacting and noninteracting compressibilities of the electron gas of density n. The first term within the bracket is the pure Hartree (RPA) contribution and the remaining two terms, which are determined by the long-wavelength limit of K_{xc}, represent the effect of exchange and correlation on the plasmon dispersion relation. For metallic densities this result is within a few per cent of what one obtains from Fermi-liquid theory.[138] It should be remembered, though, that Fermi-liquid theory does not provide the exact answer either, owing to the finite excitation energies involved.

3. Inhomogeneous Systems

In this section we turn from the linear response theory applicable to almost homogeneous systems to the truly inhomogeneous electron gases found in molecules and solids. We consider first molecular bonding.

3.1. *Aspects of Molecular Bonding*

Interest in the electronic structure of small molecules stems from several sources. The desire to understand the bonding properties and excitation spectra of these systems in terms of simple, intuitive concepts is an old and ongoing challenge. In addition to their intrinsic interest, these systems, because they have been studied by such a broad range of experimental and theoretical techniques, constitute a popular testing ground for new approaches, such as density-functional theory. An additional source of interest in small molecules is the hope that many important aspects of extended systems, such as impurities and other defects in solids, as well as chemisorption systems, can be studied using molecular clusters to simulate the portion of the extended system most relevant to the property of interest.

There are several interesting distinctions between the application of density-functional techniques to solids and to small molecules. One of these is the fact that, for small molecules there exist well-defined alternatives to the density-functional approach (the Hartree–Fock method,(117) the configuration interaction method,(140) generalized valence-bond theory,(141) for example); for extended systems, particularly metals, essentially no alternatives exist. Also different is the relative importance of the intra- and interatomic aspects of the electronic structure. In solids (certainly metals), the relative importance of interatomic interactions is reflected in the use of jellium models and calculational procedures based on plane waves. Conversely, the relative importance of intraatomic interactions in the case of molecules is reflected in their description in terms of combinations of atomic orbitals. Another important difference between solids and molecules is the relative ease with which single-particle equations can be solved. In other words, even if exchange and correlation could be neglected, the construction of molecular orbitals of sufficient accuracy for the study of bonding properties would still be a difficult task. Note, in this connection, that all the work described in the section on metallic bonding is based on the decomposition of space into regions in which the electron density and effective-single-particle potential could be assumed to be spherically symmetric. It is the fact that such "muffin-tin" and "atomic-sphere" approximations are not adequate in the context of molecular calculations that make the molecular calculations more difficult.

Fortunately, considerable progress has been made with regard to the solution of the single-particle equations, so that several calculations have now been made which can be used to calibrate the local-spin-density approximation. Gunnarsson and Johansson(142) have studied a series of molecules composed of H and He using a numerical method, which for all practical purposes is exact. These calculations are quite unambiguous, but they are restricted to molecules comprised of very light atoms, so that no assessment is provided of the ability of the local-density approximation to describe, for example, Π-bonded systems or transition-metal systems involving localized *d* electrons. The diatomic molecules formed from atoms in the first row of the Periodic Table have been studied by three groups[143–145] using successively more accurate numerical methods. These calculations taken together provide a very encouraging assessment of the accuracy of the local-density approximation.

The binding energy of the H_2 molecule [$E(H_2)-2E(H)$] as a function of internuclear separation is shown in Fig. 9. The figure compares essentially exact results for the binding curve[146,147] with those obtained using different calculational procedures (with and without the use of the muffin-tin approximation, for example) and using different approximate treatments of exchange and correlation (local-density,[142] $X\alpha$,[148] and Hartree–Fock[149]).

Three aspects of Fig. 9 warrant comment. First, the local-spin-density result is significantly more accurate than the Hartree–Fock and $X\alpha$ results. Second, the local-spin-density results differ from the exact result by only 0.1 eV. Third, as mentioned above, the binding energy is significantly corrupted by use of the muffin-tin approximation (the result in Fig. 9 labeled MS–$X\alpha$).

Consider now applications of the local-spin-density approximation to more complicated molecules, in which the presence of core states requires

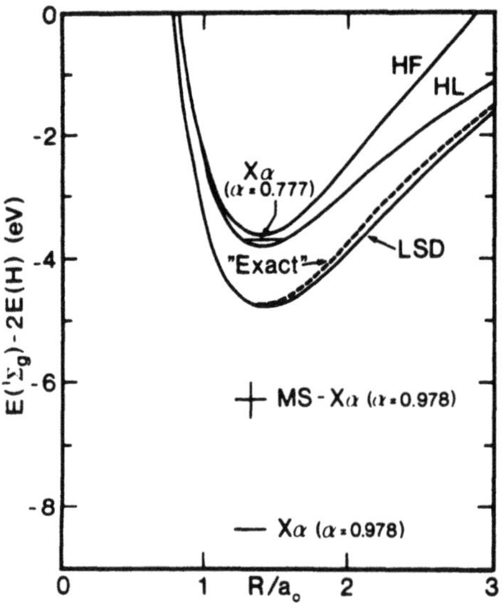

FIGURE 9. Comparison of calculated binding energies for molecular H_2. For H_2, both the exact result (Refs. 146 and 147) and the exact local-spin-density result (Ref. 142) are available both for mutual comparison and for comparison with results obtained using more approximate methods. The Hartree–Fock (HF) results are those of Ref. 149 and the $X\alpha$ results are those of Ref. 148. The MS–$X\alpha$ result (Ref. 148) was obtained using the muffin-tin approximation.

more elaborate and, unfortunately, more approximate numerical procedures than were adequate for H and He. Much of the work in this context has been performed by O. Gunnarsson, J. Harris, and R. O. Jones. A summary of their results for the first-row diatomics[143] is presented in Fig. 10. The parameter-free local-spin-density calculations are seen to describe the fundamental chemical trend across the series extremely well. The variation of the binding energy across the series exhibits the same fundamental parabolic dependence on atomic number seen in the cohesive energy in each transition series. (See Fig. 17 below.) This similarity suggests that the bonding implications of the progressive filling of the p shell are similar to those of the transition-metal d shell.

Additional insight into the accuracy of the local-density theory of molecular bonding can be gained from Table 2. Compared in Table 2 are the

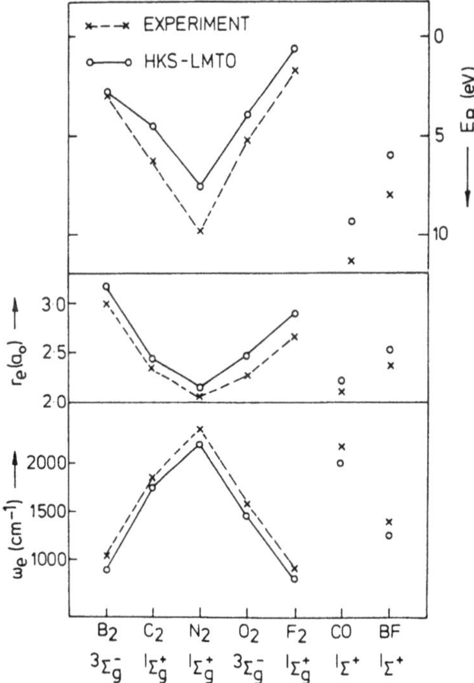

FIGURE 10. Comparison with experiment of the bonding properties of the first-row diatomic molecules calculated using the local-spin-density approximation. The properties considered are the binding energy E_B, the equilibrium nuclear separation r_e, and the vibrational frequency ω_e. The calculations (Ref. 143) employ the linear-muffin-tin-orbital technique of Ref. 150. The measured values are those of Ref. 151.

TABLE 2. Comparison of Measured and Calculated Bonding Properties of First-Row Diatomic Molecules[a]

		Molecule						
		H_2	B_2	C_2	N_2	CO	O_2	F_2
Bond energy (eV)	Exper.	4.75	2.9	6.2	9.91	11.23	5.2	1.65
	x_α(G)	3.59	3.90	5.97	9.23	11.98	7.01	3.18
	x_α(DVM)	3.6			8.4	11.5	6.6	2.9
	LSD(DVM)				11.1	12.8	7.5	3.4
	HF	3.64			5.27	6.19	1.43	−1.63
Bond length (bohr)	Exper.	1.40	3.05	2.35	2.07	2.13	2.28	2.68
	x_α(G)	1.46	3.04	2.35	2.08	2.13	2.28	2.61
	x_α(DVM)	1.46			2.13	2.15	2.36	2.67
	LSD(DVM)				2.15	2.17	2.37	2.67
	HF	1.39		2.37	2.01	2.08	2.18	2.50
Bond stiffness (cm^{-1})	Exper.	4400	1051	1857	2358	2170	1580	892
	x_α(G)	4110	1050	1920	2370	2160	1610	1090
	x_α(DVM)	4166			2362	2300	1565	1070
	LSD(DVM)				2458	2299	1429	1132
	HF	4561		1970	2730	2431	2000	1257

[a]Bond energy refers to the dissociation energy. Bond length is the internuclear separation. Bond stiffness is the vibrational frequency. Local-spin–density (LSD), $X\alpha$, and Hartree–Fock (HF) are three approximation methods which reduce the many-electron problem to a series of effective one-electron problems. Gaussian (G) and discrete variational method (DVM) are numerical methods used to solve the effective one-electron problems (from Ref. 145).

bonding properties of several small molecules, as given by (1) experiment, (2) calculations based on the local-density approximation, (3) calculations based on the Hartree–Fock approximation, (4) calculations employing the $X\alpha$ formalism, and (5) calculations employing contemporary but different numerical methods for the solution of the single-particle equations. Note first that the bond *energy* involves the total energy of the free atom, whereas the bond *length* and bond *stiffness* involve only the molecule. For the exclusively molecular properties, the local-density and $X\alpha$ calculations yield very similar results, particularly when implemented using the same numerical methods. (DVM stands for discrete variational method,[(144)] which is a numerical method.) This suggests that, if the superior numerical procedures of Ref. 145 were used in conjunction with the local-spin-density approximation, then similarly good agreement with experiment would result.

Figure 10 and Table 2 provide a very encouraging picture of the ability of calculations based on the local-density approximation to provide not only a qualitative picture of chemical trends, but a quantitative description of bonding properties as well. The agreement between calculated and measured

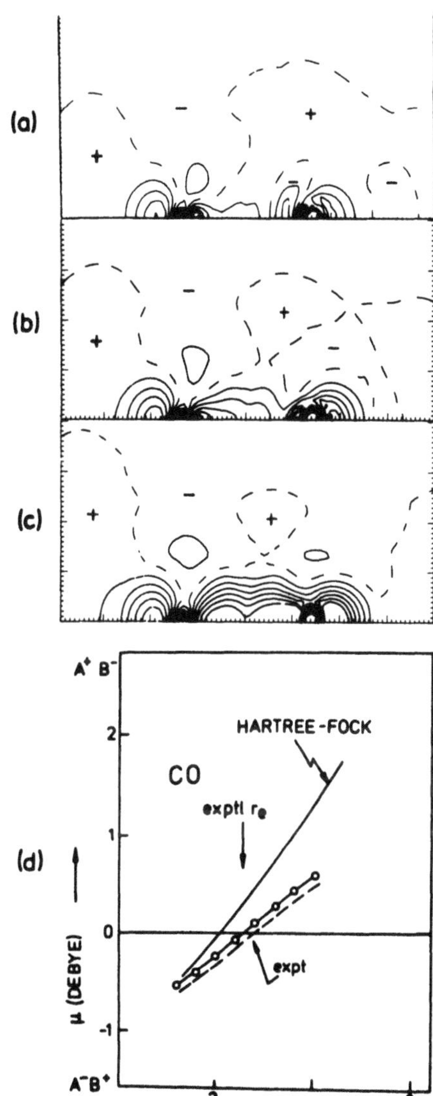

FIGURE 11. Variation of the electron density of the diatomic molecule CO with interatomic separation (from Ref. 143). Contours of constant electron density are shown for three internuclear separations, (a) 2.6, (b) 2.25, and (c) 1.9 Bohr. As indicated in (d), the internuclear separation 2.25 Bohr is close to the equilibrium value. Also indicated in (d) is the fact that the dipole moment of the molecule changes sign in this range. The contour plots therefore show the electron-density variation responsible for the dipole-moment sign change. The contour plots have been scaled so that the bond length appears to be the same in each case. The contour interval is 0.03 electrons/Bohr^3. The Hartree–Fock result for the dipole-moment variation is that of Ref. 153; the experimental result is that of Ref. 154.

atomic separations (better than 0.1 Å) shown in Table 2 is particularly significant for those interested in predicting atomic positions at solid surfaces, for example. A striking example of the potential of density-functional theory in this context is the recent work of Yin and Cohen,[152] who determine the atomic positions in bulk Si and on one of the surfaces of Si from a direct minimization of the density-functional total energy with respect to nuclear coordinates.

The calculations of Ref. 143 for the first-row diatomics also illustrate the use of local-density calculations as a microscope. The dipole moment of the heteronuclear molecule CO has interested chemists for many years. Calculations based on the Hartree–Fock formalism yield a dipole moment opposite in sign to the measured value. The value of the dipole moment at the equilibrium interatomic spacing is difficult to calculate accurately because the dipole moment varies sharply with internuclear separation and changes sign at an internuclear separation close to the equilibrium value. Local-density calculations not only describe this behavior with great accuracy; they can also be used to "observe" the charge rearrangements underlying the dipole-moment variation. Figure 11 shows the evolution of contours of constant electron density in CO as the nuclear separation is varied. The numerical accuracy with which observable quantities can be calculated and the completely parameter-free character of such calculations give credence to results, such as the contour plots in Fig. 11, which are either difficult or impossible to measure.

The calculational procedure developed by Gunnarsson *et al.*[143,155] permits another type of chemical trend to be studied. The distinguishing feature of the procedure is that the intra- and interatomic aspects of the calculations are separated. This permits molecules involving heavy atoms to be studied without a drastic increase in computational effort (with respect to calculations for molecules consisting of light atoms). Heavy atoms possess complex interiors, so that procedures in which the intra- and interatomic aspects of the molecule are determined as parts of the same calculations increase in magnitude very quickly as the constituent atomic numbers increase. Note that the linear-muffin-tin-orbital method[156] used by Gunnarsson *et al.*, like the augmented-plane-wave[157] and augmented-spherical-wave[158] methods used for solids and solid surfaces, does not merely separate the core and valence aspects of the molecular calculation, as do, for example, pseudopotential theory and the frozen-core approximation. The magnitude of the calculation in these methods is determined largely by the size of the basis set required to describe the *interatomic* variation of the molecular orbitals contributing to the valence electron density. The distinction between the separation of core and valence effects and the separation of intra- and interatomic effects is, for example, the requirement in pseudopotential techniques that the intraatomic portion of the valence wave function be produced as part of the interatomic calculation. In procedures based on augmentation, the basis set used to describe interatomic interactions is used only to describe the *interstitial* portion of the wave function and to supply boundary conditions which specify the solutions of the intraatomic problem, which is solved separately. This important calculational advantage has permitted Jones and Harris to study chemical trends corresponding to *columns* of the Periodic Table. They have calculated the bonding properties of homonuclear diatomic molecules corresponding to the first, second, and fourth valence groups. The system-independent and parameter-free character of their calculational procedure allows interesting similarities and differences to be clearly exhibited.

The most pervasive feature illuminated by these calculations is the two-sided effect of each new angular-momentum shell; that is, effects associated with the first appearance of p electrons, d electrons, and f electrons as we move down each column.

The effects are two-sided in the following sense. When the shell in question is unoccupied, it lies relatively low in energy because it is not excluded from the core by orthogonality requirements. It is therefore readily available to participate in the atomic polarization associated with bonding in the molecule. The other side of this effect is that, in the next row of the Periodic Table, the same angular momentum shell is filled. As a member of the atomic core this shell now excludes the next shell of the same angular momentum from the core, pushing it to relatively higher energies. Furthermore, the finite spatial extent of this core shell means that it imperfectly screens the now larger nuclear charge, thereby lowering the energies of the valence orbitals which penetrate somewhat the core regions. Therefore, by both lowering the energy of the occupied level and raising the energy of the unoccupied level, the entry of a new angular-momentum shell into the core reduces the polarizablity of the atom. Consider, for example, the alkalis K and Rb. In K,

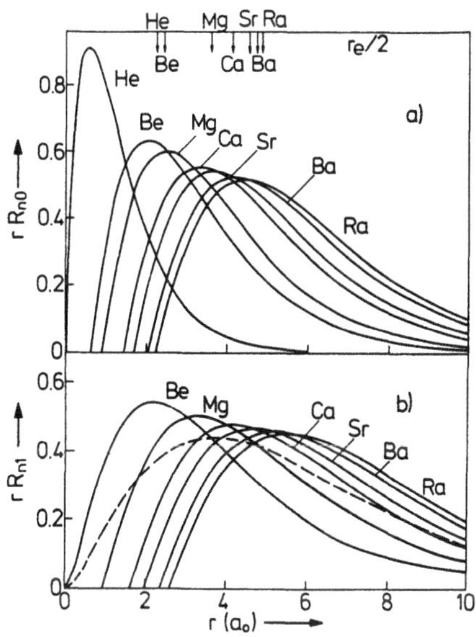

FIGURE 12. Pairing of valence wave functions for the group II atoms. The upper portion (a) of the figure shows the s wave function corresponding to the ^1S (ns^2 state; the lower portion (b) shows the radial part of the p wave function corresponding to the ^3P (ns^1np^1) state. The dashed curve is the p wave function of the 1s 2p configuration of He. The arrows at the top of the figure indicate the position of the bond center in the corresponding diatomic molecule. (From Ref. 159.)

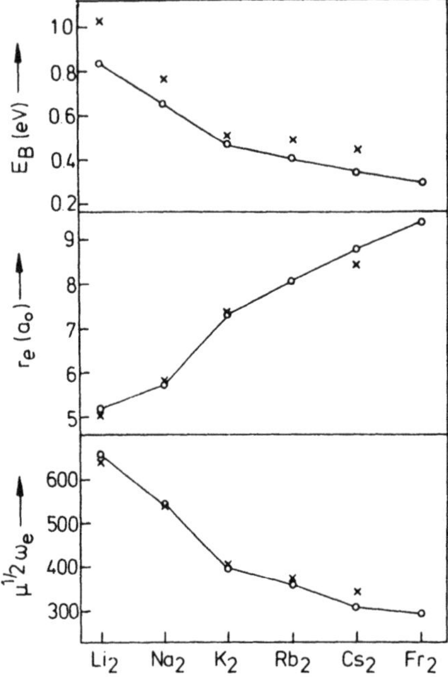

FIGURE 13. Comparison of measured and calculated molecular bonding properties of the alkali dimers. Shown are the binding energy E_B, the equilibrium internuclear separation r_e, and the weighted vibration frequency $\mu^{1/2}\omega_e$, where μ is the reduced mass. The calculated values are those of Ref. 160, where references to the experimental work can also be found.

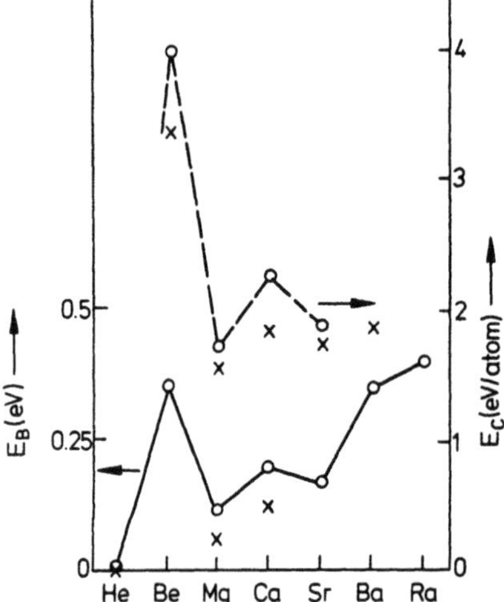

FIGURE 14. Comparison of molecular and solid-state bonding properties for the group II elements. The molecular calculations are described in Ref. 159; the cohesive-energy calculations are those of Ref. 90. The measured cohesive energies (crosses) are from Ref. 162. The measured binding energy for Mg_2 is from Ref. 163; that for Ca_2 is from Ref. 164. Both the cohesive energy and the molecular-bond energy for He can be considered to be zero in this context.

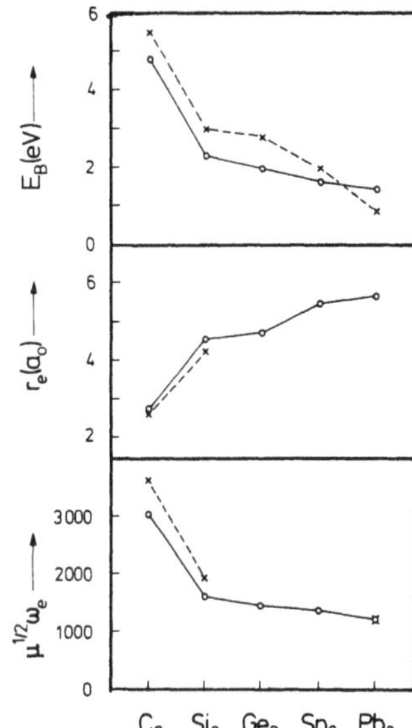

FIGURE 15. Comparison of measured (crosses) and calculated (circles) molecular bonding properties of the group IV dimers. Shown are the binding energy E_B, the equilibrium internuclear separation r_e, and the weighted vibration frequency $\mu^{1/2}\omega_e$, where μ is the reduced mass. The calculations are described in Ref. 161, where references to the experimental work can also be found.

the $3d$ levels are unoccupied and their contribution to bonding is relatively large because they lie relatively low in energy and, therefore, contribute readily to the polarization of the atom. In Rb, on the other hand, the presence of the $3d$ levels in the core pushes the unoccupied atomic $4d$ levels to relatively higher energies. At the same time, the penetration of the $3d$ core shell in Rb by the $4s$ electrons lowers their energy, further separating them from the unoccupied polarization levels.

The most direct manifestation of this effect is a pairing of the valence orbitals of the free atom. Harris and Jones have plotted these orbitals for the atoms in groups I,[(160)] II,[(159)] and IV[(161)] of the Periodic Table; we show in Fig. 12 those for group II atoms. Consider, as an example, the s orbitals of the group II atoms (Fig. 12a). Without the effect described above, the valence orbitals would become steadily more delocalized as we move down a column of the Periodic Table, because the attraction due to the increasing nuclear charge does not completely compensate for the repulsion of the orthogonality requirement. We see in Fig. 12a that this steady progression is broken by the entry of the $2p$, $3d$, and $4f$ shells into the core. The $2p$ shell makes Mg similar to Be; the $3d$ shell makes Sr similar to Ca; and the $4f$ shell makes Ra similar to Ba. The molecular bonding properties (dissociation energy, bond length, and vibration frequency) are shown for groups I[(160)] and IV[(161)] dimers in Figs. 13 and 15.

Consider the bond energy of the group II dimers shown in Fig. 14. We see that, superimposed on the slow increase in binding with valence principal quantum number is an oscillatory variation which we ascribe to the unfilled p levels in Be, to the unfilled d levels in Ca, and to the unfilled f levels in Ba. The bond length in the group IV dimers (Fig. 15) exhibits a similar oscillation, which lends itself to the same interpretation.

Another chemical trend which emerges from the comparison of bonding in the dimers of groups I, II and IV (Figs. 13, 14, and 15) is that, whereas the bond strength *decreases* with increasing valence principal quantum number in groups I and IV, it *increases* in group II. The dimers of groups I and IV are prototypical covalently bonded systems. The reversal of the bonding trend in group II suggests that the dimers of group II are held together by a different physical mechanism. The fact that the atomic polarizability tends to increase with orbital size, together with the closed-s-shell character of these systems, suggests that the relevant mechanism is the van der Waals interaction. That the local-density approximation should provide accurate results in this context is surprising, because the approximation fails to describe the physical effect with which the van der Waals interaction has come to to be identified, that is, the asymptotic power-law dependence of the interaction strength on distance. (The local-density approximation leads to an exponential distance dependence.) The resolution of this apparent paradox is the fact that the physical mechanism underlying the van der Waals interaction is simply the attraction between an electron and its exchange-correlation hole, an effect which, in many cases, is well described by the local-density approximation. The point here is that the power-law interaction strength and the local-density approximation describe complementary aspects of the *same* physical effect. The power law is appropriate when the electron and its exchange-correlation hole are well separated; the local-density approximation is appropriate when they are not. (An electron outside a metal surface and its image inside the metal constitute the simplest example of an electron well separated from its exchange-correlation hole.) Recall that the basis of the local-density approximation is the assumption that the exchange-correlation hole is centered on the electron and, even more importantly, the assumption that the size of the hole is governed by the density of the screening electron gas *at the position of the electron.* These two assumptions are responsible for the exponential distance dependence of the interaction strength. The interaction of closed-shell systems, such as the atoms of the group II dimers, involves both aspects of the electron-hole interaction. In these systems, the attraction between the electron and the small portion of the exchange-correlation hole residing on the "other" atom is the principal source of the interatomic bond. Over much of the orbit of each valence electron, the electron and this portion of the exchange-correlation hole are well separated, and their interaction is poorly described by the local-density approximation. But, in the remainder of the orbit, that in which the electron clouds of two atoms interpenetrate, the power-law description breaks down and the local-density approximation becomes a good description of the exchange-correlation hole. Lang[165] has shown, for the case of rare-gas atoms chemisorbed on the surface of a simple metal, that it is the portion of the orbit that is well described by the local-density approximation that is more important to the bond energy. It seems likely that the same argument explains the ability of calculations based on the local-density approximation to describe bonding in the group II dimers.

A final qualitative trend suggested by the molecular calculations of Harris and Jones concerns the role in chemical bonding played by relativistic effects. Because these effects are largest in the atomic core, states of low angular momentum are affected more than others. Such effects therefore tend to lower the levels occupied in the atomic ground state relative to the excited states which participate in the polarization of the atom and the formation of molecular states. Therefore, the relativistic stabilization of the free atom tends to weaken the molecular bond. This effect is most clearly seen in the optical excitations of free atoms, but is also visible in the comparison of calculated and measured binding energies of the group IV dimers (Fig. 15).

In summary, the local character of the effective-single-particle potential in the density-functional description of inhomogeneous systems has made possible molecular calculations involving a much broader spectrum of atomic constituents than was previously feasible. These calculations have been useful both in interpreting experimental measurements and in elucidating a variety of chemical trends.

3.2 *Metallic Cohesion*

One subject to which analysis based on density-functional theory has made a substantial contribution is metallic cohesion. The understanding that has resulted from the application of density-functional techniques to this problem has several aspects. First, cohesive properties are now accurately given by calculations for which the only input is the atomic number.[90] Second, theoretical methods based on density-functional theory have been developed which exhibit, in an intuitively satisfying way, the microscopic mechanisms responsible for cohesion.[166–168] Finally, these same theoretical methods exhibit explicitly, in the context of parameter-free calculations, the respective contributions of interpretive theories developed previously.[169–171]

The term "cohesive properties" usually refers to three aspects of the binding curve, the total energy of a metal E considered as a function of the volume Ω. The binding curve is shown schematically in Fig. 16. The total energy per atom $E(\Omega)$ tends to that of the free atom $E_a \equiv E(\Omega=\infty)$ as the

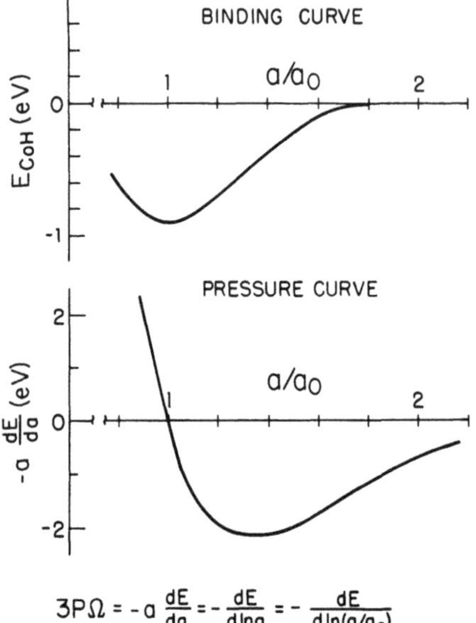

FIGURE 16. Schematic drawing of the binding curve and the equation of state. (a) The binding curve is the total energy of the solid as a function of volume. (b) The equation of state is the negative of the volume derivative of the binding curve.

volume tends to infinity, and $E(\Omega)$ takes on its minimum value when $\Omega = \Omega_0$, the equilibrium atomic volume of the solid. The three aspects of the binding curve of greatest interest are (1) the total energy gained in the condensation process, that is, the cohesive energy $E_{\text{coh}} \equiv E(\Omega = \infty) - E(\Omega = \Omega_0)$; (2) the equilibrium atomic volume Ω_0; and (3) the bulk modulus B, which is curvature of the binding curve at the equilibrium volume (multiplied by the volume),

$$B = -\Omega \frac{dP}{d\Omega} = \Omega \frac{d^2E}{d\Omega^2} \tag{69}$$

(The bulk modulus, the compressibility, the speed of sound, and the long-wavelength limit of the phonon dispersion curve are all fundamentally the same physical quantity.)

The fact that the mechanisms responsible for cohesion are contained within the expression for the total energy given by the local-density approximation was first indicated by a sequence of calculations[172,103] performed for individual systems such as Li, Cu, and Ar. A particularly clear demonstration of both the power and the limitations of the local-density description is provided by the extensive calculations published in book form by Moruzzi *et al.*[90] These authors calculated the cohesive properties of a large group of elemental metals, which included, in particular, both the 3*d* and 4*d* transition series. The calculations exhibit the ability of the theory to describe the important chemical trends, and, perhaps of greater interest, the occasional deviations of the measured values from the rather smoothly varying theoretical values indicates the presence of interestingly unusual phenomena. (Examples of the latter are discussed below.) The results of the calculations by Moruzzi *et al.*[90] are compared with experimental measurements in Fig. 17. This shows that the agreement between theory and experiment is excellent throughout the 4*d* transition series, from the the alkali Rb to the simple metals Cd and In. It should be noted in this context that the calculational procedure involved in these calculations is entirely independent of the system to which it is applied. The fact that the agreement between theory and experiment is less good for the 3*d* transition series is interesting, because it indicates that the increased localization of the 3*d* electrons, relative to their 4*d* counterparts, is significant on the scale of the domain of validity of the local-density approximation.

Two aspects of the 3*d*–4*d* comparison warrant further discussion. First, the most striking disagreement between theory and experiment seen in Fig. 17 is for the bulk modulus of magnetic members of the 3*d* transition series, Cr, Mn, Fe, Co, and Ni. The fact that the calculated bulk moduli for the 3*d* metals are similar to those for the 4*d* series suggests that it is the measured bulk moduli for the 3*d* metals which are anomalous. Janak and Williams[92] have shown that this is indeed the case. The microscopic origin of the anomalously small bulk moduli in the 3*d* series is the small lattice dilatation caused by the presence of magnetic order in conjunction with a strong dependence of the bulk modulus on volume. This effect and its implications for other systems are discussed in Section 2.2.3; the point we wish to make here is that density-functional theory makes possible the performance of "numerical experiments" which are extremely well controlled and which are therefore particularly useful in identifying unusual behavior.

The second aspect of the 3*d*–4*d* comparison worth noting is the fact that the improved description of the metal obtained by allowing for magnetic order does not improve the agreement between theoretical and experimental values for the cohesive energy in the 3*d* series; in fact, it worsens it (because the improved description results in a lower total energy for the solid). The better agreement for the cohesive energies of the 4*d* metals reflects a superior description not of the 4*d* metals, but of the free atoms in the 4*d* series. This supposition is based on the fact that the local-density approximation is more suitable for the "more nearly" homogeneous condensed phase and on the convincing description the theory provides of properties involving only the condensed phase [the atomic volume, the bulk modulus, and the magnetic properties (see Section 2.2.3]. The relative ability of the theory to describe gaseous (atomic) and condensed phases is another valuable indication of the boundary of the theory's appropriateness.

While calculations such as those which produced Fig. 17 indicate the

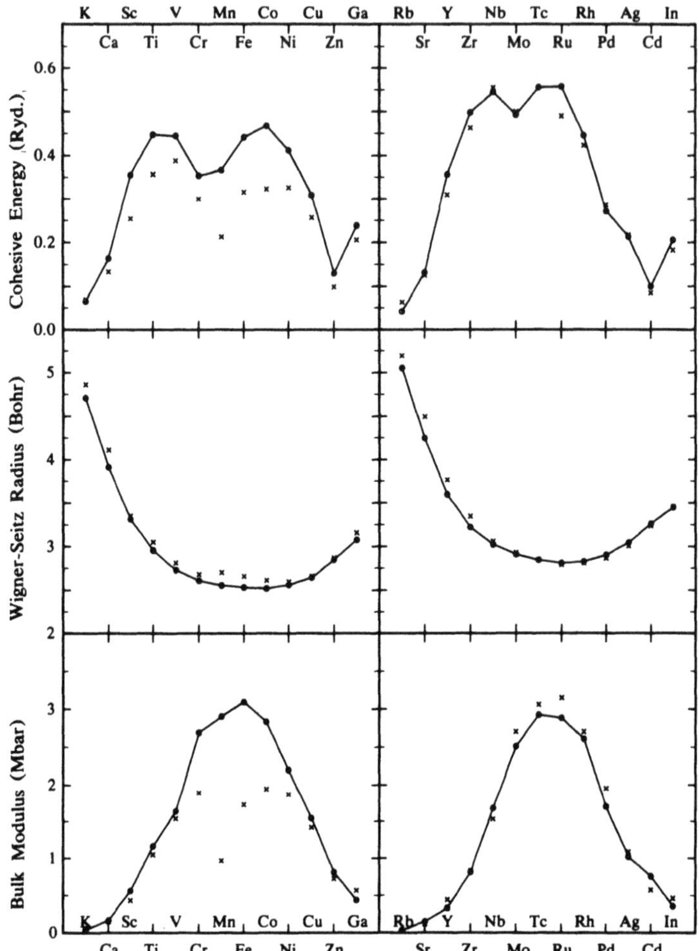

FIGURE 17. Comparison with experiment of ground-state properties given by self-consistent-field calculations based on the local-density approximation to density-functional theory. The left- and right-hand sides of the figure present results for the 3*d* and 4*d* transition series, respectively. The middle row gives equilibrium atomic volumes or lattice constants in terms of the corresponding Wigner–Seitz radii. The upper row gives cohesive energies and the bottom row gives bulk moduli. The calculations are described in Ref. 90; they employ the "muffin-tin" approximation, but are otherwise completely parameter-free. Spin-polarization effects have been included in the free-atom calculations required for the specification of the cohesive energy. These effects have not been included in the description of the solids which order magnetically. Measured values are indicated by crosses, calculated values by connected points. References to the experimental work can be found in Ref. 90.

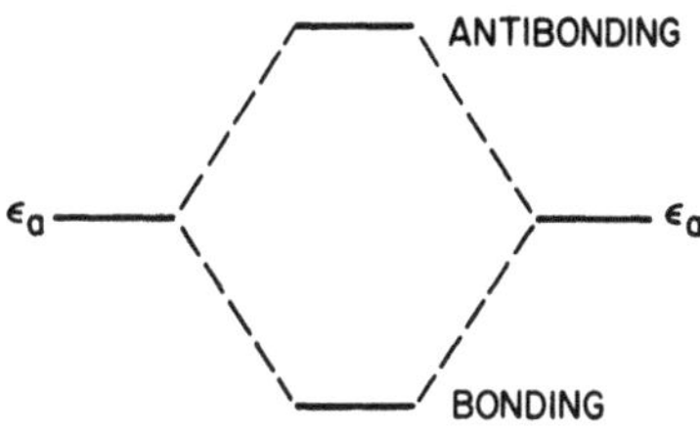

FIGURE 18. Schematic representation of the covalent interaction of localized states on neighboring atoms in a solid. The interaction leads to the formation of bonding and antibonding hybrid states. The preferential occupation of the bonding hybrid is a fundamental source of bonding.

content, accuracy, and limitations of the local-density approximation, they do not provide a transferable, intuitively satisfying picture of the microscopic mechanisms responsible for cohesion. The simplest such picture of the bonding process is shown schematically in Fig. 18, which depicts the interaction between degenerate states on neighboring atoms. This interaction results in the formation of bonding and antibonding hybrid states, and the preferential occupation of the lower-lying, more bonding level is a fundamental source of bonding. This picture of the bonding process is attractively simple, but efforts to quantify it introduce obscuring complications. In the case of solids, there are not just two atoms and two levels, but many atoms and a continuous band of states having energies lying between the bonding and antibonding levels of Fig. 18. Friedel[171] showed that, by assuming the distribution of states in this band to be uniform, the model described in Fig. 18 successfully accounts for the approximately quadratic dependence of the cohesive energy on atomic number in both transition series. (See Fig. 17.) More serious than the uncertainties associated with the distribution of states within the band, is the fact that the bonding described by Fig. 18 increases without limit as the atoms are brought closer to one another. That is, the energy separation of the bonding and antibonding levels increases with the strength of the interatomic interaction, which, in turn, increases as the interatomic separation decreases. The energy gained by preferentially occupying the bonding level therefore increases without limit. The mechanism of covalency depicted in Fig. 18 therefore does *not* contain the physical mechanism responsible for the repulsive interaction which balances the attractive force in producing the equilibrium interatomic separation of the molecule or solid. The covalent-bond mechanism of Fig. 18 presents yet another problem. Using Fig. 18 to interpret total-energy differences focuses on a particular portion of the complete expression for the total energy, namely, the sum of singe-particle eigenenergies (see Section 4.4 of Chapter 1). The single-particle energies implicitly require for their specification a reference energy. The natural impulse is to take the vacuum as the energy reference, but this cannot be correct, at least for solids, because it makes the surface dipole-layer potential barrier (see Chapter 5 for a full discussion) an important part of the cohesion process.

The dipole-layer potential rigidly shifts all the oribital eigenenergies of bulk states relative to their free-atom counterparts (i.e., relative to vacuum). This shift does not contribute to the cohesive energy, because, in adding an atom to a solid (the cohesion process), it is a neutral particle that we are bringing through the dipole-layer potential, not merely electrons. The remaining terms in the expression for the total energy (the so-called "two-electron" or "double-counting" terms) eliminate the reference-energy ambiguity, but they are difficult to incorporate in the simple, covalent-bond picture of Fig. 18.

Therefore, despite its intuitive appeal, the covalent-bond picture of Fig. 18 does not provide a complete picture of cohesion. In the pursuit of a qualitative understanding of cohesion, we might try to interpret the various terms in the complete total-energy expression directly. The problem with this approach is that, as the label suggests, the "two-electron" terms are difficult to assign to individual states or even different types of individual states (predominantly *s* or *d*, for example). Another approach, and one which eliminates the difficulty associated with interpreting the "two-electron" contributions, is based on the viral theorem. The virial theorem tells us that concomitant with the total-energy *decrease* which constitutes bonding is a kinetic-energy *increase* of precisely the same magnitude. The kinetic energy has the double advantage in this context of (1) not involving reference energies or potential-energy shifts, which make the single-particle energies ambiguous, and (2) being unambiguously assignable to particular single-particle states. Unfortunately, the interpretation of bonding provided by the kinetic energy is completely counterintuitive. (This might be expected from the fact that the kinetic energy increases with bonding.) A clear example of the counterintuitive character of this interpretation of bonding is provided by transition-metal cohesion. The important source of bonding in these systems is, as shown below, the valence *d* electrons. Nonetheless, an analysis of kinetic-energy changes shows that the kinetic energy of the valence *d* electrons decreases with solid formation. What is going on is simply that the valence *d* electrons are moving away from the nucleus, toward the neighboring atoms, but in so doing their screening of the nucleus is reduced, allowing the core electrons of the same principal-quantum-number shell to contract toward the nucleus with an increase in kinetic energy. Thus, according to the virial-theorem interpretation of bonding, it is the core electrons which are responsible for cohesion in these systems.

Some of the pitfalls associated with earlier interpretations of bonding have been discussed in order to allow the reader to appreciate the completeness of the bonding picture provided by the hydrostatic-pressure or equation-of-state construction,[166–168] which we now describe. In this analysis, the total-energy change associated with cohesion is represented simply as the volume integral of the volume derivative of the total energy. (The volume derivative of the total energy is the hydrostatic pressure; see Fig. 16.) The pressure can, of course, be calculated by numerically differentiating the total energy, but the interpretively useful representation of the pressure results from a formal manipulation which leaves the pressure described as an integral over the surface of the atomic cell of the solid.[173,174,166] For close-packed metals, the cell can be accurately approximated by a sphere, which, together with the assumption of a spherically symmetric electron density and effective single-particle potential, makes the pressure unambiguously decomposable according to electron-state type (s,p,d, etc.). When the pressure, represented in this way, is integrated with respect to volume, an intuitively satisfying picture of bonding results. The simplest application of the equation-of-state construction is illustrated in Fig. 19 for the alkali metal K.

The quantity graphed in Fig. 19 is fundamentally the equation of state (pressure vs. volume), but both the pressure and the volume have been transformed to make the interpretation even more transparent. Instead of the pressure P, we have graphed the quantity $3P\Omega$, because the latter is an energy and, as such, is easier to compare with other interaction strengths. The horizontal axis is not the volume, but rather $\ln(a/a_0)$, where a is the lattice constant (or Wigner–Seitz radius) and a_0 its value at equilibrium. [The motivation for using $\ln(a/a_0)$ is that $3P\Omega$ is the derivative of the total energy with respect to this quantity.] Graphed in this way, the equation of state provides a direct visual representation of (1) the strength of the interatomic coupling, (2) its range, (3) the cohesive energy, and (4) the bulk modulus. The bulk modulus is directly related to the slope of the pressure curve where it passes through zero (equilibrium). (The relationship is given in Fig. 19.) Figure 19 demonstrates that the strength of the interatomic interaction in an alkali metal is roughly 2 eV (taking the depth of the pressure curve as a convenient measure of strength). Figure 20 shows that the same interaction in the middle of the transition series is an order of magnitude larger. Figure 20, which shows the equation of state for Mo, also illustrates perhaps the greatest virtue of the equation of state as an interpretive tool, for we see that the contributions to the pressure of the relatively localized $4d$ electrons has been separated from that of the delocalized electrons. Figure 20 exhibits the enormous bonding strength of the d electrons, which have compressed the delocalized electrons beyond their natural equilibrium (the volume where the non-d pressure vanishes). The origin of the repulsive force which is missing from the simple covalent-bond picture of Fig. 18 is clearly indicated by Fig. 20. We see that it is the resistance of the non-d electrons to further compression that holds the lattice at its equilibrium spacing.

Much emphasis has been given here to the extraction of simple ideas from density-functional calculations. The motivation for this emphasis is that

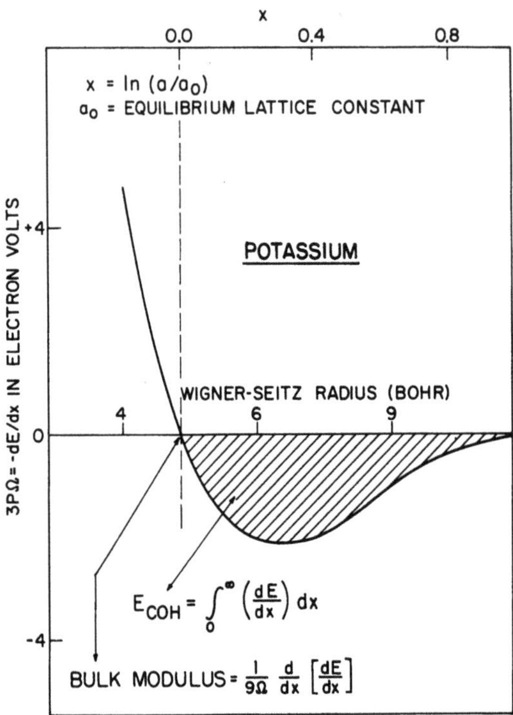

FIGURE 19. Hydrostatic pressure versus lattice constant for potassium. The quantity graphed is the logarithmic derivative of the total energy E with respect to lattice constant a. The independent variable is the logarithm of the ratio of the lattice constant to its equilibrium value. Equilibrium thus corresponds to the value zero on the upper horizontal axis and the cohesive energy corresponds to the area between the curve and the axis. (From Ref. 168.)

physical pictures, such as that presented by Fig. 20, are useful in thinking about a variety of more complex systems. For example, Fig. 20 provides an immediate explanation for the fact that the outermost atomic layers of transition-metal surfaces (apparently always) relax toward the interior of the metal. Figure 20 suggests that the surface provides a geometry in which the non-d electrons can "get out of the way," thereby allowing the d electrons to pull the surface layer of atoms toward the interior. Another geometry which permits this phenomenon to take place is the linear-chain arrangement of the transition metals in the A-15 compounds. Here, as in the case of the metal surface, the separation between the transition-metal atoms is less than it is in the corresponding bulk transition metal, presumably because the non-d elec-

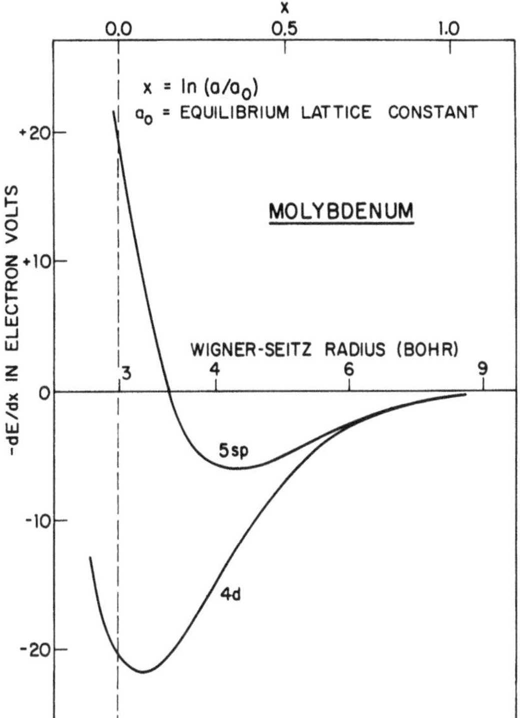

FIGURE 20. Decomposition of the pressure-versus-lattice-constant curve according to *d* and non-*d* electrons for the transition metal Mo. The quantity *a* is the lattice constant and a_0 is its equilibrium value; *E* is the total energy. Equilibrium thus corresponds to the value zero on the upper horizontal axis and the cohesive energy corresponds to the area under both curves. The curves illustrate the *d*-electron domination of bonding in the middle of the transition series. At equilibrium (zero total pressure), the attractive *d* pressure is canceled by the repulsive non-*d* pressure. (From Ref. 168.)

trons can be pushed off the axis of the chain. Similarly, the interatomic separation in transition-metal dimers is smaller than the corresponding bulk value.

The equation-of-state plot also shows clearly that, contrary to customary views, noble metals are not held apart by the repulsive interaction of their filled *d* shells. Figure 21 shows that, while the interatomic interaction due to the filled *d* shell is much smaller than that of the open *d* shell (compare Figs. 20 and 21, noting the change of vertical scale), the interaction is nonetheless *attractive*. Note (Fig. 21) that the *d*-electron pressure is attractive over the entire range contributing to the cohesive energy (from equilibrium to fully

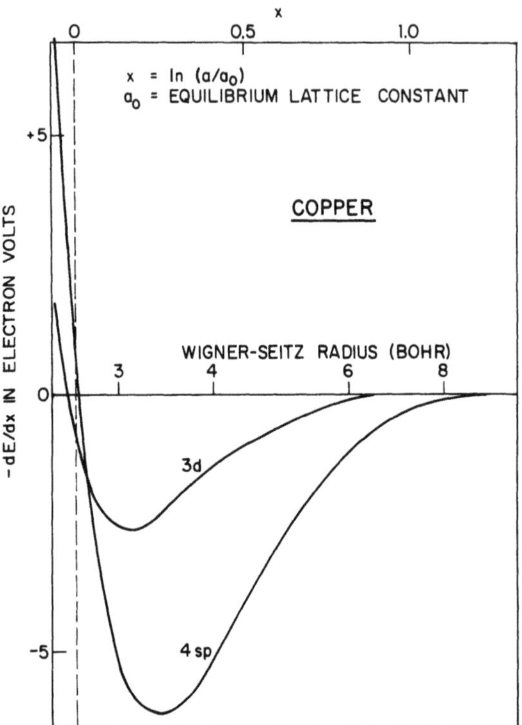

FIGURE 21. Relative contributions of *d* and non-*d* electrons to cohesion in the noble metal copper. The quantity a is the lattice constant and a_0 its equilibrium value; E is the total energy. Note the appreciable contribution of the closed *d* shell to the cohesive energy (the area underneath the pressure curve) and the fact that the *d*-shell contribution to the pressure is attractive, even at equilibrium. (From Ref. 168.)

separated atoms). As with Mo (Fig. 20), it is the resistance of the non-*d* electrons to further compression which stabilizes the lattice at its equilibrium spacing. Figure 21, therefore, unambiguously dispels the notion of hard-core repulsion, at least as it applies to the noble metals. The attractive interaction between atoms with formally closed shells is also manifest in the bonding of rare-gas solids and the group-II dimers. It is interesting that calculations based on the local-density approximation indicate such an attractive interaction, because this type of interaction is usually ascribed to van der Waals forces, which would seem to be ignored by the local-density approximation. The point is discussed in connection with the molecular calculations in Section 3, and by Lang,[165] who considers the bonding of rare-gas atoms to metal surfaces.

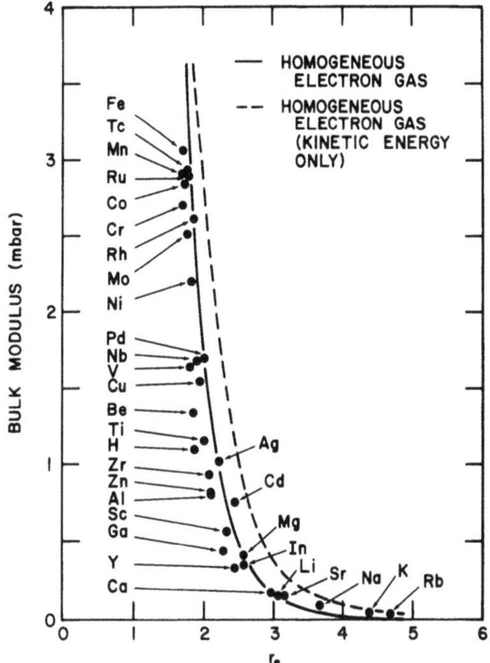

FIGURE 22. Comparison of the bulk moduli for diverse metals with that of the homogeneous electron gas. The bulk modulus of each metal is plotted against the interstitial electron density of each metal (described by the density parameter r_s). The bulk modulus of the electron gas is also plotted against r_s. The metallic bulk moduli are those calculated by Moruzzi *et al.* using self-consistent-field energy-band calculations based on the local-density approximation to exchange and correlation effects. The interstitial electron density of the metals is taken to be the spatial average of the electron density over the region external to a set of non-overlapping spheres centered on the atoms of the crystal. The calculated bulk moduli are very similar to measured values, as can be seen in Fig. 17. The dashed curve is a portion of the homogeous-gas result, that due to just the kinetic energy. We infer from the similarity of the two curves to the plotted points that the principal effect of changing the volume, even that of transition metals, is to change the volume available to the approximately homogeneous electron gas trapped in between the approximately rigid ions. (From Refs. 90 and 168.)

Another aspect of metallic cohesion which has been illuminated by density-functional calculations concerns the bulk modulus or compressibility. The calculations reveal that the variation of the bulk modulus from material to material is amenable to the extremely simple interpretation presented in Fig. 22.

Shown in Fig. 22 are the bulk moduli for a large variety of metals, plotted as a function of the electron density in the interstitial region of the metal in between the atoms. The interstitial electron density $\bar{n}$ is described by the density parameter r_s, where

$$\frac{4}{3}\pi r_s^3 = \frac{1}{\bar{n}} \tag{70}$$

The bulk moduli plotted are those given by the calculations of Moruzzi *et al.*[90] The same calculations, because they employ the so-called muffin-tin approximation, provide a precisely defined value for the average interstitial electron density $\bar{n}$. The remarkable fact exhibited by Fig. 22 is that, despite the large differences among the other aspects of the electronic structure of these materials, their bulk moduli are well approximated by a single, universal function of the interstial electron density. Figure 22 also indicates what this universal function is. The solid curve in Fig. 22 is the bulk modulus of a homogeneous electron gas graphed as a function of the density parameter r_s. The physical picture suggested by the similarity of the bulk modulus of the homogeous electron gas and those of the very inhomogeneous metals is that of an approximately homogeneous electron gas locked in between mutually attractive, but rigid, atoms. In this picture, the principal implication of atomic displacements for the total energy is the change in the volume available to the interstitial electron gas. Figure 22 provides additional support for this interpretation, by showing that most of the r_s dependence of the bulk modulus of the homogeneous electron gas is due to changes in kinetic energy.

Also consistent with this simple picture are the relative contributions to the bulk modulus from *d* and non-*d* electrons found by Pettifor.[167] That is, although the *d* electrons dominate the cohesive energy of transition metals, it is the repulsive contribution to the equation of state coming from the non-*d* electrons that dominates the bulk modulus. This dominance is consistent with the fact that the interstitial electron density is composed primarily of non-*d* electrons.

Virial Theorem in Local-Density Approximation. As a final point in this section, we note that the electronic kinetic energy T appearing in the virial theorem

$$E = 3P\Omega - T \tag{71}$$

is *not* merely the expectation value of the operator $-\nabla^2$ appearing in the single-particle equations of density-functional theory. The quantities appearing in equation (71) are the total energy E, the volume Ω, and the pressure

$P \equiv -dE/d\Omega$. The total kinetic energy T is the sum of the expectation value of $-\nabla^2$, called T_s, and the kinetic-energy portion of the exchange-correlation energy functional $E_{xc}\{n\}$, called T_{xc}. The definition of the local-density approximation to T_{xc} that is consistent with the corresponding approximation to the exchange-correlation contribution to the total energy E_{xc} is[175]

$$T_{xc} \approx \int d^3r \; n(\mathbf{r}) t^h_{xc}(n(\mathbf{r})) \tag{72}$$

where $t^h_{xc}(n)$ is the contribution of exchange and correlation to the kinetic energy of a *homogeneous,* but interacting, electron gas of density n. This contribution to the kinetic energy can be related to the corresponding contribution, $\epsilon^h_{xc}(n)$, to the total energy by noting that the virial theorem [equation (71)] is satisfied by both the noninteracting electron gas and the interacting electron gas. Therefore, subtracting the virial theorem [equation (71)] for the noninteracting gas from that for the interacting gas leaves just the contribution of exchange and correlation to each term.

$$\epsilon^h_{xc}(n) = -3\Omega \frac{d}{d\Omega} \epsilon^h_{xc}(n) - t^h_{xc}(n) \tag{73}$$

or, after solving for $t^h_{xc}(n)$, replacing $\Omega d/d\Omega$ by $-n d/dn$, and recalling that the exchange-correlation contribution to the chemical potential of the homogeneous electron gas, denoted μ_{xc}, is given by $\mu_{xc} = d/dn \, [n\epsilon^h_{xc}(n)]$, we obtain

$$t^h_{xc}(n) = 3\mu_{xc}(n) - 4\epsilon^h_{xc}(n) \tag{74}$$

When this result is combined with the fact that $\mu_{xc}(n(\mathbf{r}))$ is also the exchange-correlation potential appearing in the single-particle equations of the local-density ground-state theory, it is not difficult to demonstrate[175] that the local-density ground-state theory satisfies the virial theorem [equation (71)].

Equation (73) reveals that the $n^{1/3}$ dependence of the exchange energy for the homogeneous electron gas causes the exchange contribution to $t^h_{xc}(n)$ to vanish. This fact has two implications. First, it is the basis of the statement that the $X\alpha$ approximation obeys the virial theorem.[117] What is meant by this statement is that the virial theorem [equation (71)] is satisfied by calculations based on the $X\alpha$ approximation, if the kinetic energy is taken to be just the expectation value of $-\nabla^2$. The validity of this statement therefore rests on the simultaneous neglect of the correlation contribution to the kinetic energy and the representation of the correlation contribution to the potential energy as a scaling of the exchange contribution. (Note that the $X\alpha$ ground-state theory does not satisfy the virial theorem, even in this restricted sense, when the exchange potential is scaled by different amounts in different portions of a single physical system, that is, when different values of α are used in the vicinity of different atoms.) The second implication of the vanishing of the exchange contribution to $t_{xc}^{h}(n)$ is the fact that the exchange contribution to the kinetic energy T_{xc} for *inhomogeneous* systems, which does not in general vanish, is neglected when the local-density approximation to T_{xc} [equation (72)] is used. (The source of this contribution is the difference between the expectation value of $-\nabla^2$ taken with Hartree–Fock and density-functional orbitals; this difference is expected to be small.)

Consider for a moment the utility of the virial theorem. In calculations designed to understand the small total-energy differences associated with cohesion or compound formation, the virial theorem serves as an invaluable check against the loss of numerical significance. Because the kinetic energy is not a variational quantity in most calculations, and because the kinetic energy contains large contributions from the rapid oscillations of core orbitals, numerical agreement between the usual variational expression for the total energy [equation (7) of Chapter 5] and equation (71) is an extremely sensitive test[(158)] of the internal consistency of the calculation. The energies associated with the interior of atoms are enormously larger than cohesive energies, heats of compound formation, etc. Even the energies associated with the valence electrons, that is, the kinetic and potential energies found in the spatial region outside the atomic core but inside the atomic cell, are often much larger than the energies of interest. Thus, even if the energies associated with the atomic core are eliminated from consideration by, for example, the use of pseudopotentials or the "frozen-core" approximation, it is extremely useful to have a measure of the nonsystematic errors (e.g., numerical) involved in the calculation; the difference between the total energy given by the usual variational expression [equation (7) of Chapter 5] and by equation (71) is such a measure. Note that, when using the virial theorem within a pseudopotential formalism, a term $\int d^3r\, n(\mathbf{r})\, \mathbf{r}\cdot\nabla V_{\text{external}}$ must be included in the potential energy. While the check provided by the virial theorem is useful in connection with the valence electrons, it is even more useful in calculations that deal directly with the atomic core, e.g., calculations concerned with either the electron or spin density at the nucleus. Efforts to interpret nuclear-resonance or Mössbauer experiments must deal with subtle rearrangements of the core electrons caused (frequently indirectly) by changes in the valence electronic structure. The successful analysis of the pressure dependence of the hyperfine field (the spin density at the nucleus) by Janak[(94)] illustrates the complicated interplay of different principal-quantum-number shells of the atom in producing the effect seen experimentally at the nucleus.

3.3. Intermetallic Compound Formation

The problem of intermetallic compound formation is a particularly natural one for the application of ideas based on density-functional theory and calculations based on the local-density approximation. This is because the comparison of a compound with its constituents tends to subtract away errors introduced by the use of the local-density approximation in the atomic cores and to emphasize changes in the interatomic aspects of the electronic structure, that is, those for which the local-density approximation is most appropriate. (While the same cancellation of systematic errors is also an important part of the accurate of cohesive and molecular-bond energies, for example, the greater similarity of the atomic cores in the compound and constituents leads to a more complete cancellation in the case of compound formation.) A comparison of calculated and measured heats of compound formation is provided in Fig. 23; the agreement, particularly in view of the energy scale, is seen to be quite good.

The study of the compound-formation process is particularly interesting, because the microscopic mechanisms which cause some metals to bond together, while others do not, are still not well understood. An important aspect of this lack of understanding is the absence of a consensus as to the constituent properties that are most important to the compound-formation process. Density-functional calculations provide understanding in this context in much the same way that they are used to study molecular bonding. Calculations like those which produced Fig. 23 can be done for entire classes of compounds in which only a single variable (e.g., the valence of one of the constituents) is changing. In this way, chemical trends can be established and the relative importance of various properties of the constituents can be assessed. In addition to measureable quantities, such as the heat of formation, calculations of this type provide an array of internal quantities, such as charge transfers, electron densities, state densities etc., which can be used to identify the important microscopic mechanisms and to establish the correct simple models in terms of which to think about the formation process.

A particular challenge to theory in this context is provided by the enormous success of an empirical theory due to Miedema.[177] Miedema argues that many compound properties can be understood and predicted from a knowledge of just two crucially important properties of the constituent elements, an electronegativity and the interstitial electron density. Electronegativity is loosely defined as the ability of a particular atomic species to attract electronic charge from other atomic species. Electronegativity differences promote compound formation; the factor resisting compound formation, in the Miedema picture, is the mismatch of the electron densities of the con-

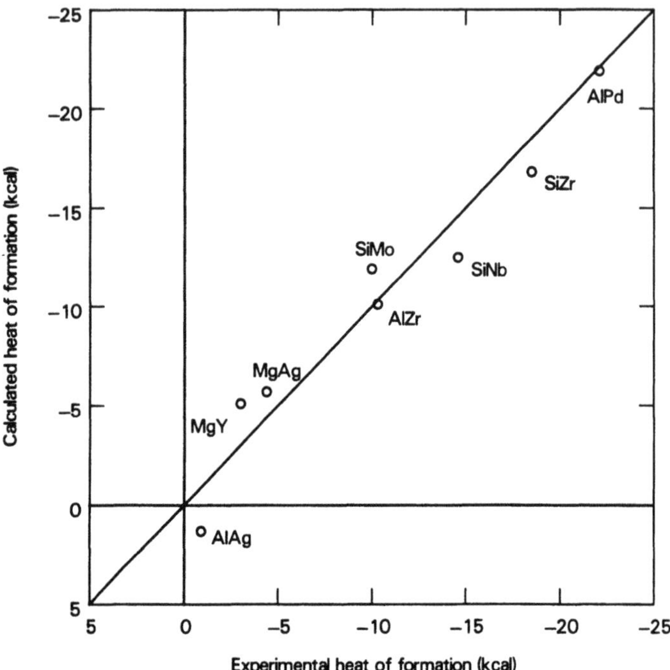

FIGURE 23. Comparison of calculated and measured heats of intermetallic compound formation. (One electron volt per atom equals 23.07 kcal.) The calculations are based on the local-density approximation used in conjunction with the augmented-spherical-wave technique (Ref. 158) for the solution of the single-particle equations. The calculations are self-consistent and entirely parameter free. (Only the atomic numbers and the crystal structure are input to the calculations.) References to the experimental work can be found in Ref. 176.

stituents at the interface separating atomic cells. Miedema envisions the formation of the compound from unit cells of the pure constituents, so that in general there is a electron-density discontiuity across the intercellular interface; Miedema ascribes resistance to compound formation to the energy required to "heal" this discontinuity. So, the challenge to theory is to determine the microscopic significance of the electronegativity notion and to understand how it and the interstitial electron density are related to the heat of compound formation. One use of density-functional calculations in this context is illustrated in Fig. 24, where the interfacial electron-density rearrangements emphasized by Miedema are shown for the compound NbMo. (Note that the large discontiuity seen in Fig. 24 at the cellular interface is not a calculational artifact, but rather an implication of the *definition* of the density difference.

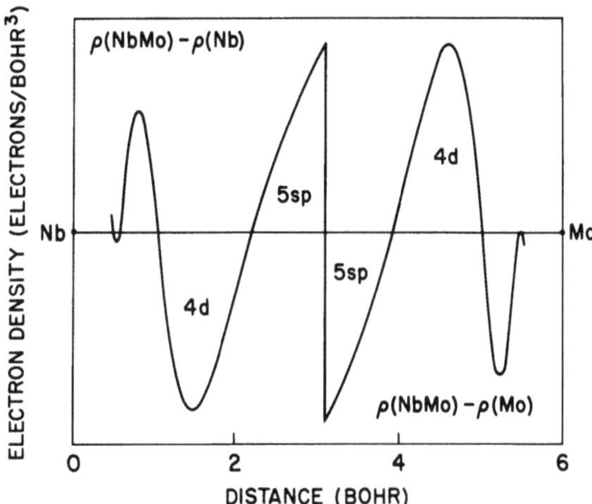

FIGURE 24. Electron-density rearrangements resulting from the compound-formation process. Shown is the electron density of the compound NbMo from which has been subtracted the electron density of pure Nb in the Nb cell (sphere) and that of pure Mo in the Mo cell (sphere). The electron-density difference, defined in this way, is in general discontinuous at the interface separating constituent atomic cells. In the empirical theory of compound formation due to Miedema, this interfacial electron-density discontinuity is one of two crucial parameters governing the formation process. For this system, local neutrality is approximately maintained by the transfer of $4d$ electrons from Nb to Mo and the transfer of $5s$ and $5p$ electrons from Mo to Nb. This system is *unrepresentative* in this regard. The electron densities are the results of self-consistent-field calculations based on the local-density approximation and the augmented-spherical-wave method (Ref. 158) of solving the single-particle equations. The quantity graphed is the difference between spherically averaged electron densities in the respective atomic spheres, graphed "back to back". (From Ref. 176.) The Wigner–Seitz radii for Nb and Mo are 3.1 and 2.9 Bohr, respectively.

That is, while the density in the compound is continuous, the constituent densities subtracted from the compound density to form the density difference are not.) Figure 24 clearly indicates that the density rearrangements emphasized by Miedema are indeed real. [The constituent electron densities used by Miedema to analyze compound properties are not obtained directly from measurements (X-ray diffraction measurements, for example), but are inferred from measured bulk moduli using the connection between bulk modulus and interstitial electron density shown in Fig. 22.]

So, the density-functional calculations confirm the *existence* of the density rearrangements emphasized by Miedema; we consider now their *importance* to the heat of compound formation. Because there are so many combinations of elements we might consider, we ask, as a preliminary, which compounds can teach us most about the fundamental factors controlling the formation process? In answering this question, we recall that Miedema's empirical theory applies in its simplest form to compounds involving only transition metals; initial efforts toward a microscopic understanding have therefore been focused on such systems. Within this restricted class of compounds, those involving only 4*d* transition metals are particularly appropriate for theoretical analysis, because they avoid the complications introduced by both magnetic effects in the 3*d* metals and relativistic effects in the 5*d* metals. There are 28 compounds even in this further restricted class, but fortunately the heats of formation of all 28 reflect a simple chemical trend. Calculations based on the local-density approximation have been useful both in establishing the importance of this trend and in elucidating its microscopic origin. The trend in question can be understood as follows: The heat of formation ΔH is a function of only the atomic numbers of the constituents z_a and z_b. It is equally correct, however, to consider the heat of formation to be a function of the average atomic number $\bar{z}$ and the atomic-number difference Δz. Considering ΔH to be a function of $\bar{z}$ and Δz has the virtue that, if we expand Δh in powers of Δz, then only the even-order terms contribute to the expansion [because $\Delta h(z_a,z_b) \equiv \Delta H(z_b,z_a)$]. Furthermore, since $\Delta H(z_a,z_a) \equiv 0$, the leading term in the expansion is second order in δz. The expansion of the heat of formation therefore has the following form:

$$\Delta H(z_a,z_b) \equiv \Delta H(\bar{z},\Delta z) = \Delta H_2(\bar{z})\, \Delta z^2 + \cdots \qquad (75)$$

The form of this expansion leads to the notion of the "normalized heat of formation," $\Delta H(z_a,z_b)/\Delta z^2$, because, to the extent that fourth- and higher-order terms in equation (75) are unimportant, we see that the normalized heat of formation is a single, universal (independent of Δz) function of $\bar{z}$. Figure 25 presents a comparison designed to show the extent of this universality. Compared in Fig. 25 are the normalized heats of formation given by Miedema's empirical theory, as tabulated in Refs. 177 and 178, and those given by calculations based on the local-density approximation. The 28 possible binary combinations of the 4*d* transition metals are shown in Fig. 25; results for compounds corresponding to the same value of Δz are connected by lines. If the higher-order terms in the expansion of ΔH [equation (75)] vanished, the curves corresponding to different values of Δz would coincide. Figure 25 shows that, although the normalized heat of formation exhibits some Δz dependence, the results of both the density-functional calculations, and Miedema's empirical formula exhibit a dominant chemical trend. That is, compounds for which $\bar{z}$ falls near the middle of the transition series tend to form,

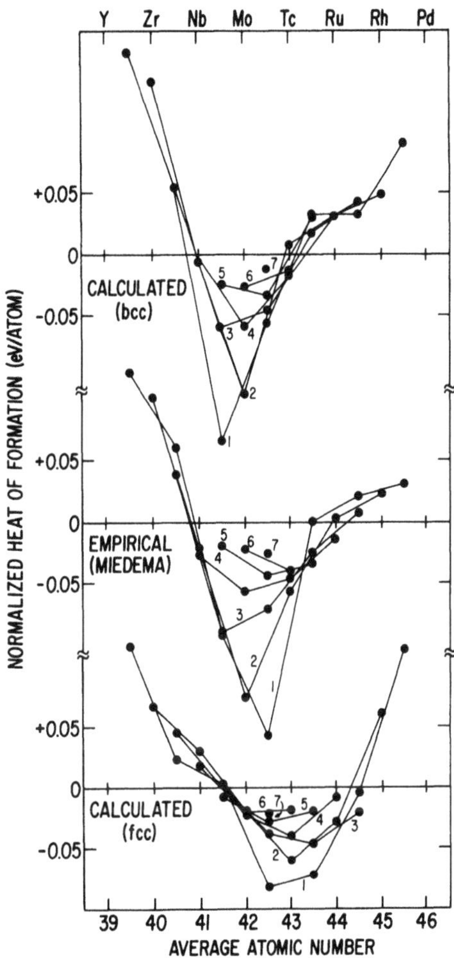

FIGURE 25. Normalized heats of formation, $\Delta H(z_a,z_b)/(\Delta z)^2$, versus the average atomic number, $\bar{z}$. The atomic numbers of the constituents and $\Delta z \equiv z_a - z_b$ are represented by z_a and z_b. Integers indicate the common value of Δz of points connected by lines. Miedema's results are tabulated in Ref. 178. Calculated results were obtained using the local-density approximation and the augmented-spherical-wave method (Ref. 158) for the solution of the single-particle equations. For the calculated results labeled fcc, the total energies of elements in the fcc structure were subtracted from those of the compounds in the CuAu (fcc-like) structure. For the results labeled bcc, total energies of elements in the bcc structure were subtracted from those of compounds in the CsCl (bcc-like) structure. Available experimental data: for RhPd, $\Delta H = +0.104$ eV/atom (Ref. 179); for ZrRu, $\Delta G/(\Delta z)^2 = -0.058$ eV/atom (Ref. 180); for NbMo, $\Delta H = 0.097$ eV/atom (Ref. 181). (From Ref. 6.)

whereas compounds with average atomic numbers near either end of the series do not. Also shown in Fig. 25 is the role of crystal structure. Two sets of theoretical results are shown: one for which the atoms in both the compound and the constituents were taken to occupy the sites of an fcc lattice; in the second set, the bcc lattice was used. The explicit structure dependence of the normalized heat of formation is consistent with the Δz dependence of both sets of calculated results. That is, crystal-structure-derived variations in the single-particle state density are the source of the Δz dependence of the normalized heat of formation. (Variations in the single-particle state density cause the sum of single-particle energies, an important component of the total energy, to vary as the Fermi energy moves through the state density with increasing valence.[182,183] The utility of systematic sets of calculations, such as those shown in Fig. 25, can be appreciated from the fact that measurements exist for only three of the 28 systems considered in Fig. 25, and it is the local-density approximation which makes such extensive sets of calculations possible.

Consider now the microscopic origin of the chemical trend exhibited by Fig. 25, that is, the approximately parabolic dependence of the normalized heat of formation on the average atomic number. Two fundamental conceptual models for the compound-formation process exist; they are illustrated schematically in Fig. 26. One of the two is much more readily identified with Miedema's empirical formula, but, before making that identification, consider the rather different physical pictures of the formation process provided by the two models. The chemical-potential model,[9–11] described in the left side of Fig. 26, emphasizes the charge transfer required to bring the chemical potentials of the constituents into equilibrium. As the figure is intended to suggest, the local electronic structure (the site-projected state densities) is envisioned in the chemical-potential model to remain unchanged by the formation process.

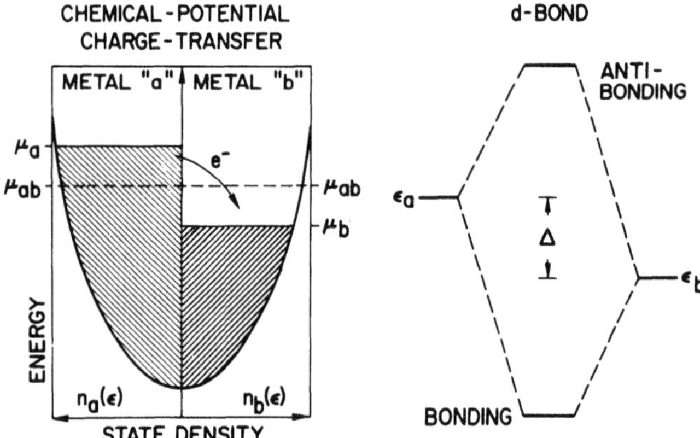

FIGURE 26. Conflicting interpretations of intermetallic compound formation. Shown schematically are the microscopic mechanisms responsible for compound formation in (a) the chemical-potential/charge-transfer model and in (b) the *d*-bond, or tight-binding, model. The fundamental difference between the two models is that the local electronic structure (state density) is assumed to be unaffected by the formation process in the chemical-potential model, whereas the change in the state density is viewed as the crucial aspect of the formation process, in the *d*-bond model. The relevance of the *d*-bond model is restricted to compounds involving transition metals.

The formation process consists in this model of charge transfer among states having energies near the Fermi level. Figure 26 is designed to contrast this conception of the formation process with that of the *d*-bond model.[184] The *d*-bond model has a long history in the context of transition-metal cohesion and alloy formation. The important components of the *d*-bond model which distinguish it from the chemical-potential model are as follows: (1) The interatomic coupling is treated wave-mechanically; (2) The mobile *s* and *p* electrons which respond most readily to chemical-potential differences are completely ignored; and (3) Changes in electronic states, even those far removed from the Fermi level, contribute to the formation process. It is important to note that the electronegativity concept plays a clearly identifiable, but interestingly different, role in the two models. In the chemical-potential model, the electronegativity of a constituent is simply its chemical-potential, and recall that it is the mobile *s* and *p* electrons that respond to chemical-potential differences. In the *d*-bond model, electronegativity is associated with the energy position of the atomic *d* level (see Fig. 26) and, clearly, it is the *d* electrons which respond to *d*-level energy differences. Because Miedema identifies his electronegativity with the work function of each constituent, and because he emphasizes the importance of the *s* and *p* electrons which contribute most of the electron density at the intercellular interface, Miedema's physical picture is more closely identified with the chemical-potential model.

Calculations based on density-functional theory are used as numerical experiments with which to determine which of the two models (chemical-potential and *d*-bond) better characterizes the compound-formation process. While such numerical experiments are never equivalent to real experiments, they are far easier to perform for large numbers of systems, and Fig. 23 indicates that the calculations model reality sufficiently closely that an understanding of the numerical experiments is likely to provide a useful guide in understanding real experiments. Note that it is the density-functional theory that makes such numerical experiments possible. The only alternative is Hartree–Fock theory which, in this context, suffers two fatal flaws: (1) the neglect of correlation leads to a poor description of the delocalized electrons and (2) the nonlocal character of the effective-single-particle potential in Hartree–Fock theory makes calculations like those underlying Figs. 23, 24, and 25 prohibitively difficult.

Consider for a moment the chemical-potential model.[9–11] This model asserts that, if the elements either occur or can be prepared with a common interstitial electron density, then charge will flow so as to equilibrate the two (in general different) chemical potentials and that this charge flow will contribute to the bonding of the constituents in two distinct ways. First, the electrons lower their energy by transferring to a region of lower chemical potential and second, the charge transfer creates a lattice of ions whose electrostatic (Madelung) energy is attractive. Because self-consistent-field calculations based on density-functional provide accurate values for the quantities characterizing this process (chemical potentials and charge transfers), the fundamental assertion of the chemical-potential model can be carefully tested. When this is done, it is found that, for compounds involving only transition metals, the chemical-potential model describes neither the chemical trend nor even the sign of the heat of formation correctly.[(176,6)]

Consider now the *d*-bond model.[(184)] When the *d*-bond model is subjected to similarly stringent tests, it passes convincingly.[(176,6)] First, when the *d*-bond model is reduced to the simplest possible formulation[184] (a tight-binding model), the model reproduces the crucial chemical trend discussed above, and it does so even when the parameters of the model (*d*-band positions and widths) are set to the values given by the corresponding density-functional calculations. The self-consistent-field calculations allow the two important assumptions of the *d*-bond model to be assessed. These are the complete neglect of the mobile *s* and *p* electrons and the assumption that the *d* bands of the two constituents merge to form a single *d*-band complex in the compound. Consider first the neglect of *s* and *p* electrons. That the neglect of the delocalized electrons should be justifiable is not at all obvious; we have seen (Fig. 24) that charge transfer among these electrons results in a dipole layer at the interface separating atomic cells of the compound. Intuition based on the electrostatic effects of dipole layers in capacitors and at metal surfaces, e.g., makes it difficult to understand why the interfacial dipole layer, such as that shown in Fig. 24, does not electrostatically shift the *d* levels of the constituents relative to one another. As Fig. 26 is intended to suggest, the relative position of the constituent *d* levels is the crucial quantity in the *d*-bond model. How then can the *d*-bond model justifiably neglect the *s* and *p* electrons which constitute this dipole layer? The self-consistent-field calculations show that the *d* levels are affected almost exclusively by charge transfer among the *d* states.[(176)] The neglect of the *s* and *p* electrons in the *d*-bond model is therefore justifiable, but *why*? An analysis of intra- and interatomic electrostatic effects reveals that the interatomic component (the Madelung potential) is electrostatically equivalent to that of charge "transferred" to a position just *inside* the surface of the atomic cell. The net result is a systematic cancellation of the effect of the interfacial dipole layer on the interior of each atomic cell. This is a particularly pleasing result, because the dipole layer makes the specification of the charge transferred among *s* and *p* states sensitively dependent on the particular decomposition of the unit cell into atomic cells. What is pleasing is to know that charge transfer is important when it is well defined, that is, when it takes place between states that are relatively localized (e.g., the *d* states); when it is poorly defined, as in the case of delocalized electrons, it is also unimportant.

The other important assumption of the *d*-bond model is that the *d* states of the compound constitute a single *d* band. The extent to which this assumption is justified is shown in Fig. 27, which depicts the evolution of the *d*-band complex of the compound as the constituents become increasingly dissimilar. We see in Fig. 27 that the *total* state density in the compound

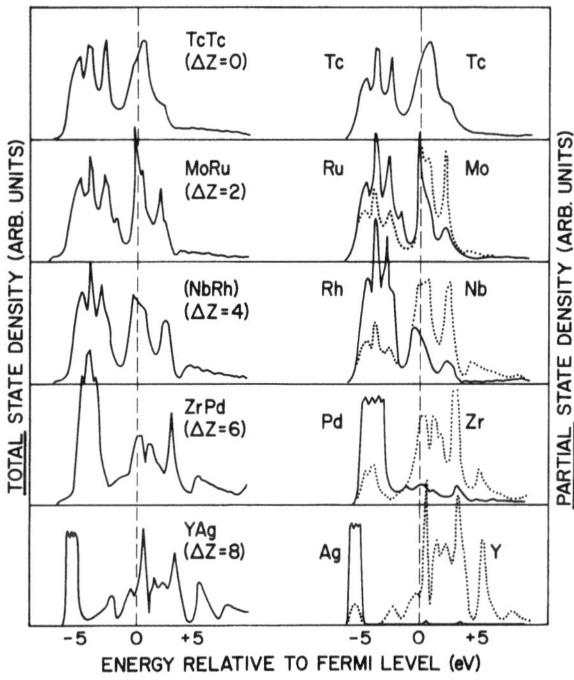

FIGURE 27. Persistence of common *d* band in transition-metal compounds of increasing valence difference. Shown on the left- and right-hand sides of the figure are the total and site projected state densities for one element (Tc) and four compounds characterized by an average valence equal to that of Tc and a valence difference that increases by two on going from one compound to the next. The comparison shows that the total state density remains similar to that of the element Tc, despite the increasing valence difference. The partial state densities reflect much more strongly the valence difference. Note that, if the site-projected state densities were not different, the common *d* band would imply a very large transfer of charge from the atom with the greater valence. The state densities were obtained from self-consistent-field calculations based on the local-density approximation and the augmented-spherical-wave method (Ref. 158) of solving the single-particle equations. (From Ref. 176.)

remains rather similar to that of a transition-metal *element* having the same crystal structure. Figure 27 shows that the common-d-band idea remains valid, even for large constituent-valence differences. Figure 27 also shows the decomposition of the total state density into site-projected state densities. The latter, unlike the total state density, strongly reflect the valence differences. Note that, if the partial state densities did not distribute the d electrons unevenly (see the right-hand side of Fig. 27), then the common d band would imply a very large charge transfer.

The density-functional calculations strongly support the d-bond model of transition-metal compound formation. We are therefore led to identify the electronegativity concept of empirical theories with the energy position of the atomic d level (for transition metals). Acceptance of the d-bond model also implies that the constituent property which resists compound formation is not the interstitial electron density, but rather the difference between the widths of the constituent d bands.[184] [Bond strength in the d-bond model is proportional to the width of the d band. Whereas the difference between the constituent d-band *centers* enhances the width of the d band in the compound (see Fig. 26), the difference between the *widths* of the constituent d bands reduces the width in the compound (see Ref. 184).] The picture of transition-metal compound formation emerging from the calculations is therefore very different from that used to justify Miedema's empirical formula. While the success of Miedema's empirical formula is understandable (see Ref. 6) for the restricted class of compounds that we have considered in detail above, the basis for the success of the formula in the broader context in which it has been found useful is not understood. If we take the microscopic picture associated with Miedema's empirical formula as the status quo, and if the picture of the compound-formation process provided by the calculations stands the test of future investigations, then density-functional theory will have succeeded in changing the way we think about an important class of chemical reactions.

4. Elements of Theory of Excitations

Density-functional theory, as described in previous sections of this chapter, is a theory of the ground state of inhomogeneous many-electron systems. The success of the theory in describing ground-state properties, together with the fact that many of the most precise experimental probes of electronic structure involve excitations, creates a strong desire to use the ground-state theory to interpret excitation spectra. In particular, the orbital eigenenergies of the ground-state theory (in the form of energy-band structures and state densities, for example) are widely used to interpret excitations such as photoemission spectra. Unfortunately, the connection between excitation spectra and the orbital eigenenergies of the ground-state theory is exact only in certain special cases. Despite its (usually) approximate character, however, this connection is immensely useful and can be discussed in terms of entirely rigorous relationships. These relationships and their relevance to the interpretation of measured spectra are the subject of this section.

We identify three distinct issues in this context. The first is the relationship of the orbital eigenenergies of the ground-state theory to total-energy differences. The second is the relationship of such total-energy differences to measurable excitation energies. The third is the effect on both total energies and orbital eigenenergies of the local-density approximation. We begin with a discussion of several exact results and definitions on which the remainder of the discussion is built. We discuss first the connection between the orbital eigenenergies and total-energy differences. We do this because the connection is amenable to a precise mathematical description and because, in the original presentation of the ground-state theory,[65] the eigenenergies were introduced as purely mathematical quantities without clear physical content. We next identify the two classes of excitations to which the ground-state theory rigorously applies. In the remaining subsections we discuss the appropriateness of the ground-state theory for a wider class of excitations and the effects of the local-density approximation on the description of localized excitations. The application of the ground-state formalism to the highly excited states corresponding to core holes brings in another aspect of the connection between the ground-state theory and the description of excitations. Finally, we apply our understanding of the connection to the meaning of the ground-state eigenenergies in a context which appears to mix all the theoretical issues, that of transition-metal valence-band spectra.

4.1. *Exact Results: Ground-State Eigenenergies and Certain Excitations*

The energy required to excite a many-electron system is the difference between the total energy of the ground state and that of another stationary or quasistationary state. We postpone for a moment the discussion of which such states can be described using the ground-state theory, in order to establish, as a preliminary, the connection between the orbital eigenenergies of the ground-state theory and total-energy differences. We will use, as an illustrative example, the ionization of a free atom, that is, the energy required to remove the outermost electron from an atom. For this type of excitation, both the initial and final states of the excitation process are ground states, so that the excitation energy can be obtained using a ground-state theory by calculating and subtracting the total energies of the neutral and ionic ground states. The ionization potential I of an atom is an example of such an excitation; it is therefore given by

$$I \equiv E(n = 0) - E(n = 1) \tag{76}$$

where $E(n)$ is the total energy considered as a function of n, the occupation of the outermost density-functional orbital of the atom. This representation of the excitation energy as a total-energy difference is called the ΔSCF representation. Whenever an excitation energy can be described by the ΔSCF total-energy difference, it is also related to the corresponding ground-state eigenenergy ϵ by the following rigorous relationship[185–187].

$$\frac{dE}{dn} = \epsilon \tag{77}$$

This relation states that the total-energy change associated with the addition or removal of charge from a density-functional orbital is the orbital eigenenergy. The only assumption required by the derivation of equation (77) is that the total-energy functional can be formally continued into the domain of fractional orbital occupation. This continuation is straightforward for all portions of the total-energy functional other than the exchange-correlation portion and, for the latter, the continuation is straightforward for all the approximate exchange-correlation functionals used in contemporary self-consistent-field calculations. It is possible, however, that the exact exchange-correlation functional's dependence on the density would lead to a nonanalytic dependence of the total energy on the orbital occupation number n. Equation (77) is the connection between the orbital eigenenergy and the total-energy change associated with an *infinitesimal* occupation-number change; the connection with the ΔSCF total-energy difference is simply the corresponding integral over the differential change. For the ionization potential considered in equation (76), we have

$$I = \int_1^0 \epsilon(n)\, dn \approx -\epsilon(\tfrac{1}{2}) \tag{78}$$

We indicate in equation (78) that the approximation of the integration over n by the midpoint formula results in Slater's "transition-state" approximation[117,188] to the ΔSCF total-energy difference. Equation (77) can be used to obtain another representation of the ΔSCF energy which, as we show below, is particularly useful in interpreting experimental spectra. This representation is obtained from the Taylor-series expansion of the ΔSCF total-energy difference in equation (76) in powers of the occupation-number change. Again using the ionization potential as an illustrative example, we obtain in this way

$$I = -\epsilon(1) - \Delta \tag{79}$$

where the quantity Δ is defined to be the remaining terms in the Taylor series

$$\Delta \equiv -\frac{1}{2}\frac{d\epsilon}{dn}\bigg|_{n=1} + \cdots \tag{80}$$

The motivation for representing the excitation energy in this way is that the excitation energy is expressed as a property of the ground state (the eigenenergy) plus a contribution describing the response of the system to the removal (or addition) of an electron. Furthermore, the representation of the excitation energy provided by equation (79) permits the contribution due to the system's response to be viewed as a shift of the orbital eigenenergy. In Hartree–Fock theory, this eigenenergy shift is called the relaxation shift. The eigenenergy shift Δ is often adequately described by the first two terms in the series representation given in equation (80); when this is the case, Δ can be written in the following compact form[189]:

$$\Delta = -\frac{1}{2}\frac{d\epsilon}{dn}\bigg|_{n=2/3} \tag{81}$$

Equations (76)–(81) describe the relationship between total-energy differences and the orbital eigenenergies. This relationship is useful both in the practical sense of providing a calculational alternative to the explicit subtraction of large total energies and in the interpretive sense of relating measurable total-energy differences to individual electron states and changes in them. We emphasize that the relationship between ΔSCF total-energy differences and orbital eigenenergies described by equations (77)–(81) is not restricted to the highest-occupied or lowest-unoccupied levels of the system; given the single assumption mentioned above, it is perfectly general. The more difficult question, to which we now turn, is: which excitation phenomena are accurately described by the ΔSCF construction?

As already mentioned in connection with our use of atomic ionization as an illustrative example, whenever both the initial and final states of an excitation process can be regarded as ground states, the ground-state theory can be used, as described above, to calculate the excitation energy. This class of excitations includes the removal of the outermost electrons from atoms and molecules, the addition of electrons (electron affinities), and, by extension to infinite systems, the work function of metals. For extended systems, the outermost orbitals are usually so delocalized that the addition or removal of an electron has a negligible effect on the electron density so that, in the language of the discussion above, the eigenenergy shift Δ is negligible. This implies, in the context of metals for example, that the eigenenergy of the highest occupied orbital equals the Fermi energy. The same argument applied to semiconductors and insulators indicates that the eigenenergies of the ground-state theory give both the top of the valence band, the bottom of the conduction band, and, therefore, the energy gap. (With regard to the gap separating the valence and conduction bands of insulators, note that, whereas the top of the valence band and the bottom of the conduction band correspond to the ground states of $(N-1)$- and $(N+1)$-electron systems, the smaller energy required to create an exciton cannot be rigorously described by the ground-state theory, because excitons represent excited states of the N-electron system.) The fact that energy-band calculations, based on the local-density approximation, usually underestimate the gap in semiconductors and insulators by a factor of $\sim 1/2$[190] might seem to contradict our statement that the gap is given by the ground-state theory. This is not a contradiction, but rather a reminder of the important distinction between the description of the ground state given by the hypothetical exact density-functional theory and that obtained in the local-density approximation.

Before ending this subsection on exact results, we recall the discussion of the possibility of defining energy functionals for the lowest nondegenerate state of each symmetry[37] (see Section 2.3) and for the sum of the energies of a finite number of the lowest eigenstates.[81] (See Section 2.3.1.)

4.2. *Delocalized Excitations*

Three aspects of the discussion above suggest that the eigenenergies of the ground-state theory might describe the delocalized single-particle excitations of extended systems, such as the one-particle spectrum of a simple metal. First, the ground-state eigenenergies give the correct Fermi energy. Second, equation (77) gives density-functional theory a strong resemblance to an exact theory of such excitations, namely, Fermi-liquid theory.[130] Third, the delocalized character of these excitations makes the eigenenergy shift, as defined above, negligible. Do these arguments imply that the ground-state eigenenergies, when calculated using the exact density-functional theory, are in fact the excitation energies, whenever Δ is negligible? Fortunately, there exists a system for which the relationship between the ground-state eigenenergies and the quasiparticle excitation spectrum can be explored in detail, and without the ambiguity usually introduced by use of the local-density approximation. This system is the homogeneous, interacting, electron gas. On the one hand, the translation symmetry of this system allows its accurate description using the techniques of traditional many-body theory; on the other hand, for this system, the local-density approximation is not an approximation. This system, therefore, allows us to compare ground-state eigenenergies which are uncorrupted by the local-density approximation with the exact quasiparticle excitation spectrum.

The exact quasiparticle excitation spectrum, $\epsilon_{\text{exact}}(k)$, is given in terms of the momentum- and frequency-dependent self-energy operator $\Sigma(k,\omega)$ by the following implicit relationship[138]:

$$\epsilon_{\text{exact}}(k) = \frac{k^2}{2m} + \Sigma(k,\epsilon_{\text{exact}}(k)) \tag{82}$$

The single-particle equations of the ground-state theory, on the other hand, contain a k-independent, translationally invariant potential; the dispersion of the ground-state eigenenergies, $\epsilon(k)$, is therefore simply proportional to k^2. As discussed above, however, the ground-state eigenenergies correctly describe the Fermi energy, and are therefore given by

$$\epsilon(k) = \frac{k^2}{2m} + \Sigma(k_F,\epsilon_{\text{exact}}(k_F)) \tag{83}$$

As the comparison of equations (82) and (83) clearly shows, the ground-state eigenenergies $\epsilon(k)$ and the quasiparticle spectrum $\epsilon_{\text{exact}}(k)$ differ as we move away from the Fermi energy, but they differ only to the extent that the self-energy varies with k and ω. In the immediate vicinity of the Fermi level, the distinction between the ground-state eigenenergies and the quasiparticle spectrum is described by the ratio of the effective mass m^* to the bare-electron mass m. The deviation of this ratio from unity is given in terms of derivatives of $\Sigma(k,\omega)$ taken at the Fermi energy,

$$\frac{m^*}{m} = \left(1 - \frac{\partial\Sigma}{\partial\omega}\right)\left(1 + \frac{m}{k_F}\cdot\frac{\partial\Sigma}{\partial k}\right)^{-1} \tag{84}$$

The difference between m and m^* provides a clear demonstration that the ground-state eigenenergies are *not* formally equivalent to excitation energies. Thus, the total-energy functional of density-functional theory differs from that of Fermi-liquid theory, and the state that corresponds to adding a density-functional orbital with wave vector k is in general not equal to the many-body state described by a quasiparticle of the same wave vector. Furthermore, while the ground-state theory, in principle, gives the correct Fermi *energy*, it remains an open question whether or not it gives the correct Fermi *surface*. Note also that the deviation of m^* from m implies that the state density at the Fermi energy, where lifetime effects are unimportant, is not given precisely by the distribution of density-functional ground-state eigenenergies. We can ask independently, as a practical matter, what is the magnitude of the difference between m and m^*? While in some special cases, such as ^{3}He, the difference can be large ($m^* \sim 3m$),[191] in the electron gas at metallic densities the difference is typically 5%.[138]

Away from the immediate vicinity of the Fermi energy, the distinction between the ground-state eigenenergies and the excitation spectrum is most easily described by the graph of the k dependence of the self-energy shown in Fig. 28. The qualitative way in which the self-energy changes as we move away from the Fermi energy is that it develops an imaginary part, reflecting the finite lifetime of the quasiparticle excitations. Beyond that, we see in Fig. 28 that the constant value of Σ appearing in the ground-state eigenenergies [equation (83)] means that the ground-state eigenenergies reflect neither the rapid variation of the self-energy for $k \sim 2k_F$, which reflects the onset of quasiparticle decay via plasmon creation, nor the gradual loss of exchange and correlation energy with increasing quasiparticle velocity. The latter reflects the relative inability of the electron gas to screen a rapidly moving disturbance. The picture of the difference between the quasiparticle spectrum and the ground-state eigenenergies provided by Fig. 28 is consistent with the known need to describe the loss of exchange and correlation energy with quasiparticle velocity in the analysis of LEED[192] and EXAFS[193] experiments, because of the large electron energies involved. Figure 28 is also consistent with the good description of simple-metal valence-band spectra provided by the ground-state eigenenergies, for we see that the variation of the Σ in the vicinity of the valence band is relatively weak. We remind the reader that, even for simple metals, the local-density approximation is not exact and its use introduces an additional source of ambiguity which is quite independent of that due to the variation of the self-energy.

4.3. *Localized Excitations: Effects of Local-Density Approximation*

Let us turn now to the effects of the local-density approximation. In order to keep this issue cleanly separated from the appropriateness of the

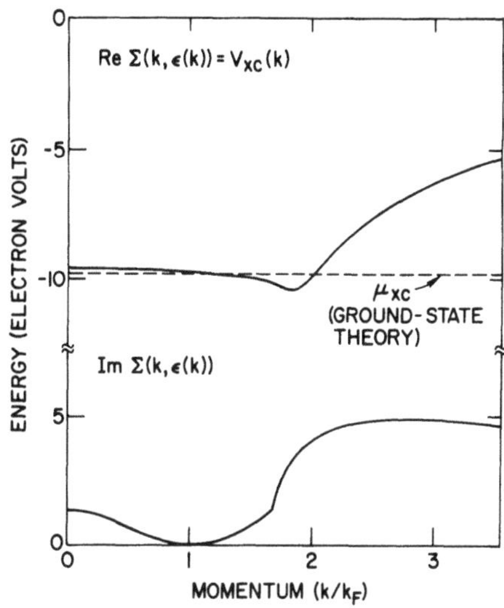

FIGURE 28. Variation of the self-energy with quasiparticle velocity. The deviation of this curve from its value at $k/k_F = 1$ is the energy error made when the eigenenergies of the density-functional description of the ground state of jellium are used to approximate the quasiparticle excitation spectrum. The error is seen to be relatively small (but finite) over the velocity range corresponding to the valence band, but to become large in the range relevant to LEED and EXAFS experiments. The self-energy depends of the value of the uniform electron density; the curve shown here is appropriate to most metals (see Fig. 22). The values of $\Sigma[k,\epsilon_{\text{exact}}(k)]$ were taken from Ref. 138.

ground-state theory, we will consider here excitations to which the ground-state theory should apply. Fortunately, here, as in the preceding discussion, special systems exist that indicate the behavior of the eigenergies of the exact density-functional ground-state theory, permitting us to focus on the effects of the local-density approximation. These special systems are atomic hydrogen and atomic helium. Recalling that, in the exact density-functional description of a system, the electron density is obtained from the solution of a single-particle Schrödinger equation containing a local potential, we see that, in the case of hydrogen, the exact density-functional description is simply the usual one-electron description. Therefore, for this system, the density-functional ground-state eigenenergy and the ionization potential are the same, that is, -1 Ry. The corresponding eigenenergy in the local-density approximation is only -0.56 Ry, even when spin polarization is accounted for. The case of atomic hydrogen has the additional virtue of clearly indicating the source of the error. The effective potential entering the density-functional description of hydrogen contains the electrostatic interaction of the electron with its own charge density. In the exact density-functional theory this self-interaction is eliminated by the contribution of the exchange term to the single-particle potential. In the local-density approximation, the elimination of the self-interaction is incomplete, leading to an elevated eigenenergy. A considerable body of evidence has been accumulated[55,56,190] which suggests that the incomplete elimination of the self-interaction in the local-density approximation is frequently a more serious source of error in the study of excitations than is the more fundamental question of the applicability of the ground-state theory to excited states.

Atomic He is another special system, with regard to understanding the implications of the local-density approximation. It is special in two ways. First, the relative unimportance of correlation (1.1 eV out of a total energy of ~79 eV) makes the Hartree–Fock description effectively exact for the present discussion. Second, because He contains only one electron of each spin, the single-particle potential entering the Hartree–Fock equations is local and state-independent. By the same argument applied above to hydrogen, the Hartree–Fock description of helium and the density-functional description, in which correlation is ignored, are the same. A particularly illuminating comparison of the local-density (exchange only) and density-functional (Hartree–Fock) results is contained in Fig. 29, where the spin-up and spin-down ground-state eigenenergies for helium are shown as a function of the occupation of the spin-down orbital.* (The spin-up orbital is fully occupied.) Note first that the incomplete cancellation of the self-interaction has elevated the local-density eigenenergies by ~ 11 eV. The difference, some 14 eV, between the spin-up eigenenergies when the spin-down orbital is empty is also purely and unambiguously a self-interaction artifact. This effect is much larger in He than in H, because the greater nuclear charge leads to more localized orbitals. That the effect of the self-interaction on the orbital eigenenergy can be large was already demonstrated by hydrogen; the new information presented by Fig. 29 is the fact that, in the full density-functional (Hartree–Fock) case, the spin-down eigenenergy *rises* as charge is removed from the orbital, whereas, in the local-density approximation, the level *falls*. That a level should rise as its occupation is reduced is the correct and intuitively reasonable behavior, because the reduced occupation of any given orbital causes the *other* orbitals in the system to contract; the change in the electrostatic potential caused by this contraction, in turn, causes the depleted level to rise. The contraction of the other orbitals is caused by the reduced screening of the nucleus. If the contraction occurs in the domain of the depleted orbital, the depleted level rises because the screening of the nucleus is increased; if the contraction occurs outside the domain of the depleted orbital, the depleted level rises because of the increased electrostatic potential in the domain of the depleted orbital. These effects constitute the usual and correct description

*The extension of the Hartree–Fock formalism to nonintegral occupation used here is analogous to that used in the local-density calculations; this extension may not be unique.

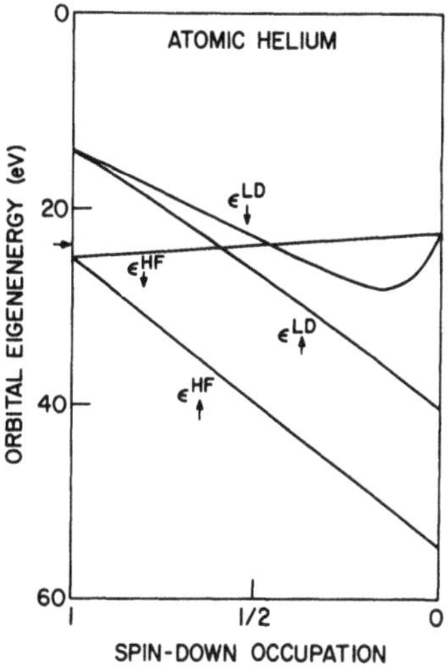

FIGURE 29. Effect of the local-density approximation on the occupation-number dependence of the ground-state eigenenergies of atomic He. The spin-up orbital contains one electron. Because the He atom contains only one electron of each spin, the Hartree–Fock potential is local. The Hartree–Fock description of this system therefore constitutes the density-functional treatment in which correlation is ignored. The Hartree–Fock results are compared to the corresponding results in the local-density approximation (i.e., correlation is ignored in both the density-functional and local-density calculations). In the density-functional description, the spin-down level is seen to rise as charge is removed from the level, whereas in the local-density approximation, the level falls. The incorrect behavior of the approximate result is an artifact of the incomplete elimination of the electrostatic self-interaction in the local-density approximation. The arrow on the energy axis indicates the value of the ΔSCF total-energy difference. The similarity of the ΔSCF result to *both* the density-functional and local-density eigenvalues at half occupation illustrates the utility of the transition-state approximation to the ΔSCF total-energy difference, even when the local-density approximation is used. The corresponding measured value is 24.6 eV, indicating the relative unimportance of correlation for this system.

of relaxation. The relaxation shift described by equations (79)–(81) should, therefore, be positive; it reflects the lowering of the final-state total energy by adjustment to the removed charge. Figure 29 indicates that, for helium within Hartree–Fock, the inconsistency between the expected behavior and that provided by calculations based on the local-density approximation is an artifact due to the approximation. The fact that the relaxation shift has the wrong sign in calculations based on the local-density approximation is a particularly direct manifestation of the incomplete elimination of the self-interaction. That is, the reason the local-density eigenenergy falls as charge is removed from the level is simply that the dominant effect of removing charge from a localized level is, in the local-density approximation, to reduce the self-interaction; this effect overshadows the true relaxation shift.

This interpretation is supported by the sign reversal of $d\epsilon/dn$ near the end of the interval, when the level is almost empty. Note that it is only the "diagonal" derivative, $d\epsilon_i/dn_i$, that has the wrong sign in the local-density approximation; the variation of the other eigenenergies ($d\epsilon_j/dn_i$) has the correct sign. That is, in the density-functional formalism, each orbital interacts electrostatically with the charges corresponding to all the occupied orbitals, through the Hartree potential. The "off-diagonal" interactions are therefore correct, as they stand. It is only the self- or "diagonal" interaction which must be eliminated by the exchange-correlation potential. The situation is summarized in Fig. 30, where we show that the relaxation shift is usually obscured in local-density calculations by the variation of the self-interaction with occupation.

Our two examples above also reveal another interesting fact with regard to the accuracy of the local-density approximation. This approximation is much more accurate for total-energies than for orbital eigenenergies. For H and He, for example, the total-energy errors are only 0.3 and 1.7 eV, whereas the corresponding errors in the eigenenergies are 6.0 and 9.0 eV. We believe this difference in accuracy to be due to the fact that total-energy errors reflect the imperfections of the exchange-correlation *functional*, whereas the eigenenergy errors are caused by the corresponding imperfections in the effective single-particle *potential*. Because the exchange-correlation energy represents an average over all the occupied states, it is relatively insensitive to the choice of exchange-correlation functional, as long as constraints such as the sum rule discussed in Section 2.1.2 are maintained. The potential, on the other hand, is obtained by functionally differentiating the energy with respect to the density; it therefore probes finer details of the density dependence of the energy functional. This idea immediately suggests that one could obtain an improved potential and improved eigenenergies by first differentiating the exact expression [equation (4)] for the exchange-correlation energy and then introducing approximations such as the local-density approximation into the resulting exact expression for the exchange-correlation potential. Such a procedure would, for example, guarantee the correct assymptotic behavior of the effective single-particle potential outside a surface or away from an isolated atom. This idea also suggests that the accuracy of the orbital eigenenergies could be used as an additional check of the quality of different approximations within density-functional theory. From the stationary property of the energy with respect to variations in the density, it could perhaps be argued that a better potential would be pointless if one does not succeed in finding also an improved functional for the exchange-correlation energy. In view of the discussion above, a more accurate exchange-correlation potential could, however, considerably improve the ability of the ground-state theory to provide an approximate description of the excitation spectrum.

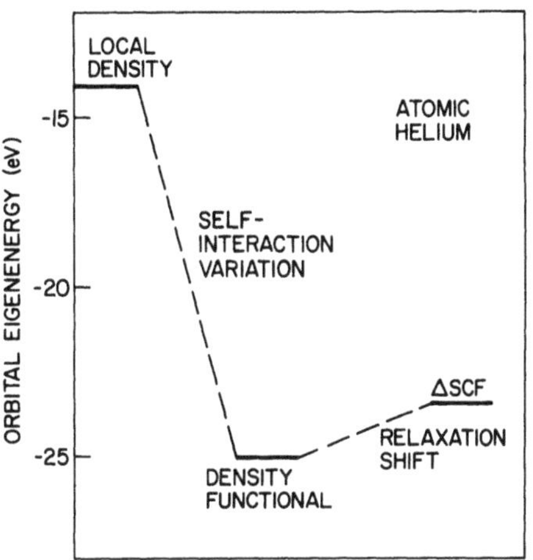

FIGURE 30. Relative energies of (a) local-density ground-state eigenenergy, (b) density-functional ground-state eigenenergy, and (c) the corresponding ΔSCF total-energy difference. The excitation considered is the ionization of atomic He. The relaxation shift (relative to the density-functional eigenenergy) is seen to be positive. The result labeled "density-functional" was obtained using Hartree–Fock calculations, which, for this system, constitute the exact density-functional treatment in which correlation is ignored. The measured ionization potential of He (24.6 eV) indicates that correlation is relatively unimportant for this system.

4.4. *Core-Level Excitations*

In terms of the preceding discussion, the application of the local-density, ground-state formalism to core-level ionization potentials should be a particularly unreliable procedure. Our discussion of the excitation spectrum of the homogeneous electron gas would suggest that core-hole states, being very far removed from the Fermi level (large excitation energy), might be poorly described by the ground-state theory. Our discussion of the effects of the local-density approximation in the preceding section would suggest that the localized character of core-hole states should introduce large errors due to the incomplete elimination of the self-interaction. Fortunately, another aspect of Fig. 29 suggests that the ground-state eigenenergies may be a useful interpretational tool in this context as well. The aspect of Fig. 29 to which we now call attention is the fact that, despite the generally large differences between the density-functional and local-density eigenenergies, the spin-down eigenvalues of the two procedures are remarkably similar when the spin-down orbital is half occupied. The significance of this fact is that the eigenenergy at half occupation is a good approximation to the corresponding ΔSCF total-energy difference [Slater's transition state,[117] equation (78)]. Thus, despite the large errors in the eigenenergies generally and despite the incorrect sign of the eigenenergy variation with occupation (see preceding section), the excitation energy, when represented as the ΔSCF total-energy difference, is reasonably well described by even the local-density approximation to the ground-state theory. Note that the ΔSCF value for the ionization potential is also indicated in Fig. 29. Note also that equations (78)–(81) allow the ΔSCF approximation to excitation energies to be constructed without the explicit evaluation and subtraction of total energies.

The ΔSCF procedure provides us with a means of avoiding some of the implications of the local-density approximation for localized excitations, but we have yet to provide any justification for using even the exact ground-state formalism to describe such highly excited states as systems containing core holes. We offer in this regard only our speculation that the crucial property of an electronic system that determines the appropriateness of the ground-state theory is the lifetime of the state to which the theory is applied. If the interactions that cause a core-hole state to decay (e.g., Auger) are neglected, then the core-hole and no-core-hole states become orthogonal subspaces of the full many-body Hamiltonian. If we think of such subspaces as possessing different symmetries, then we can appeal to the proposal by Gunnarsson and Lundqvist[37] to use the ground-state theory to describe the lowest-energy state of each symmetry subspace. In the discussion (Section 2.3) of atomic multiplets, which constitute an example of symmetry subspaces with different energies, we saw that a different energy functional is required for each symmetry subspace. In the present context, we believe that the distinguishing feature of the energy functional appropriate to the core-hole state is simply the fact that the core orbital is kept unoccupied. Note that occupying single-particle states of the energy functional other than those which lie lowest in energy represents a change in the functional, not the minimization of the ground-state functional in a restricted space of electron densities.

The screening of core holes by the valence electrons in simple metals has been studied using these ideas by Almbladh and von Barth.[186] By calculating the valence-electron response as a function of core-level occupation, it was possible to obtain the relaxation energy appropriate to the presence of either one or two core holes. This distinction permits the difference between screening in the context of photoemission (a single core hole) and Auger (two core holes) experiments to be understood. An example of the application of the ground-state theory in the context of core-hole spectroscopy that illustrates the interpretational virtues of the orbital eigenenergies and their occupation-number dependence is the analysis of core-hole screening by Williams and Lang.[189] The measurement considered is the amount by which the binding energy, or ionization potential I, of an atomic core electron is reduced when the atom is embedded in a metal. This change in the binding energy results from several physical effects: (1) the screening of the core hole by the conduction electrons of the metal; (2) the requirement that an electron photoexcited from an embedded atom traverse the surface dipole-layer potential in order to leave the solid; (3) the electrostatic elevation of the core level by the compression of the valence electron charge that occurs as part of the embedding process; and (4) changes in the electrostatic potential in the core region due to changes in the valence configuration. The binding-energy reduction is typically 5–10 eV and varies considerably from atom to atom. A principal objective of the analysis was to determine how much of this variation should be ascribed to initial-state effects (chemical shifts) and how much to final-state effects (relaxation). The interpretive value of the Taylor-series represenation of the ionization potential [equations (79)–(81)] is immediately clear, for we see that the first term in the series, $\epsilon(1)$, is manifestly a property of the ground (initial) state, whereas the remaining terms, i.e., the eigenenergy shift Δ, describe the response of the system to the creation of the core hole. If we distinguish quantities associated with the embedded atom with an asterisk, then the total binding-energy shift Δ_{tot} is simply

$$\Delta_{tot} = I - I^* \tag{85}$$

and equation (79) immediately provides the desired decomposition into initial- and final-state components, that is

$$\Delta_{tot} = \Delta_{initial} + \Delta_{final} \tag{86}$$

with $\Delta_{initial}$ given by

$$\Delta_{initial} = -\ [\epsilon(1) - \epsilon^*(1)] \tag{87}$$

and Δ_{final} given by

$$\Delta_{final} = \Delta - \Delta^* \tag{88}$$

with the eigenenergy shift Δ given by equation (81). This analysis of binding-energy shifts illustrates another practical aspect of the theory. We have shown above that, in the local-density approximation, the physical content of the eigenenergy shift Δ is often obscured by the occupation-number dependence of the artificial self-interaction. The practical point in question is that, despite the corruption of Δ by the self-interaction, the local-density approximation can be used to calculate the *change* in Δ. In other words, the large errors in the local-density approximation to Δ subtract out of the change in Δ described by the final-state shift, Δ_{final}. The total and decomposed shift obtained in this way from parameter-free self-consistent calculations are compared with measured values in Fig. 31. The close agreement between the calculated total and the measured shift lends credibility to the decomposition provided by the theory. The clear and intuitively appealing interpretation of the measurements provided by the analysis motivates the application of the ground-state formalism far outside its formally justifiable range of applicability.

The corresponding analysis of relaxation shifts, as they apply to Auger spectra, can be found in Ref. 194.

4.5. *Valence-Band Spectra of Transition Metals*

The preceding sections have presented the available theoretical information which bears on the connection between the eigenenergies of the ground-state theory and the single-particle excitation spectrum. In this subsection, we consider the relevance of the ground-state eigenenergies to a particularly important class of experiments, spectroscopic studies of transition metals and their surfaces (e.g., photoemission). These measurements constitute a large and important area of condensed-matter physics; their special significance in the context of this chapter is that many electronic-structure calculations attempt to interpret these measurements on the basis of the eigenenergies of the local-density ground-state theory. What can be said about the legitimacy and accuracy of this type of analysis? Unfortunately, the arguments of the preceding sections indicate only the issues involved (the relative ability of the electron gas to screen electrons having different velocities, the artificial self-interaction of the local-density approximation, and the lifetime of the excitation); these arguments do not indicate the quantitative importance of these effects in the context of transition-metal spectra. It is therefore necessary in this context to rely on empirical evidence.

The work of Andersen[183] and collaborators has greatly clarified the informational content of the single particle spectra of transition-metal spectra. These spectra reflect primarily[195] the crystal structure, the energy separation of the *s* and *d* bands, and the width of the *d* band. Beyond these three aspects, quantities such as the effective mass of the *s* band provide additional refinement, but the *d*-band position and width are the parameters of greatest importance. (In magnetic transition metals, the energy separation of the spin-up and spin-down bands is, of course, also an important parameter.) We therefore ask, how well does the ground-state theory predict these para-

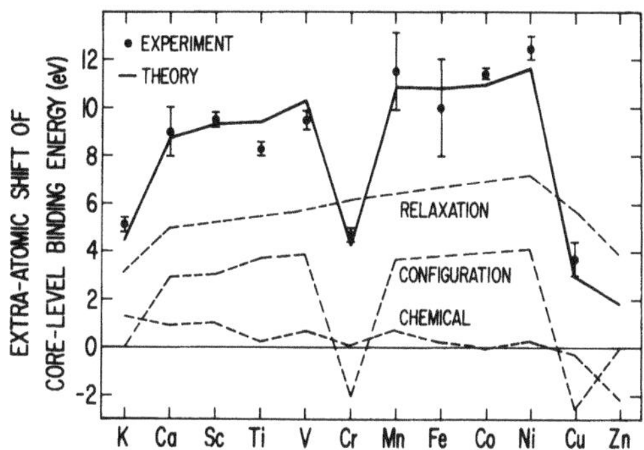

FIGURE 31. Comparison of measured and calculated core-level binding-energy shifts for $3d$ transition-metal atoms. The solid curve shows the total theoretical shift and the dashed curves indicate the contributions of initial- and final-state effects. Relaxation refers to the screening of the core hole by the metal, a final-state effect. Configuration refers to the effect on the core level of changing the electronic configuration from that of the free atom to that of the solid (a particular initial-state effect). The strong variation of the configuration component reflects the fact that only Cr and Cu have a $3d^{n-1}4s^1$ configuration as free atoms; the others have a $3d^{n-2}4s^2$ configuration. The contribution labeled chemical is the remainder of the initial-state shift; it exhibits the near cancellation of compression and dipole-layer-shift effects. (From Ref. 189.)

meters. The situation is clearest in the noble metals, where, because the Fermi level does not pass through the d band, the d-band position relative to the s band is experimentally very well determined. In the case of Cu, the d band given by the ground-state theory in the local-density approximation is ~0.4 eV too high relative to the s band.[196] The situation in zinc provides a hint as to the origin of this discrepancy. In Zn,[197] the d band lies deeper in energy (~10 eV below ϵ_F) and is narrower than the d band of Cu (1 eV as compared to 3 eV). Both the greater depth and narrowness reflect the greater spatial localization of the d states in Zn. The fact that the local-density ground-state d bands of Zn are too high by ~1.5 eV suggests that the discrepancies in the energy separation of the s and d bands of Cu and Zn result from the imperfect elimination of the self-interaction in the local-density approximation. It is the fact that the eigenenergies are too *high* and the fact that the effect increases with localization that suggest this interpretation. (Recall the discussion of atomic H and He in Section 4.3.) This interpretation is also supported by Fig. 32, where measured quasiparticle energies are compared with the results of a calculation in which the self-interaction does not enter. The calculated energy bands are seen to agree quite well with the measured spectra. In particular, the energy position of the narrow d bands relative to the Fermi level does not exhibit the 0.4 eV error seen in the corresponding local-density calculations. The effective single-particle potential used in the calculations shown in Fig. 32 is due to Chodorow,[198] and was constructed using what has come to be called the "renormalized-atom" procedure.[199] The aspect of that procedure which is relevant to the present discussion is its use of nonlocal exchange within the atomic cell. In this way, the self-interaction effect was avoided. With regard to the width of the d bands, experience with semiempirical adaptations of the ground-state theory suggest that, if the artificial elevation of localized levels by the local-density approximation could be eliminated, the eigenenergies of the ground-state theory would provide an interpretation of the excitation spectrum which is adequate for most purposes.

The excitation spectrum of copper's other Periodic Table neighbor, Ni, introduces another complication into the connection between ground-state eigenenergies and excitation spectra. Experimental measurements have shown with increasing precision over the past decade that the spectrum is significantly

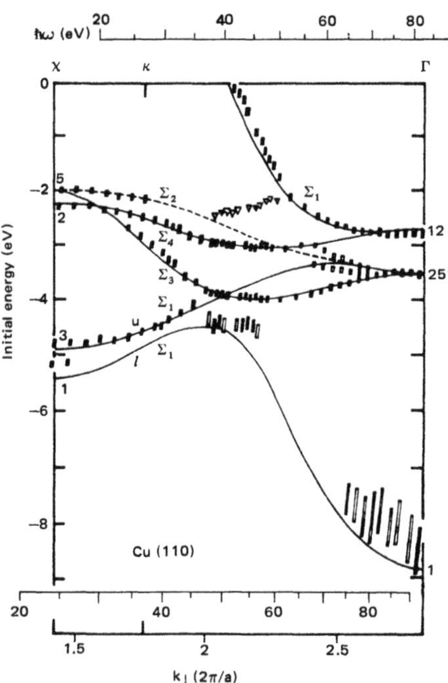

FIGURE 32. Interpretation of angle-resolved photoemmission data for Cu provided by an energy-band calculation based on an effective single-particle potential in which the self-interaction effects of the local-density approximation are not present. The experimental data are due to Thiry *et al.* (Ref. 200) and the calculation is due to Burdick (Ref. 201). The effective single-particle potential used in the calculation is due to Chodorow (Ref. 198) and is based on the use of nonlocal exchange within the atomic cell. The use of nonlocal exchange avoids the introduction of self-interaction effects which are believed to cause the d bands of similar calculations based on the local-density approximation to lie 0.4 eV closer to the Fermi level than those shown in the figure.

more narrow than the width of the ground-state energy bands. Nickel appears to be a case in which the excitation spectrum reflects a particular type of correlation which, while present in the electron gas on which the local-density description of correlation is based, is emphasized in Ni by the spatial localization of the *d* orbitals and the availability of low-energy excitations within the *d* shell. Experimental evidence[202] and theoretical analysis[203–207] both indicate that, in Ni, the state of the solid in which a single *d* hole is propagating in a state below ϵ_F is strongly coupled to other states in which a second *d* hole is created by the elevation of a *d* electron to an empty state above ϵ_F. The state consisting of two *d* holes is itself long lived and has been observed directly in several experiments. The state containing a single *d* hole may be reasonably described by the ground-state theory, but here we have a case where the discussion above of the lifetime of the excitation is probably relevant. That is, the use of the ground-state theory implicitly assumes that the single-*d*-hole state is long lived; in Ni, this assumption is not valid. Note that this interpretation of the *failure* of the ground-state energy bands to describe the excitation spectrum is consistent with the *success* of the theory in describing the ground-state properties of Ni,[87,92,94] such as the equilibrium volume, the bulk modulus, the magnetic moment, the hyperfine field etc. (See Sections 2.2 and 3.2.)

In applying the ground-state theory to other transition metals, their surfaces, etc. it would be useful to know how unusual the situation in Ni is. There is reason to believe that Ni is the exception rather than the rule. Of all the transition metals, Ni has the most localized *d* orbitals. As we proceed to the left in the Periodic Table, the decreasing nuclear charge results in less localization. As we proceed down the Ni column of the Periodic Table, localization decreases because the valence *d* orbitals must be orthogonal to the *d* orbitals of the core. The nuclear charge in Cu is, of course, greater than that of Ni; we might ask why these two-hole effects are not important in Cu. The *d* orbitals of Cu are more localized than those of Ni, but Cu lacks the high density of empty states at ϵ_F which facilitates the creation of the two-hole state in Ni. Thus, it is seems likely that the deviation of the excitation spectrum from the ground-state eigenenergy spectrum in Ni is probably unusual.

We can summarize this section by saying that, in many cases, the eigenenergies of the ground-state theory can be a useful guide to the interpretation of single-particle excitation spectra. The ambiguity introduced into the connection between the ground-state eigenenergies and the excitation spectrum by the use of the local-density approximation is frequently larger than that due to using the ground-state theory to describe excited states. But, having said that, we remind the reader of our discussion of the excitation spectrum of the homogeneous electron gas, which clearly shows that the ground-state eigenenergies are *at best* only an approximation to the true excitation spectrum. Thus, while the ground-state eigenenergies can be a useful interpretive tool, we must remain alert to the possibility, illustrated by Ni, that one or more of the assumptions implicit in their use may be invalid in particular systems.

4.6. Self-Energy Approach to Calculation of Excitation Energies

As pointed out above, the orbital eigenenergies of the density-functional ground-state theory do not, in general, give the exact single-particle excitation energies for many-electron systems. We have also seen, however, that the orbital eigenenergies are often usefully accurate approximations to excitation (quasiparticle) energies, especially when they are not corrupted by self-interaction effects associated with the local-density approximation. The most serious inadequacy of the orbital eigenenergies as a description of the excitation spectrum is the fact that they do not represent a well-defined approximation to a description which is correct in principle. There is not, at present, any systematic way in which to improve upon the description they provide. The virtue of the self-energy approach is that it provides a description of the quasiparticle excitation spectrum which is, in principle, exact. Furthermore, this approach lends itself to systematic and physically understandable levels of approximation.

The basis of the self-energy approach (see, e.g., Ref. 138) is the Dyson equation satisfied by the single-particle Green's function, or equivalently, by the quasiparticle amplitudes $\psi(\mathbf{r})$

$$\{-\tfrac{1}{2}\nabla^2 + V_c(\mathbf{r})\}\psi(\mathbf{r}) + \int d^3r'\, \Sigma\,(\mathbf{r},\mathbf{r}',\epsilon)\psi(\mathbf{r}') = \epsilon\psi(\mathbf{r}) \tag{89}$$

where $\Sigma\,(\mathbf{r},\mathbf{r}',\epsilon)$ is the self-energy operator and $V_c(\mathbf{r})$ is the Coulomb potential due to all charges in the system. The fact that the self-energy operator is not Hermitian implies that the quasiparticle energies ϵ are in general complex, as they should be, reflecting the finite lifetime of excitations in an interacting system. The fact that the self-energy depends on the quasiparticle energy means that the amplitudes $\psi(\mathbf{r})$ are not orthogonal and cannot be straightforwardly squared and summed to obtain the electron density.

The single-particle equations of the density-functional ground-state theory[65] differ from the Dyson equation only in the replacement of the self-energy $\Sigma\,(\mathbf{r},\mathbf{r}',\epsilon)$ by the exchange-correlation potential $v_{xc}(\mathbf{r})$, where

$$v_{xc}(\mathbf{r}) \equiv \frac{\delta E_{xc}\{\rho\}}{\delta n(\mathbf{r})} \tag{90}$$

We see in this way that the extent to which the orbital eigenenergies of the ground-state theory approximate excitation energies is determined by the similarity of the expectation values of $\Sigma(\mathbf{r},\mathbf{r}',\epsilon)$ and those of $v_{xc}(\mathbf{r})$. The rather different mathematical properties of these two quantities (nonlocal, non-Hermitian, and energy-dependent *versus* local, Hermitian and energy-independent), by themselves, provide little hope of a strong similarity. Figure 28 above shows that, for the homogeneous electron gas, the similarity is in fact strong, but the discussion of Fig. 28 indicates that the similarity rests on the smallness of the many-body correction to the effective mass.[138] That the similarity persists for slightly inhomogeneous systems is shown by the work of Sham and Kohn.[120] Recently, however, MacDonald[208] has suggested a mechanism by which the similarity might be considerably reduced by the presence of a lattice potential in real solids.

Consider now the calculation of the self-energy and the excitation spectrum for real systems. Most efforts to implement the self-energy approach have been based on many-body perturbation theory.[209–215] (This set of references is representative, not exhaustive.) Many-body perturbation theory is really an alternative to, rather than an aspect of, density-functional theory; we therefore restrict the present discussion to implementations of the self-energy approach which exploit the ideas of density-functional theory. The basic ideas in this context were given by Sham and Kohn[120] who, in the spirit of the local-density approximation, approximated the self-energy by that of a homogeneous electron gas,

$$\Sigma(\mathbf{r},\mathbf{r}',\epsilon) \approx \Sigma_h(\mathbf{r}-\mathbf{r}',\epsilon-\epsilon_F + \mu_h(n);n) \tag{91}$$

The physical idea here is that the response of the inhomogeneous electron gas to the passage of a given electron or hole should be similar to that of a homogeneous electron gas. In particular, the two important physical parameters characterizing the response should be (i) the energy of the quasiparticle relative to the Fermi level (recall that the imaginary part of the true self-energy vanishes at the Fermi level); and (ii) electron density in the vicinity of the electron or hole considered. The approximation described by equation (91) clearly reflects this intuition; its specification is completed by taking the density parameter n appearing in the right-hand side of equation (91) to be the density of the inhomogeneous system at the position midway between $\mathbf{r}$ and $\mathbf{r}'$. In addition to its intuitive appeal, this approximation has the virtue of becoming exact for systems in which the density varies slowly. Because, however, the density variations in real systems are not slow, justification for the application of this approximation to real systems remains empirical.

It should be noted that the local-density approximation to the self-energy [equation (91)] retains the nonlocal and energy-dependent character of the exact self-energy. To eliminate the practical difficulties associated with the nonlocality, Sham and Kohn also suggested a second, more approximate, form for the self-energy:

$$\Sigma(\mathbf{r},\mathbf{r}',\epsilon) \approx \Sigma_h(\mathbf{p}(\mathbf{r});\epsilon-\epsilon_F + \mu_h[n(\mathbf{r})];n(\mathbf{r})) \cdot \delta(\mathbf{r}-\mathbf{r})' \tag{92}$$

This approximation is based on arguments similar to those leading to the WKB approximation and they are valid for slowly varying electron densities. Here, the self-energy is approximated again by that of a homogeneous electron gas, but in this case for a quasiparticle of momentum $\mathbf{p}(\mathbf{r})$. This local momentum is in turn specified by the implicit relationship

$$\epsilon-\epsilon_F + \mu_h = \tfrac{1}{2}\,|\,\mathbf{p}\,|^2 + \Sigma_h(\mathbf{p};\epsilon-\epsilon_F + \mu_h;n) \tag{93}$$

with the chemical potential μ_h and the electron density n taking on local values, $n = n(\mathbf{r})$ and $\mu_h = \mu_h(n)$. (Reference 60 contains a discussion of the choice of quasiparticle momentum.) An important aspect of this approximation [equation (93)] is that at the Fermi energy ($\epsilon = \epsilon_F$), the local momentum $\mathbf{p}(\mathbf{r})$ is given by the Fermi momentum of a homogeneous electron gas (possessing the local density). The approximate self-energy given by equation (92) then reduces to the exchange-correlation potential of the ground-state theory in the local-density approximation. In other words, *in the approximation of equation (92),* the Fermi *surface* given by the single-particle equations of the ground-state theory is that of the quasiparticle spectrum. In contrast to the ground-state theory, however, the quasiparticle spectrum implied by equation (92) contains the many-body correction to both the effective mass and the state density. (See Section 4.2.) The local approximation to $\Sigma(\mathbf{r},\mathbf{r}',\epsilon)$ given by equation (92) has been tested in Al where it was found to give a quasiparticle band structure quite similar to that given by the local-density ground-state theory.[(68)] Similar conclusions were reached by Watson *et al.*[(216)] for Ni and, more recently, by MacDonald[(208)] for the alkali metals.

The more elaborate, nonlocal approximation to the self-energy [equation (91)] has been applied by Vosko, Rasolt, and co-workers. These tests have been carried out for model systems,[(217,218)] for Cu,[(219)] and for the alkali metals.[(208,220)] These applications indicate, quite interestingly, that the implications of the nonlocality can be substantial. In particular, Fermi-surface distortions for the alkali metals were found to be much reduced for the nonlocal self-energy relative to those given by the local-density ground-state theory. The reduced distortions given by the nonlocal theory are much closer to the measured values. In the case of Cu, the nonlocal theory was shown[(219)] to lead to corrections of sufficient magnitude to remove the deviation from measured Fermi-surface "neck" radii found by Janak *et al.*[221] using the local-density ground-state theory.

As mentioned above, MacDonald[208] has pointed out another interesting effect which has its origin in the nonlocal character of the self-energy. When interactions with the lattice of ions produce a substantial "band mass," the many-body contribution to the effective mass can be enhanced relative to its value without scattering by the lattice. For Li and Cs the corrections to the effective mass were estimated to be 10% and 12%.

The examples considered here suggest that the superior description of the excitation spectrum offered by the nonlocal self-energy approach provides just the level of refinement required to understand many-body contributions to Fermi surface distortions and effective masses. It should be noted, however, that this approach still employs a "local-density" approximation, and it has been tested primarily on systems in which the electron density is relatively slowly varying at least with respect to transition metals and surfaces. We have seen above that there are two principal sources of error in using the local-density ground-state theory to describe excitations: (i) the use of the local-density approximation, and (ii) the use of the ground-state theory. The non-local self-energy approximation of equation (91) eliminates only the second of these, and we have seen that the first can be the more important source of error in systems containing relatively localized states. Further tests are needed to determine the range of validity of this approximation scheme.

5. Physical Models Based on Density Functional Theory

In the other sections of this chapter we have considered the physical content of the local-density approximation, the reason for its success and several examples of its utility. Here we consider another aspect of the theory, its use as a basis for schemes which, while unabashedly approximate, attempt to abstract from relatively complicated systems the variables and quantities responsible for their behavior.

The common theme to this section is the attempt to understand systems possessing relatively little symmetry in terms of systems which permit a relatively detailed understanding because their greater symmetry facilitates their description. Thus, for example, free atoms are often adequately described in the central-field approximation, the basis of which is the assumption of full rotation symmetry. Similarly, many properties of metal surfaces are adequately described by the semi-infinite jellium model, the basis of which is the assumption of full translation symmetry in two dimensions. Rotation and translation symmetry permit the reduction of the partial differential equations of density-functional theory (Schrödinger's and Poisson's equations) to one-dimensional (ordinary) differential equations. If the solutions of these highly symmetric models can be used to understand the interactions of atoms with one another and with metal surfaces, for example, a great simplification has been accomplished. Furthermore, the accomplishment is not merely one of practicality, of facilitating calculations, because the description of complex systems in terms of easily characterized components is often what it means to understand the complex system.

5.1 *Chemical-Potential Model of Intermetallic Compound Formation*

One example of this general approach was mentioned in Section 3.3, that introduced by Hodges and Stott[10,11] and developed by Alonso and Girifalco[9] for the description of intermetallic-compound formation. Here the theory tries to exploit the smaller number and greater translational symmetry of the pure-metal constituents in an effort to understand the process of compound formation. The fundamental assumption which makes the scheme approximate, on the one hand, but powerfully simple, on the other, is the idea that the electronic structure of each atomic cell is unchanged on going from the element to the compound (except for rigid electrostatic shifts caused by charge transfer). This conception of the formation process is illustrated schematically in the left-hand side of Fig. 26. In terms of Fig. 26, the fundamental assumption is that the state densities on the left and right sides of the figure are unchanged by the formation process and that bonding results from charge transfer among states near the Fermi level. The assumption that the electronic structure of an atomic cell is unaffected by the change of environment between element and compound is implicitly a Thomas–Fermi-like (local) approximation to the kinetic-energy operator. (The kinetic-energy operator is the essential source of interatomic coupling in calculations which accurately describe compound formation.) This type of approximation is least justified when the interatomic bonding results from the formation of bonding and antibonding states by the interaction of relatively localized atomic states on neighboring atoms, as in the right-hand side of Fig. 26. (Covalency is an intrinsically wave-mechanical effect, that one derived from the nonlocality of the kinetic-energy operator and, therefore, beyond the scope of a Thomas–Fermi description.) Since covalent bonding is known to dominate the cohesive properties of transition-metal elements, it is not surprising that the formation process described by Fig. 26 does not adequately describe the formation of transition-metal compounds. Available evidence suggests, however, that the charge-transfer mechanism of Fig. 26 is responsible for the formation of simple-metal compounds.

5.2. *Kim–Gordon Description of Interacting Closed-Shell Systems*

A successful example of the use of calculations for high-symmetry systems to describe bonding in lower-symmetry systems is that proposed by Kim and Gordon.[7,8] The basic idea is most easily described in the context of colliding rare-gas atoms. Here a local-density approximation to the total-energy functional $\mathscr{E}\{\rho\}$ is used (including a local-density approximation to the kinetic energy), but the functional is used to calculate only the difference between the total energy of the composite two-atom system and those of the individual atoms. In this regard it is similar to the approach of Hodges and Stott[10,11] and Alonso and Girifalco[9] described above. It differs from that approach in taking the electron density of the composite system to be a simple superposition of the constituent electron densities and in restricting the use of the approximation to systems where the superposition approximation is expected to be valid. The superposition approximation fails when open-shell systems interact to form covalent bonds (as in transition metals). The Kim–Gordon procedure has been shown to accurately describe the interaction of rare-gas atoms and closed-shell ionic systems, such as alkali halide crystals.[229]

5.3. *Effective-Medium Theory of Chemical Bonding*

As the last example of the use of density-functional theory as the basis of simplified conceptual models, we shall discuss the very promising "effective-medium" formalism proposed recently by Norskov and Lang[12] and by Stott and Zaremba.[13] Here the high-symmetry system whose properties are brought to bear on lower-symmetry systems is not the free atom, but an atomic "impurity" in otherwise homogeneous jellium. The prototypical system to which this procedure is applied is atomic chemisorption on metals. Here an atom, as part of a polyatomic system, is viewed as an atom embedded in an electron-density distribution. This distribution is in general *nonuniform*, but the binding energy of the atom in its true nonuniform environment can be rigorously represented as a series, the zero-order term of which corresponds to the atom embedded in a *uniform* electron gas of an appropriately chosen[12] average density. The fundamental idea of the procedure is illustrated schematically in Fig. 33. The most important quantity in this procedure is the heat of solution of an atom in jellium, considered as a function of the jellium density. This quantity is shown for two representative atoms in Fig. 34. Perhaps the greatest virtue of this representation of the chemical bond, aside from its simplicity, is that it identifies the strength of the chemical bond as a property of the atom. Thus, Fig. 34, by itself, indicates that oxygen bonds are typically 4 eV in strength, whereas hydrogen bonds are approximately half as strong.

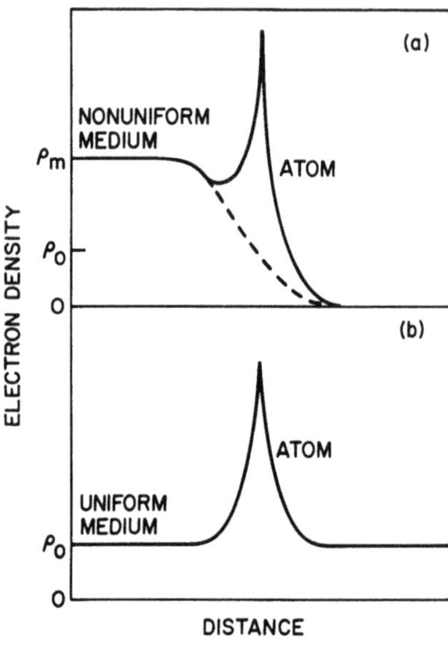

FIGURE 33. Schematic representation of the "effective-medium" description of chemical bonding. An atom embedded in a *nonuniform* electron gas, such as that at a metal surface, is described by a series, the first term of which describes the same atom embedded in a *uniform* electron gas.

To estimate the energy with which an atom is chemisorbed on a metal surface, using this formalism, all that is required is the electron-density distribution at the surface of the *bare* metal. This density profile, together with the information in Fig. 34, gives the total-energy variation of the combined system as a function of metal–adatom separation. The minimum in such a curve indicates the energetically preferred metal–adatom separation.

What we have just described is only the zero-order contribution to the effective-medium theory; the first- and second-order correction terms have

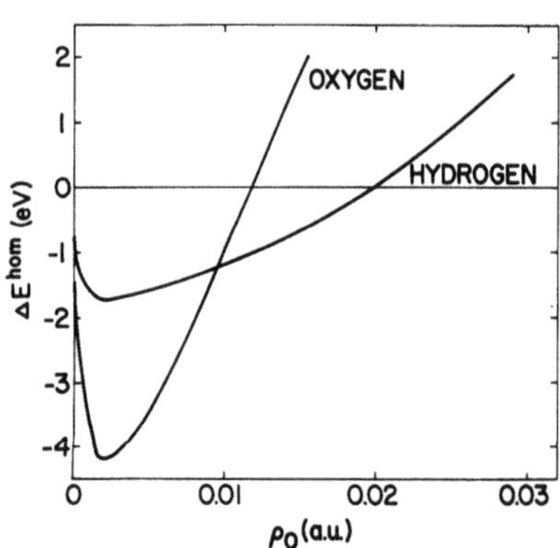

FIGURE 34. Zero-order "effective-medium" energy, the energy of atomic hydrogen and oxygen embedded in jellium, as a function of the jellium density. The energy zero is the energy of the free atom. Calculations (non-spin-polarized) were done for densities down to 0.001 a.u., after which each curve was joined to a point at zero density corresponding to the negative of the measure electron affinity of the free atom. (After Ref. 12.)

been derived and the first-order correction has been tested numerically.[12] The test consisted of a comparison of the energies of atomic chemisorption as given by the effective-medium theory with those given by elaborate wave-mechanical calculations[3] based on the local-density approximation. Both sets of calculations considered atomic chemisorption on jellium, but, it should be emphasized, not because the effective-medium theory is restricted to or more appropriate to such a model, but rather because reliable energies of chemisorption are only available for such simple systems. The results of these tests of the effective-medium formalism are shown in Fig. 35. It is quite remarkable that, even in the zero-order approximation, the effective medium calculations reasonably describe both the length and strength of the metal–adatom bond. When the first-order correction terms are included, the agreement between the two sets of calculations is astonishing.

It should be noted that the domain of validity of this promising technique is not firmly established. The fundamental criterion for validity may again involve the notion of closed and open atomic valence shells. In the context of the effective-medium theory, the atoms considered in Fig. 34 are closed-shell systems in the sense that when these atoms are embedded in jellium, the resonance states corresponding to the valence shell fall below the Fermi

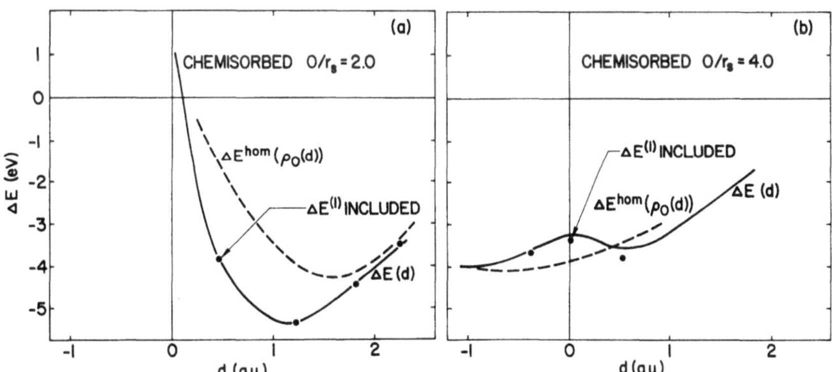

FIGURE 35. Test of zero- and first-order "effective-medium" theory for the chemisorption of atomic oxygen on simple-metal (jellium) surfaces. The horizontal coordinate *d* indicates the distance from the edge of the positive background (of the jellium model) to the adatom nucleus. Solid curves indicate the first-principles results the dots indicate, respectively, the results of the zero- and first-order "effective-medium" theory. Upper set of curves (a) describe results for a substrate density similar to that of Al; lower curves (b) correspond to Na. (From Ref. 12.)

energy and can therefore be regarded as filled or closed. When the chemisorption of atomic Si is studied using this technique, the results of the first-order theory are somewhat less accurate than those shown in Fig. 35. They differ from those of the full calculations[223] by ~ 10%. The significance of Si in this context is that its valence ($3p$) shell does not fill when the atom is embedded in jellium. The significance of the open shell is, in turn, the fact that it provides a high density of low-energy excitations which facilitate the polarization of the atom when it is embedded in a nonuniform electron density. The second-order contribution to the effective-medium result involves just such a polarizability, but this contribution remains to be tested. Even if this theory proves to be of greatest use for atoms and molecules with effectively closed shells, it should be noted that the definition of "closed shell" is significantly more general than that appropriate to the formalism of Kim and Gordon.[7,8] It should also be noted that, in the effective-medium description of chemisorption, the substrate entered the description only through its surface-density profile. The formalism therefore does not appear to be sensitive to the existence of high substrate state density characteristic of transition metals. It would indeed be a remarkable fact if a transition metal, such as tungsten, and a simple metal, such as aluminum, differed, from the point of view of chemisorption, only in their surface electron-density profiles. Studies in progress suggest that the contribution of localized d electrons can be included in the theory using only perturbation theory.

Acknowledgments

We gratefully acknowledge very helpful discussions with our colleagues Drs. N. D. Lang, C. D. Gelatt, Jr., J. F. Janak, C. O. Almbladh, and M. Scheffler. We also wish to thank the authors of the other chapters of this book and especially the editors, Professors S. Lundqvist and N. H. March for their generous patience.

References

1. O. Gunnarsson, M. Jonson, and B. I. Lundqvist, *Phys. Rev. B* **20,** 3136 (1979).
2. J. A. Appelbaum and D. R. Hamann, *Rev. Mod. Phys.* **48,** 479 (1976).
3. N. D. Lang and A. R. Williams, *Phys. Rev. B* **18,** 616 (1978).
4. G. A. Baraff and M. Schlüter, *Phys. Rev. Lett.* **41,** 892 (1979).
5. J. Bernholc, N. O. Lipari, and S. T. Pantelides, *Phys. Rev. Lett.* **41,** 895 (1979).
6. A. R. Williams, C. D. Gelatt, Jr., and V. L. Moruzzi, *Phys. Rev. Lett.* **44,** 429 (1980).
7. R. G. Gordon and Y. S. Kim, *J. Chem. Phys.* **56,** 3122 (1972).
8. M. J. Clugston, *Adv. Phys.* **27,** 893 (1978).

9. J. A. Alonso and L. A. Girifalco, *J. Phys. F* **8,** 2455(1978); *Phys. Rev. B* **19,** 3889 (1979).
10. C. H. Hodges and M. J. Stott, *Phil. Mag.* **26,** 375 (1972).
11. C. H. Hodges, *Phil. Mag.* **38,** 205 (1978).
12. J. K. Norskov and N. D. Lang, *Phys. Rev. B* **21,** 2131 (1980).
13. M. J. Stott and E. Zaremba, *Phys. Rev. B* **22,** 1564 (1980).
14. U. von Barth and L. Hedin, *J. Phys. C* **5,** 1629 (1972).
15. P. Hohenberg and W. Kohn, *Phys. Rev.* **136,** B864 (1964).
16. L. J. Sham, in: *Computational Methods in Band Theory*, P. J. Marcus, J. F. Janak, and A. R. Williams, eds., p. 458 (Plenum, New York, 1971).
17. D. J. W. Geldart, M. Rasolt, and R. Taylor, *Solid State Commun.* **10,** 279 (1972).
18. L. Kleinman, *Phys. Rev. B* **10,** 2221 (1974).
19. D. J. W. Geldart, M. Rasolt, and C. O. Almbladh, *Solid State Commun.* **16,** 243 (1975).
20. A. K. Rajagopal and S. Ray, *Phys. Rev. B* **12,** 3129 (1975).
21. A. K. Gupta and K. S. Singwi, *Phys. Rev. B* **15,** 1801 (1977).
22. A. K. Rajagopal and S. P. Singhal, *Phys. Rev. B* **16,** 601 (1977).
23. D. J. W. Geldart and M. Rasolt, *Phys. Rev. B* **13,** 1477 (1976).
24. M. Rasolt, *Phys. Rev. B* **16,** 3234 (1977).
25. F. Herman, J. P. Van Dyke, and I. B. Ortenburger, *Phys. Rev. Lett.* **22,** 807 (1969).
26. K. S. Singwi, A. Sjölander, M. P. Tosi, and R. H. Land, *Phys. Rev. B* **1,** 1044 (1970).
27. D. J. W. Geldart and R. Taylor, *Can. J. Phys.* **48,** 155–165 (1970); **48,** 167 (1970).
28. J. H. Rose, H. B. Shore, D. J. W. Geldart, and M. Rasolt, *Solid State Commun.* **19,** 619 (1976).
29. K. H. Lau and W. Kohn, *J. Phys. Chem. Solids,* **37,** 99 (1976).
30. J. S–Y. Wang and M. Rasolt, *Phys. Rev. B* **13,** 5330 (1976).
31. M. Rasolt, J. S–Y. Wang, and L. M. Kahn, *Phys. Rev. B* **15,** 580 (1977).
32. D. C. Langreth and J. P. Perdew, *Solid State Commun.* **31,** 567 (1979).
33. D. C. Langreth and J. P. Perdew, *Phys. Rev. B* **15,** 2884 (1977).
34. J. E. Inglesfield and E. Wikborg, *Solid State Commun.* **16,** 335 (1975).
35. J. P. Perdew, D. C. Langreth, and V. Sahni, *Phys. Rev. Lett.* **38,** 1030 (1977).
36. O. Gunnarsson, B. I. Lundqvist, and J. W. Wilkins, *Phys. Rev. B* **10,** 1319 (1974).
37. O. Gunnarsson and B. I. Lundqvist, *Phys. Rev. B* **13,** 4274 (1976).
38. C. O. Almbladh, Technical Report, University of Lund (1972).
39. B. Y. Tong, *Phys. Rev. A* **4,** 1375 (1971).
40. O. Gunnarsson, M. Jonson, and B. I. Lundqvist, *Phys. Lett.* **59A,** 177 (1976).
41. O. Gunnarsson, M. Jonson, and B. I. Lundqvist, *Solid State Commun.* **24,** 765 (1977).
42. J. A. Alonso and L. A. Girifalco. *Phys. Rev. B* **17,** 3735, (1978).
43. U. von Barth, unpublished notes.
44. A. K. Rajagopal, in: *Advances in Chemical Physics*, I. Prigogine and S.A. Rice, eds., Vol. 41 (Addison-Wesley, New York, 1980).
45. O. Gunnarsson and R. O. Jones, *Phys. Scripta* **21,** 394 (1980).
46. U. von Barth and C. D. Gelatt, *Phys. Rev. B* **21,** 2222 (1980).
47. V. L. Moruzzi, A. R. Williams, and J. F. Janak, *Phys. Rev. B* **15,** 2854 (1977).
48. J. C. Kimball, *Phys. Rev. B* **14,** 2371 (1976).
49. C. O. Almbladh, U. von Barth, Z. D. Popovic, and M. J. Stott, *Phys. Rev. B* **14,** 2250 (1976).
50. K. Schwarz, *Chem. Phys. Lett.* **57,** 605 (1978).
51. D. Pines, *Elementary Excitations in Solids*, p. 75 (Benjamin, New York, 1963).

52. A. K. Rajagopal, J. C. Kimball, and M. Banerjee, *Phys. Rev. B* **18,** 2339 (1978).
53. H. Stoll, C. M. E. Pavlidou, and H. Preuss, *Theoret, Chim. Acta* **49,** 143 (1978).
54. D. R. Hartree, *Proc. Cambridge Phil. Soc.* **24,** 111 (1927).
55. I. Lindgren, *Int. J. Quant. Chem.* **5,** 411 (1971).
56. I. Lindgren and A. Rosen in : *Case Studies in Atomic Physics*, M.R.C. McDowell and E.W. McDaniel, eds., Vol. 4, p. 97 (North Holland, 1974).
57. B. Y. Tong and L. J. Sham, *Phys. Rev.* **144,** 1 (1966).
58. J. P. Perdew, *Chem. Phys. Lett.* **64,** 127 (1979).
59. A. Zunger and M. L. Cohen, *Phys. Rev. B* **18,** 5449 (1978).
60. L. Hedin and B. I. Lundqvist, *J. Phys. C* **4,** 2064 (1971).
61. T. L. Gilbert, *Phys. Rev. B* **12,** 2111 (1975).
62. M. Rasolt, G. Malmström, and D. J. W. Geldart, *Phys. Rev. B* **20,** 3012 (1979).
63. M. Rasolt and D. J. W. Geldart, *Phys. Rev. B* **21,** 3158 (1980).
64. W. Kohn and W. Hanke, "Non-local Corrections in the Exchange and Correlation Energy of an Inhomogeneous Electron Gas," in CECAM Report of Workshop on Ab Initio One Electron Potentials, University of Paris, Orsay (1976).
65. W. Kohn and L. J. Sham, *Phys. Rev.* **140,** A1133 (1965).
66. A. K. Rajagopal and J. Callaway, *Phys. Rev. B* **7,** 1912 (1973).
67. J. G. Zabolitzky, *Phys. Rev. B* **22,** 2353 (1980).
68. G. Arbman and U. von Barth, *J. Phys. F* **5,** 1155 (1975).
69. M. Gell-Mann and K. A. Brueckner, *Phys. Rev.* **106,** 364 (1957).
70. E. P. Wigner, *Trans. Faraday Soc.* **34,** 678 (1938).
71. D. M. Ceperly, *Phys. Rev. B* **18,** 3126 (1978); and D. M. Ceperly and B. J. Alder, *Phys. Rev. Lett.* **45,** 566 (1980).
72. R. F. Bishop and K. H. Lührmann, *Phys. Rev. B* **17,** 3757 (1978); and to be published.
73. U. von Barth, *Phys. Scripta* **21,** 585 (1980).
74. J. Kübler, *J. Magnetism Magn. Mater.* **20,** 277 (1980).
75. H. L. Skriver, *J. Phys. F* **11,** 97 (1981).
76. S. H. Vosko, L. Wilk, and M. Nusair, *Can. J. Phys.* **58,** 1200 (1980).
77. A. K. Rajagopal, *J. Phys. C* **11,** L943 (1978).
78. A. H. MacDonald and S. H. Vosko, *J. Phys. C* **12,** 2977 (1979).
79. N. D. Mermin, *Phys. Rev.* **137,** A1441 (1965).
80. V. Peuckert, *J. Phys. C* **11,** 4945 (1978).
81. A. K. Theophilou, *J. Phys. C* **12,** 5419 (1979).
82. T. Ziegler, A. Rauk, and E. J. Baerends, *Theoret. Chim. Acta.* **43,** 261 (1977).
83. U. von Barth, *Phys. Rev. A* **20,** 1693 (1979).
84. D. G. Pettifor, *J. Magnetism Mag. Mater.* **15–18,** 847 (1980).
85. S. H. Vosko and J. P. Perdew, *Can. J. Phys.* **53,** 1385 (1975).
86. O. Gunnarsson, *J. Phys. F: Metal Phys.* **6,** 587 (1976).
87. J. F. Janak, *Phys. Rev. B* **16,** 255 (1977).
88. U. K. Poulsen, J. Kollar, and O. K. Andersen, *J. Phys. F: Metal Phys.* **6,** L241 (1976).
89. O. K. Andersen, J. Madsen, U. K. Poulsen, O. Jepsen, and J. Kollar, *Physica* **86–88B,** 249 (1977).
90. V. L. Moruzzi, J. F. Janak, and A. R. Williams, *Calculated Electronic Properties of Metals* Pergamon Press, New York, 1978).
91. C. S. Wang and J. Callaway, *Phys. Rev. B* **15,** 298 (1977); *Phys. Rev. B* **16,** 2095 (1977).

92. J. F. Janak and A. R. Williams, *Phys. Rev. B* **14,** 4199 (1976).
93. J. F. Janak, *Solid State Commun.* **25,** 53 (1978).
94. J. F. Janak, *Phys. Rev.* **20,** 2206 (1979).
95. J. Kübler, Proceedings of First Intl. Conf. on Phys. of Magnetism and Mag. Mater., Poland (1980).
96. J. Kübler, *J. Magnetism Mag. Mater.* **20,** 107 (1980).
97. N. A. Cade, *J. Phys. F* **10,** 1187 (1980).
98. D. Hackenbracht and J. Kübler, *J. Phys. F: Metal Phys.* **10,** 427 (1980).
99. R. Zeller, R. Podloucky, and P. H. Dederichs, *Z. Phys. B* **38,** 165 (1980).
100. R. Podloucky, R. Zeller, and P. H. Dederichs, *Phys. Rev. B* **22,** 5777 (1980).
101. O. Jepsen, J. Madsen, and O. K. Andersen, *J. Magnetism Mag. Mater.* **15,** 867 (1980).
102. C. S. Wang and A. J. Freeman, *Phys. Rev. B* **21,** 4585 (1980).
103. J. F. Janak, V. L. Moruzzi, and A. R. Williams, *Phys. Rev. B* **12,** 1257 (1975).
104. J. F. Janak and A. R. Williams, *Phys. Rev. B* **23,** 6301 (1981).
105. J. Harris and R. O. Jones, *J. Chem. Phys.* **68,** 3316 (1978).
106. S. H. Vosko, J. P. Perdew, and A. H. MacDonald, *Phys. Rev. Lett.* **35,** 1725 (1975).
107. E. C. Stoner, *Proc. R. Soc. London, Ser. A* **169,** 339 (1939).
108. J. C. Slater, *Phys. Rev.* **49,** 537 (1936).
109. E. P. Wohlfarth, *Rev. Mod. Phys.* **25,** 211 (1953); *J. Inst. Math. Applic.* **4,** 359 (1968).
110. O. K. Andersen, H. L. Skriver, H. Nohl, and B. Johansson, *Pure Appl. Chem.* **52,** 93 (1979).
111. H. L. Skriver, O. K. Andersen, and B. Johansson, *Phys. Rev. Lett.* **44,** 1230 (1980).
112. H. L. Skriver, O. K. Andersen, and B. Johansson, *Phys. Rev. Lett.* **41,** 42 (1978).
113. R. A. de Groot, D. D. Koelling, and F. M. Mueller, *J. Phys. F.*
114. K. J. Tauer and R. J. Weiss, *Bull, Am. Phys. Soc.* **6,** 125 (1961).
115. D. M. Roy and D. G. Pettifor, *J. Phys. F: Metal Phys.* **7,** L183 (1977).
116. T. M. Hattox, J. B. Conklin, Jr., J. C. Slater, and S. B. Trickey, *J. Phys. Chem. Solids* **34,** 1627 (1973).
117. J. C. Slater, *Quantum Theory of Molecules and Solids:* Vol. 4 (McGraw-Hill, New York, 1974).
118. J. H. Wood, *J. Phys. B* **13,** 1 (1980).
119. O. Gunnarsson and R. O. Jones, *J. Chem. Phys.* **72,** 5357 (1980).
120. L. J. Sham and W. Kohn, *Phys. Rev.* **145,** 561 (1966).
121. S. Chakravarty, M. B. Fogel, and W. Kohn, *Phys. Rev. Lett.* **43,** 775 (1979).
122. A. K. Rajagopal, *Phys. Rev. B* **17,** 2980 (1978).
123. O. Gunnarsson, B. I. Lundqvist, and S. Lundqvist, *Solid State Commun.* **11,** 149 (1972).
124. J. C. Stoddart, *J. Phys. C* **8,** 3391 (1975).
125. J. Callaway and C. S. Wang, *J. Phys. F* **5,** 2119 (1975).
126. S. P. Singhal and J. Callaway, *Phys. Rev. B* **14,** 2347 (1976).
127. D. M. Edwards and M. A. Rahman, *J. Phys. F* **8,** 1501 (1978).
128. K. L. Liu and S. H. Vosko, *J. Phys. F* **8,** 1539 (1978).
129. J. Callaway and A. K. Chatterjee, *J. Phys. F* **8,** 2569 (1978).
130. D. Pines and P. Nopiéres, *The Theory of Quantum Liquids:* Vol. I, (Benjamin, New York, 1966).
131. G. Niklasson, *Phys. Rev. B* **10,** 3052 (1974).
132. W. Hanke and L. J. Sham, *Phys. Rev. Lett.* **33,** 582 (1974).
133. A. Zangwill and P. Soven, *Phys. Rev. A* **21,** 1561 (1980).

134. M. J. Stott and E. Zaremba, *Phys. Rev. A* **21,** 12 (1980).
135. C. P. Slichter, *Principles of Magnetic Resonance* Springer-Verlag, Berlin, 1978).
136. E. Zaremba and D. Zobin, *Phys. Rev. Lett.* **44,** 175 (1980).
137. B. Goodman and A. Sjolander, *Phys. Rev. B* **8,** 200 (1973).
138. L. Hedin and S. Lundqvist, in *Solid State Physics,* F. Seitz, D. Turnbull, and H. Ehrenreich, eds., Vol. 23, p. 1 (Academic Press, New York, 1969).
139. D. M. Edwards and B. Fisher, *J. Phys. Paris* **32,** C1, 697 (1971).
140. A. C. Wahl and G. Das, *Modern Theoretical Chemistry,* Vol. 3, H. F. Schaefer, III, ed. (Plenum Press, New York, 1976).
141. A. Goddard, *Ann. Rev. Phys. Chem.* **29,** 363 (1978).
142. Gunnarsson and P. Johansson, *Int. J. Quantum Chem.* **10,** 307 (1976).
143. O. Gunnarsson, J. Harris, and R. O. Jones, *Phys. Rev. B* **15,** 3027 (1977); *J. Chem. Phys.* **67,** 3970 (1977).
144. E. J. Baerends, D. E. Ellis, and P. Roos, *Chem. Phys.* **2,** 41 (1973).
145. B. I. Dunlap, J. W. D. Connolly, and J. R. Sabin, *J. Chem. Phys.* **71,** 4993 (1979).
146. W. Kolos and C. J. Roothan, *Rev. Mod. Phys.* **32,** 219 (1960).
147. W. Kolos and L. J. Wolniewicz, *J. Chem. Phys.* **43,** 2429 (1965).
148. K. H. Johnson, J. G. Norman, and J. W. D. Connolly, in *Computational Methods for Large Molecules and Localized States in Solids,* (F. Herman, A. D. McLean, and R. K. Nesbet, eds.) (Plenum Press, New York, 1973).
149. P. E. Phillipson and R. S. Mulliken, *J. Chem. Phys.* **28,** 1248 (1958).
150. O. Gunnarsson, J. Harris, and R. O. Jones, *Int. J. Quant. Chem. Symp.* **11,** 71 (1977).
151. K. P. Huber, in *American Institute of Physics Handbook,* D. E. Gray, ed., Sec. 7g (McGraw-Hill, New York, 1972).
152. M. T. Yin and M. L. Cohen, *Phys. Rev. Lett.* **45,** 1004 (1980).
153. W. M. Huo, *J. Chem. Phys.* **43,** 624 (1965).
154. C. Chakerian, Jr., *J. Chem. Phys.* **65,** 4228 (1976).
155. O. Gunnarsson, J. Harris, and R. O. Jones, *J. Phys. C: Solid State Phys.* **9,** 2439 (1976).
156. O. K. Andersen, *Phys. Rev. B* **12,** 3060 (1975).
157. L. F. Mattheiss, J. H. Wood, and A. C. Switendick in *Methods in Computational Physics,* B. Adler, S. Fernbach, and M. Rotenberg, eds., Vol. 8 (Academic Press, New York, 1968).
158. A. R. Williams, J. Kübler, and C. D. Gelatt Jr., *Phys. Rev. B* **19,** 6094 (1979).
159. R. O. Jones, *J. Chem. Phys.* **71,** 1300 (1979).
160. J. Harris and R. O. Jones, *J. Chem. Phys.* **68,** 1190 (1978).
161. J. Harris and R. O. Jones, *Phys. Rev. A* **18,** 2159 (1978).
162. K. A. Gschneider, Jr., *Solid State Phys.* **16,** 276 (1976).
163. W. J. Balfour and A. E. Douglas, *Can. J. Phys.* **48,** 901 (1970); C. R. Vidal and H. Scheingraber, *J. Mol. Spectrosc.* **65,** 46 (1977).
164. W. J. Balfour and R. F. Whitlock, *Can. J. Phys.* **53,** 472 (1975).
165. N. D. Lang, *Phys. Rev. Lett.* **46,** 842 (1981).
166. D. G. Pettifor, *Comm. Phys.* **1,** 141 (1976).
167. D. G. Pettifor, *J. Phys. F: Metal Phys.* **8,** 219 (1978).
168. A. R. Williams, C. D. Gelatt, Jr., and J. F. Janak, in *Theory of Alloy Phase Formation,* L. H. Bennett, ed., Conference Proceedings: The Metallurgical Society of AIME (1979).
169. E. P. Wigner and F. Seitz, *Phys. Rev.* **43,** 804 (1933); **46,** 509 (1934).
170. H. Brooks, *Trans. Metal. Soc. AIME* (Am. Inst. Min. Metal. Pet. Eng.). **227,** 546 (1963).

171. J. Friedel, *The Physics of Metals*, J. M. Ziman, ed., Chap. 8 (Cambridge V.P., London,) (1969).
172. F. W. Averill, *Phys. Rev. B* **6,** 3637 (1972); J. B. Conklin, Jr., F. W. Averill, and T. M. Hattox, *J. Phys. (Paris) Suppl. C* **3,** 213 (1972); *J. Phys. Chem. Solids* **34,** 1627 (1973); S. B. Trickey, F. R. Green, Jr., and F. W. Averill, *Phys. Rev. B* **8,** 4822 (1973).
173. D. A. Liberman, *Phys. Rev. B* **3,** 2081 (1971).
174. J. F. Janak, *Phys. Rev. B* **9,** 3985 (1974).
175. U. von Barth, unpublished notes (1975).
176. A. R. Williams, C. D. Gelatt, Jr., and V. L. Moruzzi (to be published).
177. A. R. Miedema, P. F. de Chatel, and F. R. de Boer, *Physica* **100B,** 1 (1980); A. R. Miedema, R. Room, and F. R. de Boer, *J. Less-Common Metals* **41,** 283 (1975).
178. A. R. Miedema, *Philips Tech. Rev.* **36,** 217 (1976).
179. K. M. Myles, *Trans. Metall. Soc. AIME* **242,** 1523 (1968).
180. L. Brewer and P. R. Wengert, *Metall. Trans.* **44,** 83 (1973).
181. S. C. Singhal and W. L. Worrell, *Metall. Trans.* **44,** 1125 (1973).
182. D. G. Pettifor, *J. Phys. C* **3,** 367 (1970).
183. A. R. Mackintosh and O. K. Andersen, in *Electrons at the Fermi Surface*, M. Springford, ed., (Cambridge Univ. Press, Cambridge, 1980).
184. D. G. Pettifor, *Phys. Rev. Lett.* **42,** 846 (1979); *Solid State Commun.* **28,** 621 (1978).
185. J. C. Slater and J. H. Wood, *Int. J. Quantum Chem. Suppl.* **4,** 3 (1971).
186. C. O. Almbladh and U. von Barth, *Phys. Rev. B* **33,** 3307 (1976).
187. J. Janak, *Phys. Rev. B* **18,** 7165 (1978).
188. A. R. Williams and R. A. deGroot, *J. Chem. Phys.* **63,** 628 (1975).
189. A. R. Williams and N. D. Lang, *Phys. Rev. Lett.* **40,** 954 (1978).
190. A. Zunger, J. P. Perdew, and G. L. Oliver, *Solid State Commun.* **34,** 933 (1980).
191. J. C. Wheatley, *Rev. Mod. Phys.* **47,** 415 (1975).
192. J. E. Demuth, P. M. Marcus, and D. W. Jepsen, *Phys. Rev. B* **11,** 1460 (1975).
193. P. Lee and G. Beni, *Phys. Rev. B* **15,** 2862 (1977).
194. N. D. Lang and A. R. Williams, *Phys. Rev. B* **20,** 1369 (1979).
195. D. G. Pettifor, *J. Phys. C* **3,** 367 (1970).
196. J. F. Janak, A. R. Williams and V. L. Moruzzi, *Phys. Rev. B* **11,** 1522 (1975).
197. F. J. Himpsel, D. E. Eastman, E. E. Koch, and A. R. Williams, *Phys. Rev.* **22,** 4604 (1980).
198. M. I. Chodorow, Ph. D. thesis (MIT, 1939), unpublished.
199. C. D. Gelatt, Jr., H. Ehrenreich, and R. E. Watson, *Phys. Rev.* **15,** 1613 (1977).
200. P. Thiry, D. Chandesris, J. Leconte, C. Guillot, R. Pinchaux, and Y. Petroff, *Phys. Rev. Lett.* **43,** 82 (1979).
201. G. A. Burdick, *Phys. Rev.* **129,** 138 (1963).
202. C. Guillot, Y. Ballu, J. Paigne, J. Lecante, K. P. Join, P. Thirty, R. Pinchaux, Y. Petroff, and L. M. Falicov, *Phys. Rev. Lett.* **19,** 1632 (1977).
203. D. R. Penn, *Phys. Rev. Lett.* **42,** 921 (1979); *J. Appl. Phys.* **50,** 7480 (1979).
204. A. Liebsch, *Phys. Rev. Lett.* **43,** 1431 (1979).
205. G. Treglia, F. Ducastelle, and D. Spanjaard, *Phys. Rev. B* **21,** 3729 (1980).
206. D. M. Edwards, *Inst. Phys. Conf. Ser.* **39,** 279 (1978).
207. L. C. Davis and L. A. Feldkamp, *J. Appl. Phys.* **50,** 1944 (1979).
208. A. H. MacDonald, *J. Phys. F* **10,** 1737 (1980).
209. L. Hedin, *Arkiv. Fysik* **30,** 231 (1965).
210. W. Brinkman and B. Goodman, *Phys. Rev.* **149,** 597 (1966).
211. A. B. Kunz, *Phys. Rev. B* **6,** 606 (1972).

212. J. C. Inkson, *J. Phys. C* **6,** L181 (1973).
213. L. Dagens and F. Perrot, *Phys. Rev. B* **8,** 1281 (1973).
214. E. O. Kane, in *Proceedings of the Twelfth International Conference on the Physics of Semiconductors,* M. H. Pilkuhn, ed., p. 169 (B. G. Teubner, Stuttgart, 1974).
215. G. Strinati, H. J. Mattausch, and W. Hanke, *Phys. Rev. Lett.* **45,** 290 (1980).
216. R. E. Watson, J. F. Herbst, L. Hodges, B. I. Lundqvist, and J. W. Wilkins, *Phys. Rev. B* **13,** 1463 (1976).
217. M. Rasolt and S. H. Vosko, *Phys. Rev. B* **10,** 4195 (1974).
218. S. B. Nickerson and S. H. Vosko, *Phys. Rev. B* **14,** 4399 (1976).
219. J. S–Y. Wang and M. Rasolt, *Phys. Rev. B* **15,** 3714 (1977).
220. M. Rasolt, S. B. Nickerson, and S. H. Vosko, *Solid State Commun.* **16,** 827 (1975).
221. J. F. Janak, A. R. Williams, and V. L. Moruzzi, *Phys. Rev. B* **6,** 4367 (1972).
222. L. L. Boyer, *Phys. Rev. Lett.* **45,** 1858 (1980).
223. N. D. Lang and A. R. Williams, *Phys. Rev. Lett.* **37,** 212 (1976).

CHAPTER 5

DENSITY FUNCTIONAL APPROACH TO THE ELECTRONIC STRUCTURE OF METAL SURFACES AND METAL–ADSORBATE SYSTEMS

N. D. LANG

1. Introduction

We consider in this chapter the use of the density functional theory to study the electronic structure of metal surfaces and atoms chemisorbed on these surfaces. We refer to the review of the density functional formalism by Kohn and Vashishta[1] in Chapter 2 of this book, and mention in addition the articles by Hedin and Lundqvist,[2] March[3] and Rajagopal.[4] There is a variety of other theoretical reviews[5–21] on the electronic structure problems at metal surfaces, many emphasizing points of view different from that of the present article.

We refer to Chapter 2 for the presentation of the general results of the density functional approach.[22,23] In the wave-mechanical formulation of this approach, discussed fully in Chapter 2, a local potential v_{eff} is defined as the sum of the total electrostatic potential and the exchange-correlation potential. The density distribution $n(\mathbf{r})$ can then be determined by solving the set of one-particle equations

N. D. LANG • IBM Thomas J. Watson Research Center, Yorktown Heights, New York 10598.

$$\{-\tfrac{1}{2}\nabla^2 + v_{\text{eff}}[n;\mathbf{r}]\}\Psi_i(\mathbf{r}) = \epsilon_i\Psi_i(\mathbf{r}) \tag{1a}$$

together with

$$n(\mathbf{r}) = \sum_i |\Psi_i(\mathbf{r})|^2 n_i \tag{1b}$$

where the Ψ_i are orthonormal solutions of equation (1a) and n_i is 1 for the N lowest-lying solutions and zero otherwise. The total energy is given by

$$\begin{aligned}\tilde{E}[n] = \sum_i \epsilon_i n_i - \int v_{\text{eff}}[n;\mathbf{r}]n(\mathbf{r})d\mathbf{r} + \int v(\mathbf{r})n(\mathbf{r})d\mathbf{r} \\ + \frac{1}{2}\int \frac{n(\mathbf{r})n(\mathbf{r}')}{|\mathbf{r}-\mathbf{r}'|} d\mathbf{r}\, d\mathbf{r}' + E_{xc}[n]\end{aligned} \tag{2}$$

where as usual $E_{xc}[n]$ denotes the exchange and correlation energy functional. [This does not include the self-energy of the static charge distribution which gives rise to $v(\mathbf{r})$, e.g., the lattice self-energy; when this self-energy is included, we will write E instead of $\tilde{E}$.] Approximations to v_{eff} can be obtained by expanding $E_{xc}[n]$ in a series of gradients of the density, and dropping certain of the terms. The simplest ("local-density") approximation consists in keeping only the first (nongradient) term:

$$E_{xc}[n] \doteq \int g_{xc}(n(\mathbf{r}))d\mathbf{r} \tag{3}$$

where $g_{xc}(n)$ is the exchange-correlation part of the energy density of a homogeneous electron gas of density n. The wave-mechanical approach treats the single-particle kinetic-energy part $T_s[n]$ exactly, in contradistinction to the Thomas–Fermi method dealt with in Chapter 1.

Sometimes it is convenient for purposes of discussion to study the density of eigenstates in equation (1a). In this connection, we define a local density of eigenstates as

$$n(\epsilon,\mathbf{r}) = \sum_i |\Psi_i(\mathbf{r})|^2 \delta(\epsilon - \epsilon_i) \tag{4}$$

and a total density of eigenstates as

$$n(\epsilon) = \int n(\epsilon,\mathbf{r})d\mathbf{r} = \sum_i \delta(\epsilon - \epsilon_i) \tag{5}$$

1.1. Work Function

Consider a crystal with several different macroscopic faces labeled by the index i. The work function Φ_i of face i is the minimum energy required to remove an electron from the crystal to any point $\mathbf{r}_i$ which is a distance outside of face i small compared with the face dimensions but large compared with the lattice spacing (and is at a macroscopic distance from the face edges). There are electric fields extending well outside the crystal such that moving an electron from $\mathbf{r}_i$ to $\mathbf{r}_j$ requires an amount of work $\Phi_j - \Phi_i$. The work function of a metal is therefore given by

$$\Phi_i = [\phi(\mathbf{r}_i) + E_{N-1}] - E_N \tag{6}$$

with E_N the ground-state energy of the neutral N-electron crystal, E_{N-1} the ground-state energy of the singly ionized crystal, and $\phi(\mathbf{r}_i)$ the electrostatic potential at $\mathbf{r}_i$. Since the chemical potential μ is $E_N - E_{N-1}$,

$$\Phi_i = \phi(\mathbf{r}_i) - \mu \tag{7}$$

The two terms in equation (7) are each referred to the arbitrary zero of potential; it is convenient to choose this reference to be the mean electrostatic potential in the metal,

$$\bar{\phi} \equiv \Omega^{-1} \int_{\text{metal}} \phi(\mathbf{r}) d\mathbf{r} \tag{8}$$

where Ω is the volume of the metal. Thus we write

$$\bar{D}_i \equiv \phi(\mathbf{r}_i) - \bar{\phi} \tag{9a}$$

$$\bar{\mu} \equiv \mu - \bar{\phi} \tag{9b}$$

so that

$$\Phi_i = \bar{D}_i - \bar{\mu} \tag{10}$$

If, as in Chapter 2, equation (42), we denote by $G[n]$ the sum of the single-particle kinetic energy $T_s[n]$ and the exchange and correlation energy $E_{xc}[n]$, then we can average both sides of equation (45), Chapter 2, over the volume of the metal, and we find that

$$\bar{\mu} = \mu - \bar{\phi} = \Omega^{-1} \int_{\text{metal}} \frac{\delta G[n]}{\delta n(\mathbf{r})} \, d\mathbf{r} \tag{11}$$

We see therefore that $\bar{\mu}$ is a bulk property of the metal. It is the bulk chemical potential, relative to the mean electrostatic potential in the metal; its independence of this potential follows from the definition of $G[n]$ as the total nonelectrostatic energy. Equation (10) thus divides the work function into bulk $(-\bar{\mu})$ and surface $(\bar{D}_i)$ components.*[27,28]

All many-body effects are contained in the exchange and correlation contributions to $\bar{\mu}$ and in their effect on the electrostatic surface barrier $\bar{D}_i$. In particular, the image force effect on Φ_i may be regarded as contained in the disappearance of part of the correlation energy when one electron is moved away from the surface.

Often it is convenient to use a reference other than the mean potential $\bar{\phi}$ for the two terms in equation (10). In a Wigner–Seitz calculation,[26] for example, a natural reference is the electrostatic potential at the Wigner–Seitz sphere, ϕ^{WS}. Thus we would write†

$$D_i^{\text{WS}} \equiv \phi(\mathbf{r}_i) - \phi^{\text{WS}} \tag{12a}$$

$$\mu^{\text{WS}} \equiv \mu - \phi^{\text{WS}} \tag{12b}$$

and hence

$$\Phi_i = D_i^{\text{WS}} - \mu^{\text{WS}} \tag{13}$$

A form of equation (7) that is particularly convenient in wave-mechanical calculations has been derived by Schulte.[30] He shows that $\mu = \epsilon_F$, the highest occupied eigenvalue of equation (1a).[31] Thus

$$\Phi_i = \phi(\mathbf{r}_i) - \epsilon_F \tag{14}$$

This formula corresponds to the intuitive definition of the work function as the energy required to remove an electron from the Fermi level to the vacuum level. Just as before, we can introduce a reference such as ϕ^{WS}.

*Starting with a similar equation, Tong,[24] Hodges and Stott,[25] and Nieminen and Hodges[26] have discussed the work function for positrons in metals.

†Care must be taken in reviewing the literature not to compare values of D_i defined using different references. See the discussion by Lang and Kohn.[29]

1.2. *Surface Energy*

The surface energy σ_i of a crystal face i is the work required, per unit area of new surface formed, to split the crystal in two along the ith crystal plane. We consider a macroscopic crystal, and take the two fragments into which the crystal is split to be identical. Let A_i be the area of the newly exposed face on each fragment, E_i the total energy of each fragment, and E the total energy of the unsplit crystal. Then

$$\sigma_i = (2A_i)^{-1}(2E_i - E) \tag{15}$$

Note that the energies E differ from the energies $\tilde{E}$ discussed above in that they include the electrostatic self-energy of the lattice of nuclei.*

1.3. *Linear Response*

In this section we discuss the response of a metal surface to a small static perturbation. We take this perturbation to consist of a distribution of positive charge of (number) density $\delta n_+(\mathbf{r})$, which causes a change in electron (number) density $\delta n(\mathbf{r})$ and a possible shift in chemical potential $\delta\mu$. Specific results for $\delta n(\mathbf{r})$ will be presented later; here we simply obtain the general expression for the change δE in total energy that occurs as a result of the metal-perturbing charge interaction.

Let

$$\delta v(\mathbf{r}) = -\int \frac{\delta n_+(\mathbf{r}')}{|\mathbf{r}-\mathbf{r}'|} d\mathbf{r}' \tag{16}$$

(we always define electrostatic potentials to be those seen by a negative test charge), and

$$\delta\phi(\mathbf{r}) = \delta v(\mathbf{r}) + \int \frac{\delta n(\mathbf{r}')}{|\mathbf{r}-\mathbf{r}'|} d\mathbf{r}' \tag{17}$$

Then δE is given by

$$\delta E = -\int v(\mathbf{r})\delta n_+(\mathbf{r})d\mathbf{r} + \tilde{E}_{v+\delta v}[n + \delta n] - \tilde{E}_v[n] \tag{18}$$

*We omit the contribution to the surface energy from changes of the zero-point lattice vibrations.

Using the explicit form for $\tilde{E}[n]$, namely,

$$\tilde{E}\,[n] = \int v(\mathbf{r})n(\mathbf{r})\,d\mathbf{r} + \frac{1}{2}\int \frac{n(\mathbf{r})n(\mathbf{r}')}{|\mathbf{r}-\mathbf{r}'|}\,d\mathbf{r}\,d\mathbf{r}' + G[n] \tag{19}$$

we can write equation (18) to second order as

$$\begin{aligned}\delta E = &\int \delta v(\mathbf{r})\delta n(\mathbf{r})d\mathbf{r} + \frac{1}{2}\int \frac{\delta n(\mathbf{r})\delta n(\mathbf{r}')}{|\mathbf{r}-\mathbf{r}'|}\,d\mathbf{r}\,d\mathbf{r}' \\ &+ \int \phi(\mathbf{r})[\delta n(\mathbf{r}) - \delta n_{+}(\mathbf{r})]d\mathbf{r} \\ &+ \int \frac{\delta G[n]}{\delta n(\mathbf{r})}\delta n(\mathbf{r})d\mathbf{r} + \frac{1}{2}\int \frac{\delta^2 G[n]}{\delta n(\mathbf{r})\delta n(\mathbf{r}')}\delta n(\mathbf{r})\delta n(\mathbf{r}')d\mathbf{r}\,d\mathbf{r}'\end{aligned} \tag{20}$$

The first three terms in equation (20) are the electrostatic components of the energy change. From equation (45), Chapter 2, we can readily show that the fourth and fifth (nonelectrostatic) terms are equal, respectively, to

$$-\int \phi(\mathbf{r})\delta n(\mathbf{r})d\mathbf{r} + \mu\delta N$$

and

$$-\frac{1}{2}\int \delta\phi(\mathbf{r})\delta n(\mathbf{r})d\mathbf{r} + \frac{1}{2}\delta\mu\delta N$$

where $\delta N \equiv \int \delta n(\mathbf{r})d\mathbf{r}$. Thus we may write (to second order)

$$\delta E = \tfrac{1}{2}\int \delta v(\mathbf{r})\delta n(\mathbf{r})d\mathbf{r} - \int \phi(\mathbf{r})\delta n_{+}(\mathbf{r})d\mathbf{r} + (\mu + \tfrac{1}{2}\delta\mu)\delta N \tag{21}$$

Note the appearance of the factor ½ in the first term and the disappearance of the terms in equation (20) corresponding to the electrostatic interaction of the screening charge distribution $\delta n(\mathbf{r})$ with itself and with the unperturbed metal. The second term in equation (21) is simply the electrostatic energy of interaction between the perturbing charge distribution and the unperturbed metal. Note that if the perturbing charge is spatially localized, $\delta\mu$ will vanish, since such a perturbation will not change the chemical potential of a macroscopic system. [The derivation given here of equation (21) is clearly not restricted to the case of a surface.] For a point perturbing charge far from the surface, the first term in equation (21) has the image-potential form (see Section 2.3.3d).

2. Uniform-Background Model of Metal Surface

2.1. General Discussion

We turn now to consider the uniform-background (jellium) model of a metal surface. This is the surface analog of a model that is widely used in the study of the bulk properties of simple (s–p bonded) metals. We discuss the case of a semi-infinite crystal, imagining that the charge on the ionic lattice has been smeared out into a homogeneous positive background of density $\bar{n}$ that terminates abruptly at a plane. This charge density [which gives rise to the external potential $v(\mathbf{r})$ discussed earlier] is thus taken to have the form

$$n_+(z) = \begin{cases} \bar{n}, & z \leq 0 \\ 0, & z > 0 \end{cases} \tag{22}$$

The density $\bar{n}$ is the mean density of positive charge in the ionic lattice; a convenient measure of this density is the radius r_s defined by $(4/3)\pi r_s^3 \equiv \bar{n}^{-1}$ (r_s ranges from about 2 to 6 bohrs for simple metals). This is probably the most elementary model of a metal surface that yields quantitatively accurate information on such basic properties as the work function.

The electron density distribution $n(z)$ that is obtained by solving equation (1) self-consistently for the uniform-background model[32,33] has the form shown in Fig. 1. Note the Friedel oscillations in the density already referred to in Chapter 1. These are characteristic of any disturbance in an electron gas (in this case the surface acts as the disturbance), and arise as a consequence of the sharpness of the Fermi surface (not of the positive background). The electric double layer formed by the spreading out of the electron distribution past the edge of the positive background means that in general the electrostatic potential in the vacuum [$\phi(\infty)$] will be higher than it is in the metal interior [$\phi(-\infty)$]. Thus, an electron trying to leave the metal encounters an electrostatic barrier of height

$$D = \phi(\infty) - \phi(-\infty) = 4\pi \int_{-\infty}^{\infty} z[n(z) - n_+(z)]dz \tag{23}$$

Since $\bar{\phi} = \phi(-\infty)$ for this case, this is just the quantity defined in equation (9a). By equations (10) and (11), the work function is given by

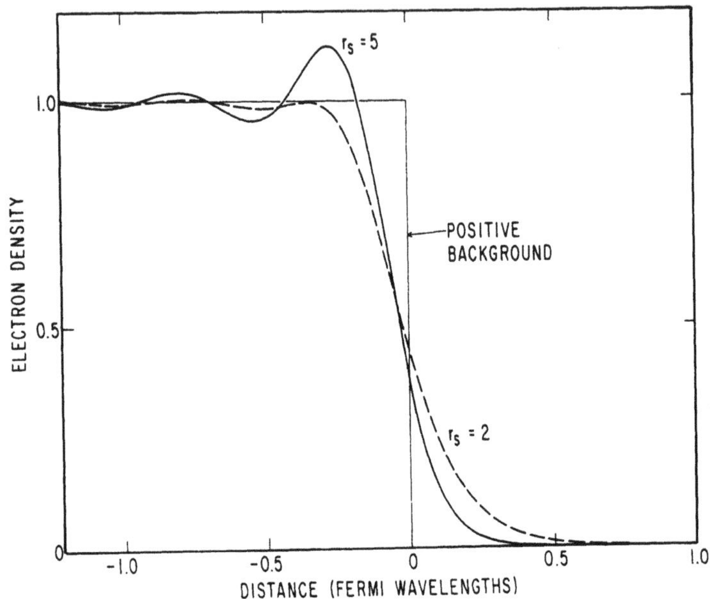

FIGURE 1. Electron density in surface region of uniform-background model. One Fermi wavelength is equal to $2\pi/k_F$. (From Ref. 32.)

$$\Phi = D - \bar{\mu} \tag{24}$$

with

$$\bar{\mu} = \frac{d\bar{n}\epsilon(\bar{n})}{d\bar{n}} \tag{25}$$

here $\epsilon(\bar{n})$ is the average energy per particle of a uniform electron gas of density $\bar{n}$.

Often the entire potential v_{eff} in equation (1a), and not just its electrostatic component, is described as the surface barrier potential. This reflects the fact that the formation of an exchange-correlation hole about an electron in the metal interior lowers the energy of the electron relative to the vacuum level. It is only the electrostatic component of the total surface barrier height, however, that depends on the details of the surface region.

2.2. Two Sum Rules

We discuss in this section two sum rules derived by Budd and Vannimenus[34,35] for the uniform-background model of a surface: one for the surface potential, the other for the rate of change of surface energy with bulk density. Several other sum rules relating to linear-response properties and to the surface phase shift are discussed in later sections.

Consider a slab of positive charge of density $\bar{n}$ extending from $z = -L$ to $z = 0$, with the faces normal to the z axis each having area A, to which $N = AL\bar{n}$ electrons have been added, making a neutral system. We take periodic boundary conditions for the faces other than the two normal to the z axis.

We now consider the change in energy associated with stretching the slab so that it extends from $z = -L$ to $z = \delta L$, while holding the areas A and the total charge of the slab constant. We can view the stretching simply as the addition of a perturbation which, within the range $-L < z < \delta L$ and laterally within the slab, has the form

$$\delta n_+(z) = \begin{cases} -\delta\bar{n}, & -L < z < 0 \\ \bar{n} - \delta\bar{n}, & 0 < z < \delta L \end{cases} \tag{26}$$

with $L\delta\bar{n} = (\bar{n} - \delta\bar{n})\delta L$. Using equation (21) and retaining only terms of first order shows that the change in energy is given by

$$\begin{aligned} \delta E &= -A\int_{-L}^{\delta L} [\phi(z) - \phi(-\tfrac{1}{2}L)]\delta n_+(z)dz \\ &= -AL\delta\bar{n}\left\{[\phi(0) - \phi(-\tfrac{1}{2}L)] - \frac{2}{L}\int_{-L/2}^{0} [\phi(z) - \phi(-\tfrac{1}{2}L)]dz\right\} \end{aligned} \tag{27}$$

where we introduce a potential reference in the center of the slab.

Now the total slab energy can be expressed (for large N and A) as

$$E = N\epsilon(\bar{n}) + 2A\sigma \tag{28}$$

with σ the surface energy. The energy change due to the stretching of the slab can therefore be written

$$\begin{aligned} \delta E &= -N\frac{d\epsilon}{d\bar{n}}\,\delta\bar{n} - 2A\frac{d\sigma}{d\bar{n}}\,\delta\bar{n} \\ &= -AL\delta\bar{n}\left\{\bar{n}\frac{d\epsilon}{d\bar{n}} + \frac{2}{L}\frac{d\sigma}{d\bar{n}}\right\} \end{aligned} \tag{29}$$

We now equate the two expressions for δE. If we assume* that $\phi(z)$ tends toward its value in the slab interior faster than $1/z$, and that as $L \to \infty$, $\phi(-\tfrac{1}{2}L)$

*See the discussion in the Appendix of Ref. 35.

tends toward its asymptotic value [which we will denote $\phi(-\infty)$] faster than $1/L$, then we see that for the limiting case of the semi-infinite uniform-background model,

$$\phi(0)-\phi(-\infty) = \bar{n}\frac{d\epsilon(\bar{n})}{d\bar{n}} \tag{30a}$$

$$\frac{d\sigma(\bar{n})}{d\bar{n}} = \int_{-\infty}^{0} [\phi(-\infty)-\phi(z)]dz \tag{30b}$$

Our derivation of these sum rules of Budd and Vannimenus follows the derivation given in Ref. 35 with a few changes in notation. Note that equation (30a) provides an exact relation between surface and bulk properties.*

If we combine equation (30a) with equations (23)–(25), we see immediately that

$$\Phi = \phi(\infty)-\phi(0)-\epsilon(\bar{n}) \tag{31}$$

This form for the work function has been discussed by Mahan and Schaich[41]; the proof they gave for it is specific to the local-density approximation, and assumes a particular form for the surface charge deficiency present after an electron has been removed from the metal. Note that it is the average energy per particle that appears here, in contrast to the maximum energy (i.e., the bulk Fermi energy) appearing in equation (24).

2.3. Solutions

2.3.1. Thomas–Fermi. As discussed in Chapter 1, the Thomas–Fermi approximation[42] consists of taking $G[n]$ in equation (19), e.g. to be given by

$$G[n] \doteq \int g_s(n(\mathbf{r}))d\mathbf{r} \tag{32}$$

where $g_s(n)$ is the kinetic energy density of a noninteracting electron gas of density n; the resulting equation is then solved simultaneously with Poisson's equation. By evaluating equation 45 in Chapter 2 directly at $z = \pm\infty$, we find using equations (24) and (25) [note $g(n) = n\epsilon(n)$] that the Thomas–Fermi work function vanishes for all r_s:

*We also mention several papers on analogous relations for bimetallic interfaces,[36–38] as well as the derivation of a relation between the kinetic and interaction parts of the surface energy (virial theorem).[39,40]

$$\Phi_{TF} = 0, \qquad \forall r_s \tag{33}$$

Solving the Thomas–Fermi equation for the uniform-background model[43] shows that

$$n(z) \propto z^{-6} \qquad (z \to \infty) \tag{34a}$$

$$1 - n(z)/\bar{n} \propto e^{z/\lambda_{TF}} \qquad (z \to -\infty) \tag{34b}$$

with λ_{TF} the Thomas–Fermi length $[0.640\sqrt{r_s}]$.The Thomas–Fermi approximation, as expected, yields a density distribution inside the metal without Friedel oscillations. In the vacuum, the density shows a power-law decrease, just as in the Thomas–Fermi atom treated in Chapter 1,[44] while the true density at a surface is expected to decay exponentially.

The inclusion of exchange and correlation in equation (32) (still omitting terms in gradients of the density, however) yields a density distribution that has the peculiar property of dropping discontinuously to zero at a certain distance from the surface[45–47] (just as in the corresponding treatment of atoms[48]). The work function for this case again has a value independent of r_s[45,46] (in the range for which the density discontinuity occurs outside the positive-background region[46]).*

The surface energy for the uniform-background model in the Thomas–Fermi approximation is found to be[21,43]

$$\sigma = -0.0763 r_s^{-9/2} \tag{35}$$

The fact that σ is negative (i.e., that the crystal cleaves spontaneously) reflects a deficiency not only of the Thomas–Fermi approximation, but of the use of a model without a discrete ionic lattice, as will be seen below.

2.3.2. Extended Thomas–Fermi. We saw above that omitting all gradient terms from the density-gradient series for $G[n]$ precluded obtaining a good account of surface electronic structure. Smith[49] considered the question of whether this defect would be remedied when at least some gradient terms were included. He retained the first gradient term in the series and used the simplest approximation for its coefficient, in which only kinetic-energy effects are included. Thus he took

$$G[n] \doteq \int \left[g(n(\mathbf{r})) + \frac{1}{72 n(\mathbf{r})} |\nabla n(\mathbf{r})|^2 \right] d\mathbf{r} \tag{36}$$

*For example, if only exchange is included (Thomas–Fermi–Dirac), this value is 1.3 eV.

where $g(n)$ is the energy density for the homogeneous electron gas of density n, including exchange and correlation. He parametrized the electron density as

$$n(z) = \begin{cases} \bar{n} - \frac{1}{2}\bar{n}e^{\beta z}, & z < 0 \\ \frac{1}{2}\bar{n}e^{-\beta z}, & z > 0 \end{cases} \tag{37}$$

and minimized the expression for the total energy $E[n]$ with respect to β. [The resulting surface energy is negative for $r_s < 2.6$, a shortcoming which is only remedied when discrete lattice effects are included.]

The work function Φ can be computed from equation (24), with $D = 4\pi\bar{n}\beta^{-2}$. The results showed approximate agreement with experimental work functions for a large number of metals. While exchange-correlation effects were found to be important in all cases, ordinary electrostatic effects were seen to be quite strong at higher densities. This success indicated the utility of the density-functional formalism in the study of surface electronic structure.*

2.3.3. Wave-Mechanical. A wave-mechanical treatment is most directly effected by self-consistently solving the set of equations (1). For the uniform-background model of a surface, the potential v_{eff} in equation (1a) is a function only of z, and is constant everywhere in the metal except the surface region. The eigenfunctions of this equation can therefore be written

$$\Psi_{\mathbf{k}}(\mathbf{r}) = \Psi_{k_z}(z)e^{i(k_x x + k_y y)} \tag{38}$$

where for $z \to -\infty$,

$$\Psi_{k_z}(z) = \sin[k_z z - \gamma(k_z)] \tag{39}$$

with γ determined uniquely by the requirements that $\gamma(0) = 0$ and $\gamma(k_z)$ be continuous. (The origin for z is taken at the positive background edge, as before, in defining γ.)

Langreth[65] has presented a proof of the fact that the charge neutrality of the system implies the phase-shift sum rule†

*We note a number of later calculations of the extended-Thomas–Fermi type for surfaces, both spin-unpolarized[50–52] and spin-polarized[53] interfaces or interacting surfaces,[54–58] cavities in metals,[59] and electron–hole droplets.[60–64] Ma and Sahni[52] explore in particular the consequences of including the second gradient term in the kinetic energy for the surface calculation.

†See also the papers of Appelbaum and Blount,[66] Paasch and Wonn,[67] and Inglesfield.[68]

$$\frac{2}{k_F^2}\int_0^{k_F} k\gamma(k)dk = \frac{\pi}{4} \tag{40}$$

due originally to Sugiyama,[69] where $k_F \equiv (3\pi^2\bar{n})^{1/3}$. In analogy with equation (28) (written for a slab with two faces each of area A and with periodic boundary conditions for the remaining faces), we can define a surface density of eigenstates $\eta^s(\epsilon)$ by writing

$$n(\epsilon) = V\eta^b(\epsilon) + 2A\eta^s(\epsilon) \tag{41}$$

where $n(\epsilon)$ is the total eigenstate density for the crystal [cf. equation (5)], $\eta^b(\epsilon)$ is the bulk eigenstate density per unit volume, and V is the positive background volume. By generalizing a result in Ref. 21, Kenner and Allen[70] have shown that for the semi-infinite case,

$$\eta^s(\epsilon) = \frac{1}{\pi^2}\left\{\gamma\left(\sqrt{2\epsilon}\right) - \frac{\pi}{4}\right\} \tag{42}$$

where the bottom of the metal band is at $\epsilon = 0$ (i.e., the electrostatic potential reference is chosen so that $v_{\text{eff}}[n; -\infty] = 0$). Thus the phase shift sum rule can be written as[67]

$$\int_0^{\epsilon_F} \eta^s(\epsilon)d\epsilon = 0 \tag{43}$$

this form exhibits its equivalence to the charge neutrality condition ($\epsilon_F = \frac{1}{2}k_F^2$ here).

Lang and Kohn[32,33] have given details of the self-consistent solution of equations (1) for the uniform-background surface model,* using the local-density approximation for exchange and correlation. The local-density ap-

*We note a number of other calculations of this general class, including the very early calculation of Bardeen,[71] for surfaces, both spin-unpolarized[71–77] and spin-polarized,[78] interfaces or interacting surfaces,[79–82] cavities in metals[83–85] (see also Ref. 86), and electron–hole droplets.[87,88] The calculations of Sahni and co-workers[76] for surfaces have been done largely analytically by using simple parametric forms (such as a step or a truncated linear potential) for the surface barrier in conjunction with theoretical constraints such as equation (30a). These studies lead to results for surface properties close to those of fully self-consistent calculations. (If the surface potential parameters are varied, this analysis permits study of the variation of the surface properties as the density profile is changed from slowly varying to rapidly varying.) van Himbergen and Silbey[77] have also given analytic solutions for a variety of surface properties in the square-barrier potential.

proximation has given good results in the study of atoms, molecules, and solids,[89–92] as discussed in the previous chapter. Its adequacy in the treatment of metal surfaces is considered further in the sections below, particularly in the section on surface energy.

(*a*) *Charge Density and Potentials.* The charge density obtained by Lang and Kohn is shown in Fig. 1. Deep in the metal, the density has the Friedel oscillation form

$$n(z) = \bar{n} + \frac{A \cos(2 k_F z + \alpha)}{z^2} + O\left(\frac{1}{Z^3}\right) \tag{44}$$

a result which can be obtained by inserting the form for the wave function given in equations (38) and (39) into equation (1b). For low bulk densities ($r_s \sim 5$), the Friedel oscillations are large, but for high bulk densities ($r_s \sim 2$), they are greatly reduced, and the density begins to resemble the monotonically decreasing Thomas–Fermi form, except in the tail region. There, the wave-mechanical density decreases exponentially,* in contrast to the z^{-6} decrease of the Thomas–Fermi solution.

Figure 2 gives an example of the computed potentials. Note that in this case ($r_s = 4$), the electrostatic potential barrier is only a small part of the total effective barrier (as given by v_{eff}). This is generally true for low and intermediate bulk densities: the exchange-correlation part of the surface barrier is of major importance.†[71]

When one of the metal electrons passes out through the surface region, its exchange-correlation hole stays behind, flattening out on the surface and assuming an image-charge form.[95–98] This clearly nonlocal effect is not reproduced by the local-density approximation, according to which the exchange-correlation potential has an exponential, rather than an image, form. This circumstance, however, mainly affects the details of the electron density well out in the vacuum tail of the distribution and is of little importance[33,98] to the calculation of quantities such as the work function. Note also that the total height of the exchange-correlation part of the potential barrier is a bulk property [it is equal to $g_{xc}'(\bar{n})$].

(*b*) *Work Function.* The work function associated with the density distri-

*For a simple step potential barrier, $n(z) \propto z^{-2} \exp[-2z\sqrt{(2\Phi)}$ for $z \to \infty$, as shown by Gupta and Singwi.[93] (Φ is nonzero in the wave-mechanical calculation.)

†This does not, however, mean that if exchange and correlation are omitted the electrons will no longer be confined at the surface. In this case, the surface electron distribution can be expected to spread out further, but this raises the electrostatic potential barrier in a self-consistent way (cf. Ref. 71). Peuckert[94] indicates that the Hartree work function is small but positive (high-density limit).

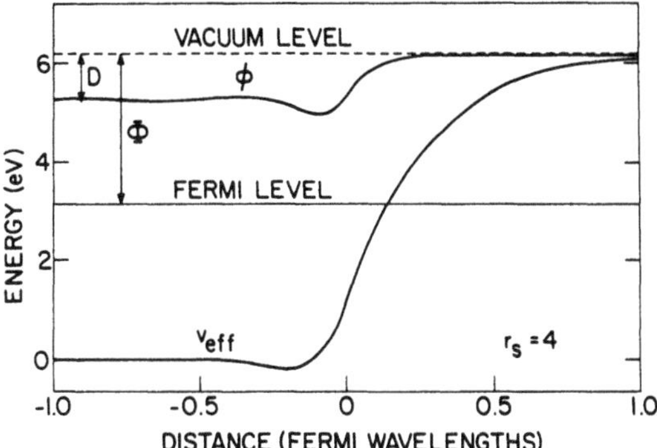

FIGURE 2. Effective one-electron potential v_{eff}, with electrostatic part ϕ, in surface region of uniform-background model ($r_s = 4$). One Fermi wavelength is equal to $2\pi/k_F$. (Data from Ref. 33.)

butions discussed above is obtained by adding, according to equation (24), the electrostatic barrier height (the surface part of Φ) and the negative of the bulk chemical potential (the bulk part of Φ). The computed work function[28] and its two components are shown as a function of r_s in Fig. 3. As r_s decreases, Φ reaches a maximum and begins to decrease. The components of Φ diverge as $r_s \to 0$, but Φ itself, according to Peuckert,[94] tends to the finite limit of ~1.3 eV. This latter type of behavior was seen in the Thomas–Fermi case (with Φ

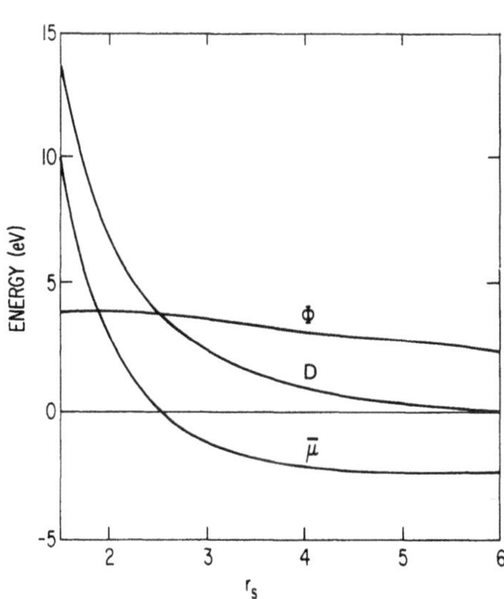

FIGURE 3. Components of the work function in the uniform-background model. $\Phi = D - \bar{\mu}$. (From Ref. 143.)

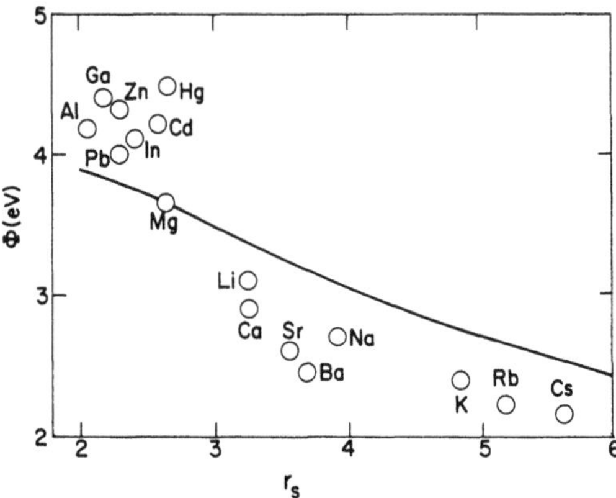

FIGURE 4. Solid line: work function for the uniform-background model. Circles: measured work function for polycrystalline samples. (From Ref. 21.)

$\equiv 0$ for all r_s). We remind the reader of the arbitrary character of the split-up of Φ into components (see Section 1.1).

Figure 4 compares the $\Phi(r_s)$ computed by Lang and Kohn[28] with experimentally measured* work functions for polycrystalline simple metals. (The uniform-background model is not intended to describe transition or noble metals.) The agreement between theory and experiment is quite reasonable, and is improved when discrete lattice effects are taken into account using perturbation theory (see Section 3.1).

(*c*) *Surface Energy*. It will be recalled that the total energy was divided into electrostatic and nonelectrostatic parts, with the latter, denoted $G[n]$, further divided into $T_s[n]$ (noninteracting kinetic) and $E_{xc}[n]$ (exchange-correlation). This division leads to a corresponding division in the surface energy. For the uniform-background model, it yields†

$$\sigma_{es} = \frac{1}{2}\int_{-\infty}^{\infty} \phi(z)[n(z) - n_+(z)]dz \tag{45a}$$

$$\sigma_s = \int_0^{\epsilon_F} \epsilon\eta^s(\epsilon)d\epsilon - \int_{-\infty}^{\infty} v_{\text{eff}}[n;z]n(z)dz \tag{45b}$$

$$\sigma_{xc} = \int_{-\infty}^{\infty} [g_{xc}(n(z)) - g_{xc}(\bar{n}\theta(-z))]dz \tag{45c}$$

*Sources of data are given in Ref. 21.

†σ_s has been written in this form by Paasch and Wonn.[67] It is given in terms of phase shifts by Huntington[99] and in Appendix A of Ref. 21; the equivalence of the two forms is seen using equations (42) and (40) [or (43)].

where we use the local-density form for the exchange-correlation term, and where (in σ_s) we continue to choose our zero of energy so that $v_{eff}[n; -\infty] = 0$. The values of these three terms,[33] and their total, are shown as a function of r_s in Fig. 5. As $r_s \to 0$, kinetic-energy effects dominate, and σ becomes negative. This behavior was seen earlier in the Thomas–Fermi case, where σ was negative for all r_s. It should be noted that σ need not be expressed as a sum of large contributions of different signs, but can be found directly from the sum rule (30b) by inserting the electrostatic potentials computed for the model at a variety of r_s values, and integrating the sum rule over r_s (i.e., over $\bar{n}$). This has been done by Vannimenus and Budd.[35]

Now the surface energy of a high-density metal is positive (otherwise the metal would cleave spontaneously), in contrast to the result shown in Fig. 5. It therefore must be determined which of the two approximations employed is responsible for this failure: the smearing out of the discrete ionic lattice into a uniform background, or the use of the local-density approximation for exchange-correlation. The question has been a controversial one; however, it has been demonstrated in a variety of ways that it is the absence of the discrete lattice which leads to a negative surface energy, and that only a much smaller part of the underestimate of σ is ascribable to the use of the local-density approximation. A review of this question has been given by Langreth.[100]

In the paper of Lang and Kohn,[33] the discrete lattice was reintroduced into the calculation of σ using first-order pseudopotential perturbation theory, with the Madelung energy taken into account exactly. This yielded a positive

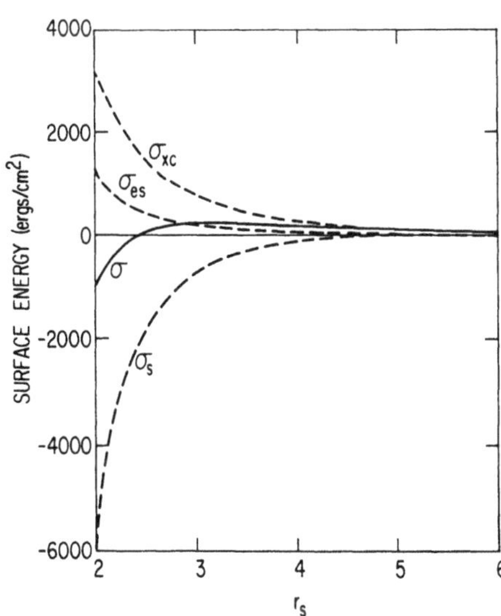

FIGURE 5. Components of the surface energy in the uniform-background model. $\sigma = \sigma_s + \sigma_{es} + \sigma_{xc}$. (From Ref. 143.)

σ in reasonable agreement with experiment for a number of simple metals, except Pb (which has a particularly strong pseudopotential). A similar calculation in which part of the z axis variation of the ionic pseudopotential is taken into account nonperturbatively has been presented by Monnier and Perdew.[101],* The results for σ are generally fairly close to those of Lang and Kohn except for the case of Pb, where the calculated value is brought into agreement with experiment. Finally, Appelbaum and Hamann,[102] still using the local-density approximation for exchange-correlation, have performed a calculation of σ for Cu$\langle 111 \rangle$, in which the discrete ionic lattice is taken into account completely nonperturbatively, and have obtained good agreement with experiment (see Section 3.2). All of this work suggests therefore that the nonlocal exchange-correlation effects, which were omitted, are not extremely large.

A complementary way of exploring this question is to obtain and evaluate forms for σ_{xc} that go beyond the local-density approximation, avoiding arguments based on agreement with experiment.

Schmit and Lucas,[103] Craig,[104] and Peuckert[105] discussed a particular nonlocal contribution to σ_{xc} that can be expected not to be well represented in the local-density approximation. When a crystal is cleaved, the bulk plasmon modes of longest wavelength (along the cleaved surface normal) become unallowed, but new surface plasmon modes are introduced. The contribution to σ_{xc} discussed by these authors is the difference in zero-point energies of these modes. This work represented a very interesting conceptual point in the discussion of surface energies; but the actual magnitude of the contribution became a matter of controversy. In addition to general discussion of the question,[106–111] several relevant calculations have been presented.[112–119] One group of these calculations[116–118] obtain σ_x exactly and σ_c in the random-phase approximation (thereby as it can be shown[118] including the effects of Schmit and Lucas), for the case in which the surface potential is taken to be an infinite square barrier. A comparison between σ_{xc} computed by Wikborg and Inglesfield[118] for this model and a local-density calculation of σ_{xc} for the infinite-barrier density distribution shows a difference of $\sim 10\%$. (A comparison of the exchange or correlation parts separately shows a much larger difference, but as pointed out in this context by Lang and Sham[120] and by Langreth and Perdew,[119] it is not proper to separate exchange and correlation energies in the local-density approximation, since a regular power series in density gradients does not exist for these energies separately, but only for their sum.) These model calculations thus suggest that the local-density calculation provides a reasonable estimate of σ_{xc}.

*These authors have also presented calculations in which nonlocal exchange-correlation effects are taken into account—see Section 3.1.

Another, model-independent approach was used by Langreth and Perdew.[119] They take the Coulomb part of the Hamiltonian to be a function of a dimensionless coupling constant λ varying between 0 and 1, with the electron–electron interaction simply proportional to λ, and the external potential taken to depend on λ in such a way that the electron density distribution does not change as λ is varied. If the equal-wave-vector Fourier transform of the density–density correlation function is defined as follows:

$$S_\lambda(\mathbf{k}) = \frac{1}{N}\int e^{i\mathbf{k}\cdot(\mathbf{r}'-\mathbf{r})}\langle[n(\mathbf{r})-\langle n(\mathbf{r})\rangle][n(\mathbf{r}')-\langle n(\mathbf{r}')\rangle]\rangle_\lambda\, d\mathbf{r}\, d\mathbf{r}' \tag{46}$$

(N is the number of electrons), then it is seen that σ_{xc} can be written

$$\sigma_{xc} = \int \sigma_{xc}(\mathbf{k})\frac{d\mathbf{k}}{(2\pi)^3} \tag{47}$$

with

$$\sigma_{xc}(\mathbf{k}) = \frac{2\pi N}{Ak^2}\int_0^1 [S_\lambda(\mathbf{k}) - S_\lambda^B(\mathbf{k})]d\lambda \tag{48}$$

Here S_λ^B is the density–density correlation function appropriate to a bulk uniform electron gas, and A is the surface area. It is most convenient to discuss the spherical average of $\sigma_{xc}(\mathbf{k})$, denoted $\sigma_{xc}(k)$; only this average enters σ_{xc}. Langreth and Perdew are able to show that the long-wavelength limit of $\sigma_{xc}(k)$ is given by*

$$\lim_{k\to 0} \sigma_{xc}(k) \to \frac{\pi}{4k}\left[\omega_s(\bar{n}) - \frac{1}{2}\omega_p(\bar{n})\right] \tag{49}$$

it is dominated by the change in zero-point energies of the plasmons,[103] and is independent of the details of the surface.

Note that these authors use a three-dimensional wave-vector decomposition of the surface energy, even though it is only the two-dimensional wave vector parallel to the surface that is a conserved quantity. They demonstrate that the two-dimensional decomposition of σ_{xc} is not dominated by surface plasmons in the long-wavelength limit, and is sensitive to details of the surface; this is the reason for using the less obvious three-dimensional wavevector

*$\omega_p(n) = (4\pi n)^{1/2}$ is the bulk plasmon frequency and $\omega_s = \omega_p/\sqrt{2}$ is the surface plasmon frequency, discussed fully in Chapter 3.

decomposition. They do note that on the average the intuitive notion of "short wavelength" and "long wavelength" can still be expected to hold in this case.

Now σ_{xc} in the local-density approximation can also be decomposed according to wave-vector. In the long-wavelength limit,

$$\lim_{k\to 0} \sigma_{xc}^{\mathrm{LDA}}(k) \to \frac{1}{2}\int_{-\infty}^{\infty} [\omega_p(n(z)) - \omega_p(\bar{n})\theta(-z)]dz \tag{50}$$

While this is dominated by shifts in plasmon zero-point energies, it sharply underestimates the importance of these effects (it does not diverge as $k\to 0$, as the exact limit does), and, again in contrast to the exact limit, depends on the form of the surface density profile. On the other hand, Langreth and Perdew show that $\sigma_{xc}^{\mathrm{LDA}}(k)$ is exact in the large-k limit, which can be described by saying that $\sigma_{xc}(k)$ represents the energy of a packet of excitations whose spatial extent is $O(k^{-1})$ and thus for large k the excitations have a small extent and do not "know" the density is varying.[100]

Given the electron density distribution, $\sigma_{xc}^{\mathrm{LDA}}(k)$ is straightforward to evaluate (an example is shown in Fig. 6). Now let it be assumed that $\sigma_{xc}^{\mathrm{LDA}}(k)$ (which

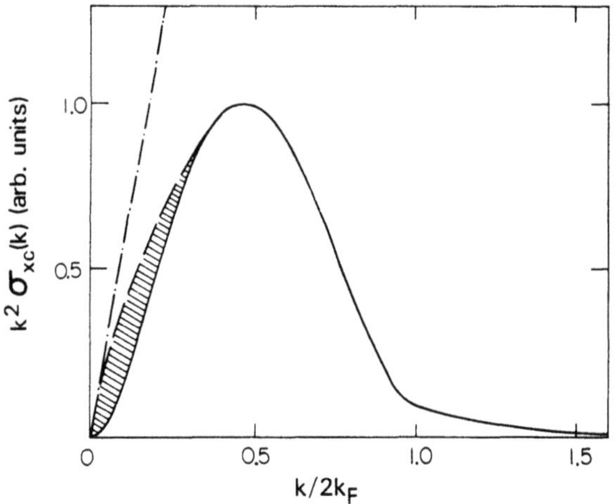

FIGURE 6. Wave-vector analysis of the exchange-correlation part of the surface energy in the uniform-background model ($r_s = 4$). Solid curve: local-density approximation; dash-dot line: exact $k\to 0$ asymptote [equation (49)]; dashed arc: interpolation. Value of σ_{xc} is proportional to the area under the curve; shaded area represents the correction for effects beyond the local-density approximation. (After Ref. 119.)

is exact for $k \to \infty$) represents a good approximation to $\sigma_{xc}(k)$ even down to $k \gtrsim k_F$ (the region of the maximum in the curve shown). Langreth and Perdew[119] employ a simple interpolation between the exact $k \to 0$ limit (dot-dash straight line) and the maximum in the $k^2\sigma_{xc}^{\mathrm{LDA}}(k)$ curve (Fig. 6). Using this interpolation, together with $\sigma_{xc}^{\mathrm{LDA}}(k)$ for larger k values, should therefore provide accurate values of $\sigma_{xc}(k)$ for all k. This in turn will yield an accurate value for σ_{xc}, which is proportional to the area under the curve of $k^2\sigma_{xc}(k)$. We see in particular that the error in the local-density approximation is not large; this is a direct result of the presence of the phase-space factor k^2, which has the effect of minimizing the actual numerical importance of collective effects [that dominate $\sigma_{xc}(k)$ at small k]. In addition to testing the adequacy of the local-density approximation, Langreth and Perdew have thus devised a method for going beyond the local-density approximation in the calculation of surface energies.

Another possible method is to use terms in the gradient series for $E_{xc}[n]$ beyond the local-density term.[93,121–123] The proper coefficient for the first gradient term in this series has been obtained by Rasolt and Geldart[124] for metallic densities within the random-phase approximation; but it has been shown by Perdew, Langreth, and Sahni[125] that for density profiles similar to those actually found for the surface, the use of this term overestimates σ_{xc} by about as much as the local-density approximation underestimates it. Langreth and Perdew[126] have decomposed this gradient term (and hence its contribution to the surface energy) by wave vectors (as above) and have shown that the gradient contribution to a curve such as that of Fig. 6 rises rapidly as $k \to 0$, instead of decreasing to zero as required by the exact $k \to 0$ result. It is this error which leads to the overestimate of σ_{xc}. The surface density variation would have to be far slower than that of any realistic profile for this spurious $k \to 0$ contribution to become negligible, and for the use of this gradient term to become valid.[126]

The work of Langreth and Perdew,[126] and that of Rasolt, Malmström, and Geldart,[127] also shows that the gradient term introduces a noticeable correction to $\sigma_{xc}^{\mathrm{LDA}}(k)$ for $k \gtrsim k_F$, which indicates the approximate character of restricting the interpolation scheme to the $k \lesssim k_F$ region (the circular interpolation of Fig. 6). The correction has an oscillatory character, however, and as a result does not represent a large contribution to σ_{xc}.*

*Langreth and Perdew[126] have suggested a procedure for correcting the $k \to 0$ behavior of the gradient term, in which (at each k) it is evaluated not using the local slope of the density profile but rather a sort of average slope, with the averaging done over a length $\sim k^{-1}$. In this modified form the gradient correction represents a replacement for the interpolation scheme mentioned above. See also the paper of Peuckert.[128]

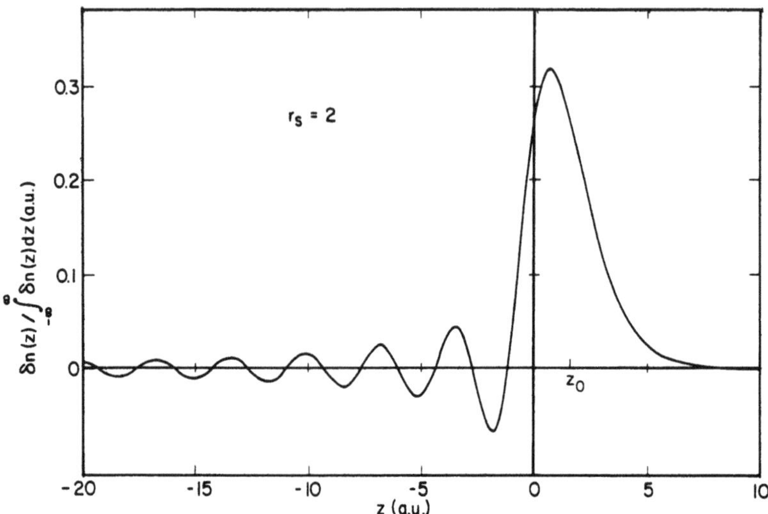

FIGURE 7. Screening charge in the surface region of the uniform-background model induced by a weak uniform normal electric field. The position of the center of gravity of the screening charge is given by z_0. (The background occupies the $z < 0$ half-space.) (After Ref. 132.)

(*d*) *Linear Response.* We consider in this section density-functional analyses of the response of the metal surface to a static perturbing charge distribution.[28,129–137] Studies of magnetic response,[78,138–140] and of response at nonzero frequency,* are not discussed.

We describe first the simple case of the charge density $\delta n(z)$ induced by a uniform electric field $\mathscr{E}$ normal to the surface. Obtaining the induced density for this case by calculating the surface response function is rather difficult. It is, however, straightforward to obtain $\delta n(z)$ by repeating the calculation for the unperturbed uniform-background model with the potential due to the field included as part of the external potential (the one-dimensional symmetry remains), and subtracting from the resulting density distribution that for the unperturbed case. If the field is small, $\delta n(z)$ will be the distribution appropriate to the linear-response regime. An example of such a distribution is shown in Fig. 7.[132] In our discussion below, we take $\delta n(z)$ to be normalized to unity: $\int_{-\infty}^{\infty} \delta n(z) dz = 1$.

*The density-functional formalism, in the form discussed here, is not designed to treat the case of time-dependent external potentials. Possible extensions of the formalism to treat this case have been discussed by Peuckert[141] and by Ying.[142]

One of the most important features of the induced density distribution is the position of its center of gravity

$$z_0 = \int_{-\infty}^{\infty} z\delta n(z)dz \tag{51}$$

This determines the effective location of the metal surface in the sense that if we take the total electrostatic potential inside the metal to be fixed, then well outside the surface (with the field pointing out of the metal),

$$\delta\phi(z) = \mathscr{E}(z-z_0) \tag{52}$$

Using the local-density approximation for exchange-correlation, it is found[132] that z_0 (relative to the positive background edge) ranges from 1.9 bohr for $r_s = 1.5$ to 1.2 bohr for $r_s = 6$.*

A relation between the rate of change of the work function with bulk density and the field-induced distributions $\delta n(z)$ has been obtained by Budd and Vannimenus[137]:

$$\frac{d\Phi(\bar{n})}{d\bar{n}} = 4\pi \int_{-\infty}^{0} dz\, z \int_{-\infty}^{z} dz' \delta n(z') \tag{53}$$

where $\delta n(z)$ is the distribution appropriate to $\bar{n}$. This is an analog of equation (30b) for the surface energy.

We now consider the linear response of the metal to a point charge of strength q at a coordinate z well outside the surface. The change in energy of the system due to the presence of the charge is

$$\delta E = \frac{1}{2} \int \delta v(\mathbf{r}) \delta n(\mathbf{r}) d\mathbf{r}$$

where we have taken $\mu = 0$ in equation (21), where $\delta v(\mathbf{r})$ is the potential due to the bare point charge, and where $\delta n(\mathbf{r})$ is the induced density. It has been shown by Lang and Kohn[132] that

*Note that (see Section 3.1) the semi-infinite uniform-background model represents an ionic lattice whose outermost lattice plane lies half an interplanar spacing behind the background edge. Thus, if we denote the interplanar spacing by d and if we express z_0 relative to the background edge, then the effective surface in a metal will be a distance $z_0 + \frac{1}{2}d$ in front of the outermost lattice plane. (We neglect here any small distortion of the metal lattice at the surface.) This result determines, for example, the relation between the electrical spacing of a parallel-plate condenser and its physical spacing.

$$\delta E = -\frac{q^2}{4(z - z_0)} + O((z - z_0)^{-3}) \tag{54}$$

i.e., the z_0 of equation (52) also gives the location of the image plane.

It is straightforward to see that, within the context of linear response, a point charge and a sheet of charge (parallel to the surface) that are at the same coordinate $z = d$ will induce electron density distributions whose centers of gravity are both at the same coordinate $z = z_0(d)$. Thus to determine the center of gravity of the electron density induced by a point charge, we can do the much simpler calculation for a sheet (which preserves the one-dimensional symmetry of the unperturbed problem), just as we earlier calculated $\delta n(z)$ and its center of gravity z_0 for the uniform normal field. Note in particular that (since the source of a uniform normal field can be taken to be a sheet of charge far outside the surface), the center of gravity of the image charge distribution induced by a distant point charge is at $z_0(\infty) \equiv z_0$, the position of the image plane (reference plane for the image potential). As the point charge moves closer in, the center of gravity of the charge it induces at first remains fixed and then it too starts to move in; eventually $z_0(d)$ and d become equal when the point charge is inside the metal. This behavior is shown in Fig. 8.[143] Note that the dipole moment of the point charge q together with its induced charge is just $\mu = q[d - z_0(d)]$. Instead of computing $z_0(d)$ directly as described above, we can use a formula of Budd and Vannimenus[137] which expresses it in terms of the density $\delta n(z)$ induced by a uniform normal field:

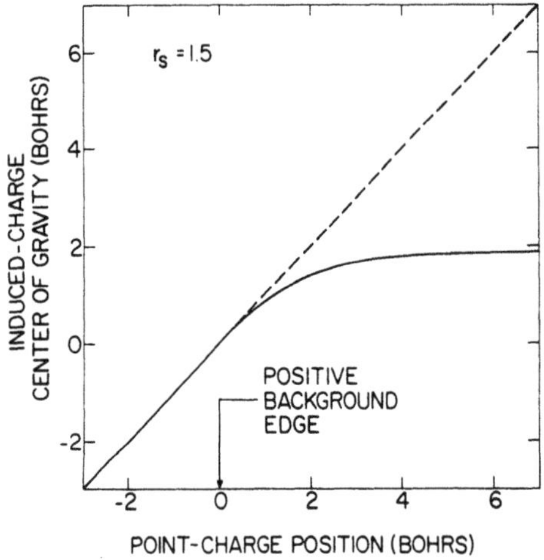

FIGURE 8. Solid curve gives position of center of gravity of screening charge in the uniform-background model induced by a weak point charge, as a function of point-charge position. (Dashed line is bulk asymptote.) (From data in Ref. 143.)

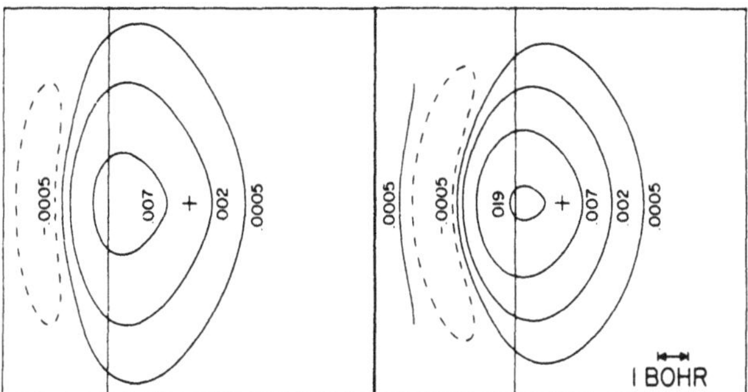

FIGURE 9. Contours of constant screening charge density in the uniform-background model ($r_s = 2$) induced by a weak point charge at two different distances (2.6 and 1.5 bohrs) from the surface. In each box, vertical line indicates positive-background edge, metal is to left-hand side, and vacuum is to right-hand side. Point-charge position indicated by +. For computational convenience, contours are not shown outside inscribed circle in box. Contour values are in electrons/bohr^3, and have been scaled up to correspond to a unit point charge. (From Ref. 145.)

$$z_0(d) = \int_{-\infty}^{\infty} \delta n(z)[z\theta(z) - (z - d)\theta(z-d)]dz \tag{55}$$

When a point charge is far from the surface [at (0,0,d)], the electron density it induces can be shown to be given by[21]

$$\delta n(\mathbf{r}) \approx \frac{q(d-z_0)}{2\pi[x^2 + y^2 + (d-z_0)^2]^{3/2}} \delta n(z) \tag{56}$$

Short of this asymptotic region, the chemisorption analysis of Refs. 144–146 (see Section 4.2.1b below) can be used to obtain the induced density; two examples are shown in Fig. 9.

3. Lattice Models of Metal Surfaces

In this section we will discuss both studies of the electronic structure of metal surfaces which at the outset take the external potential to be that due to a lattice of nuclei (or a lattice of ionic pseudopotentials) and studies in which the lattice is perturbatively introduced into the analysis of the uniform-background model. The presence of the discrete lattice permits the consideration of the anisotropy of quantities such as the work function and surface energy (and leads to a positive surface energy for high interior electron densities, in contrast to the uniform-background model). Analysis of such features as surface states, and the quantitative study of the surfaces of transition metals (and semiconductors and insulators) are possible using nonperturbative treatments of the lattice.

3.1. *Perturbative and One-Dimensional Variational Solutions*

We begin by discussing in a qualitative way the effect of the discrete lattice on the work function and surface energy, using the construct of the classical neutralized lattice.

Consider a semi-infinite crystal consisting of a cubic lattice (for convenience) of point positive charges (in place of ions) and a rigid uniform distribution of negative charge that terminates abruptly along a plane at the surface (in place of conduction electrons). Denote the spacing of the lattice planes that are parallel to the surface by d. The boundary of the negative charge must be a distance $\frac{1}{2}d$ in front of the outermost lattice plane (in order that the electric field averaged over a macroscopic region vanish both inside and outside the crystal). This classical model crystal is shown schematically on the left-hand side of Fig. 10.

The potential difference $\bar{D}_i$ in equation (9a) depends only on the charge averaged parallel to the surface. To discuss this quantity, therefore, we can

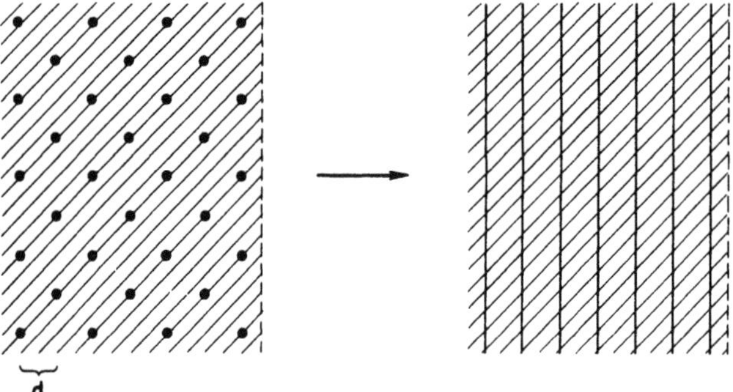

FIGURE 10. Left-hand side: classical neutralized lattice—shaded region represents negative charge distribution which terminates abruptly at a plane (dashed line), dots represent point positive charges. Right-hand side: each lattice plane of point charges in the classical neutralized lattice is smeared out into a thin sheet of positive charge. (From Ref. 143.)

imagine that each lattice plane of point charges in the model crystal has been smeared out into a thin positive sheet, as in Fig. 10. A simple electrostatic calculation shows that $\bar{D}_i$ is positive and proportional to d^2. Since the more closely packed a crystal face is, the larger is the associated interplanar spacing, we see that the more closely packed crystal faces have higher work functions. This is in fact observed for the majority of metals for which experiments have been performed.

A discussion of the surface energy for various faces of the classical neutralized lattice involves a more complicated electrostatic calculation. When this crystal is cleaved and the two halves are separated slightly there are electric fields in the gap that cause the halves to attract each other (Fig. 11). These fields would not be present if the lattice were smeared into a homogeneous positive distribution; this is one of the chief reasons why the uniform-background model of the surface is not satisfactory for the computation of surface energies.

For simple metals, the effect of the ionic lattice on the electrons can be described by $v_{\text{pseudo}}(\mathbf{r})$, a superposition of ionic pseudopotentials. With respect to the potential $v_+(\mathbf{r})$ due to the uniform background, this represents a perturbation

$$\delta v(\mathbf{r}) = v_{\text{pseudo}}(\mathbf{r}) - v_+(\mathbf{r}) \tag{57}$$

Consider now a neutral crystal with N electrons, all of whose dimensions are macroscopic but whose surface consists almost entirely of two parallel faces (normal to the z axis) of a given orientation. We can therefore omit the index i used in Section 1.1, since only these faces will enter our discussion. [It makes no difference in measuring the work function of these faces whether

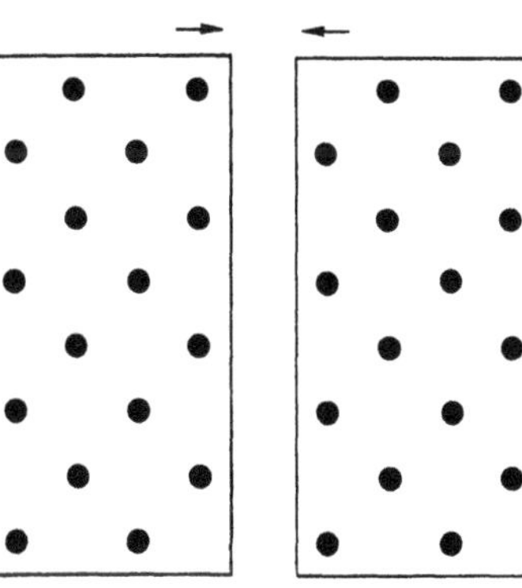

FIGURE 11. A bcc classical neutralized lattice split along the ⟨110⟩ plane. Dots represent point charges lying on a selected plane through the crystal. Electric fields in the gap cause the two halves to attract. [The halves would repel for the simple cubic lattice split along the ⟨100⟩ plane.] (After Ref. 143.)

an electron is removed from the crystal to a point $\mathbf{r}_i$ of the type defined in Section 1.1 or to infinity.] Equation (6) may be written

$$\Phi = E'_N - E_N \tag{58}$$

where E_N is the ground-state energy of the neutral slab and E'_N is the energy of an excited state in which $N-1$ electrons reside in the lowest possible state in the metal, while one electron is at rest at infinity. The density distributions associated with these two states of the system are denoted here $n_N(\mathbf{r})$ and $n'_N(\mathbf{r})$, respectively.

By standard perturbation theory, the first-order* change of the work function Φ due to $\delta v(\mathbf{r})$ (i.e., to reintroduction of the lattice into the uniform-background model) is

$$\delta\Phi = \int \delta v(\mathbf{r}) n'_N(\mathbf{r}) d\mathbf{r} - \int \delta v(\mathbf{r}) n_N(\mathbf{r}) d\mathbf{r} \tag{59}$$

Taking $\delta v(\infty) \equiv 0$, we may write this equation as

$$\delta\Phi = \int \delta v(\mathbf{r}) n_\sigma(\mathbf{r}) d\mathbf{r} \tag{60}$$

with

$$n_\sigma(\mathbf{r}) \equiv n_{N-1}(\mathbf{r}) - n_N(\mathbf{r}) \tag{61}$$

Here $n_{N-1}(\mathbf{r})$ is the density distribution of the $N-1$ electrons in their ground state [i.e., $n_{N-1}(\mathbf{r})$ is equal to $n'_N(\mathbf{r})$ minus the density distribution associated with the electron at infinity]. Because the two faces normal to the z axis are so large relative to the other faces, $n_\sigma(\mathbf{r})$ in the slab can be treated as a function of z alone. Since this function is symmetric about the slab center, we may write[28]

$$\delta\Phi = -\int_a^\infty \delta v(z) n_\sigma(z) dz \Big/ \int_a^\infty n_\sigma(z) dz \tag{62}$$

where $\delta v(z)$ is the $x-y$ average of $\delta v(\mathbf{r})$ and a is a location in the slab interior that is a distance large compared with a screening length from the surfaces. Because (62) is homogeneous in n_σ, $n_\sigma(z)$ can be taken to be the surface-charge

*The studies of Finnis[147] and of Lert and Weare[148] represent perturbative treatments of surface properties that take second-order effects into account by using an approximation for the relevant surface linear-response function.

TABLE 1. Work Function Φ for the Three Principal Crystal Faces of Na, as Computed in Ref. 28[a]

Face	Φ(eV)
⟨110⟩	3.09
⟨100⟩	2.77
⟨111⟩	2.67

[a] Value for uniform-background model is 3.06 eV.

density in the semi-infinite uniform-background model induced by a weak perpendicular electric field, denoted $\delta n(z)$ in Section 2.3.3d.

Values for the work-function change $\delta\Phi$ were computed in this way by Lang and Kohn[(28)] for the principal crystal faces of a variety of simple metals. To obtain $\delta v(\mathbf{r})$, the individual ionic pseudopotentials were taken to have a form proposed by Ashcroft.[(149)] An example of the results of this computation is given in Table 1; it shows the trend discussed above of higher work function with closer packing.* For the alkalis, the average $\delta\Phi$ was found to be negative, bringing the computed work functions into closer agreement with polycrystalline data (cf. Fig. 4).

We consider now the effect of the reintroduction of the lattice on the surface energy.† When the uniform background is replaced by the lattice, a shift $\delta\sigma$ in the surface energy occurs as a result of changes in the total energies of the cleaved and uncleaved crystals. These energies can each be written as the sum of the electrostatic self-energy of the associated positive charge configuration (background or lattice), and a term of the form $\tilde{E}_v[n]$ (cf. Section 1) giving the self-energy of the electron gas plus its energy of interaction with the positive charges. The contribution of the surface-energy shift that is related to the electrostatic self-energy of the positive charge configurations can be treated exactly, while that related to the $\tilde{E}_v[n]$ terms can be calculated using perturbation theory.

The stationarity of $\tilde{E}_v[n]$ described by $\delta\tilde{E}_v[n] = 0$ means that the associated contribution to $\delta\sigma$ can be computed correctly to first order in $\delta v(\mathbf{r})$ by using the electron density distributions of the uniform-background model. To this order, therefore, the surface-energy components σ_s and σ_{xc} are not affected by δv and only σ_{es} is shifted. It is easily seen that

*Slight reversals of this order were found in Ref. 28 for the ⟨111⟩ and ⟨100⟩ faces of Al and Pb; recent experiments[(150)] for Al differ on this reversal.

†A study of metallic adhesion along these lines has been given by Ferrante and Smith,[(80,151)] and of the positions of the outermost lattice planes by Allen and Rice[(152)] (see also Ref. 33).

$$\delta\sigma = \sigma_{cl} + \int_{-\infty}^{\infty} \delta v(z)[n(z) - n_+(z)]dz \tag{63}$$

with $\delta v(z)$ defined as above and $n_+(z)$ given in equation (22). The quantity σ_{cl} is the surface energy of a classical neutralized lattice (discussed above), having the same structure and cleavage plane as the crystal of interest.[33,153]*

A dimensional argument shows that $\sigma_{cl} \propto Z\bar{n}$ (with Z the ionic charge), and thus this quantity increases rapidly as r_s decreases. The second term in equation (63) is also positive for the cases of interest, increases rapidly as r_s decreases, and turns out to be comparable in magnitude to σ_{cl}.[33] Together the two contributions cause the calculated surface energy to be positive even at high densities (small r_s), as it ought to be. Table 2 compares some of the calculated values with the rather incomplete experimental data now available.

As the Miller indices of the surface plane considered are raised, the surface energy computed in this way increases quickly. The surface-energy anisotropy of metals is found experimentally to be rather small, however. The origin of this discrepancy, as pointed out by Manninen and Nieminen,[154] is the fact that as the surface becomes sparser, the use of an electron density that is constrained to be uniform in the surface plane (first-order perturbation theory applied to the uniform-background model) becomes less and less appropriate.

We turn now to consider the particular treatment of Monnier and Perdew,[101,155] in which a variational solution for the full discrete-lattice problem is sought among a class of electron distributions which are both one-dimensional in symmetry and become constant in the metal interior. A treatment similar to this in some respects, within the extended-Thomas–Fermi context, has also been presented by Paasch and Hietschold.[156,157],† These analyses lie in a certain sense between the perturbative approach just discussed and the nonperturbative approaches described in Section 3.2 below.§

The study of Monnier and Perdew first obtains the electron density $n(z)$ for the uniform-background model in which a step potential (with the step taken in the surface region) has been added to the external potential; the

*The cleavage energy constant for the bcc⟨111⟩ case given in Ref. 33 is in error; a corrected value is given in Ref. 153.

†In connection with Ref. 156, we note that in computing changes in σ_{cl} due to lattice relaxation, only the Fourier component of the lattice potential with zero wave-vector parallel to the surface is taken into account (in contrast to Appendix E of Ref. 33), and that surface dipole moments defined in two different ways are compared (cf. Sec. 1.1 above).

§Other studies of this type are given in Refs. 154, 158, and 159.

TABLE 2. Calculated Surface Energies for the Most Densely Packed Faces of Nine Simple Metals.[a] Single-crystal experimental data are not in general available (with Zn an exception), so to permit a crude comparison, measured values for polycrystalline samples or liquid metals (zero-temperature extrapolations) are listed. $\sigma_{\text{pert}}^{\text{LDA}}$ gives values calculated in the local-density approximation for the uniform background model[b] plus the correction of equation (63). Values for $\Delta\sigma_{\text{var}}$ and $\Delta\sigma_{xc}$ are extracted from the paper of Monnier and Perdew.[101] $\Delta\sigma_{\text{var}}$ is defined to be the difference between the surface energy calculated using the variational procedure of Monnier and Perdew, within the local-density approximation, and that using the perturbation procedure of Lang and Kohn[33] ($\Delta\sigma_{\text{var}} = \sigma_{\text{var}}^{\text{LDA}} - \sigma_{\text{pert}}^{\text{LDA}}$). $\Delta\sigma_{xc}$ is the nonlocal correction to the variationally calculated surface energy using the "wave vector analysis" method of Langreth and Perdew[119] (as shown, e.g., in Fig. 6 of the present chapter). σ_{tot} is then the total of the entries in the three previous columns[c]: $\sigma_{\text{tot}} = \sigma_{\text{pert}}^{\text{LDA}} + \Delta\sigma_{\text{var}} + \Delta\sigma_{xc}$.

Metal	Face	r_s	$\sigma_{\text{pert}}^{\text{LDA}}$ (ergs/cm^2)	$\Delta\sigma_{\text{var}}$ (ergs/cm^2)	$\Delta\sigma_{xc}$ (ergs/cm^2)	σ_{tot} (ergs/cm^2)	σ_{expt} (ergs/cm^2)
Al fcc	⟨111⟩	2.07	835	−85	150	900	1169[d]
Pb fcc	⟨111⟩	2.30	1200	−775	90	515	692[d]
Zn hcp	⟨0001⟩	2.30	565	−5	110	670	1040[d]
							575[e]
Mg hcp	⟨0001⟩	2.65	590	−5	75	660	720[f]
Li bcc	⟨110⟩	3.28	390	−15	35	410	470[f]
Na bcc	⟨110⟩	3.99	235	0	20	255	273[d]
K bcc	⟨110⟩	4.96	145	0	10	155	135[f]
Rb bcc	⟨110⟩	5.23	125	−15	10	120	115[f]
Cs bcc	⟨110⟩	5.63	105	−20	10	95	85[f]

[a]All σ and $\Delta\sigma$ values are rounded to the nearest 5 ergs/cm^2, except the experimental data of Wawra.
[b]The calculation is done as described in Ref. 33, except that the exchange-correlation energy density is taken to be the parametrization by L. Hedin and B. I. Lundqvist [*J. Phys. C* **4,** 2064–2083 (1971)] of the results of K. S. Singwi, A. Sjölander, M. P. Tosi, and R. H. Land [*Phys. Rev. B* **1,** 1044–1053 (1970)], rather than the Wigner form [as given in D. Pines, *Elementary Excitations in Solids* (Benjamin, New York, 1963)]. I am grateful to John Perdew for pointing out that this could have a noticeable [$O(100)$ ergs/cm^2] effect on the values for the high-density metals. In addition, the calculations were done for the actual r_s of the metal, rather than by using an interpolation between σ values for half-integral r_s values as in Ref. 33.
[c]We neglect the fact that those calculations that were done with the local-density approximation by Monnier and Perdew (Ref. 101) used Wigner's form for the correlation, rather than that of Hedin and Lundqvist. This fact should have a negligible effect on the values of $\Delta\sigma_{\text{var}}$ and $\Delta\sigma_{xc}$.
[d]From ultrasonic measurements on polycrystalline solid metals by H. Wawra, *Z. Metallkunde* **66,** 395–401 and 492–498 (1975).
[e]Measurement by cleavage along ⟨0001⟩ face, A. H. Maitland and G. A. Chadwick, *Phil. Mag.* **19,** 645–651 (1969).
[f]Zero-temperature extrapolations of liquid-metal surface tensions, as given in Ref. 33.

density is found by solving self-consistently the equations of Kohn and Sham [equations (1)]. The step can be imagined to simulate crudely the difference between the mean ionic pseudopotential in the crystal and the vacuum. Depending on its sign, it can make the electron density profile more or less spread out relative to that of the pure uniform-background model. The step height (or alternatively its position) is varied; and the variational solution to the problem is obtained by selecting from among the density profiles corresponding to the different steps that one which gives the lowest calculated total energy, where the energy expression evaluated includes the full discrete-lattice pseudopotential.

Monnier and Perdew distinguish their procedure (which they call "variational-self-consistent") from those variational approaches in which either the density or the entire effective potential is parametrized. The usual implementation of the density parametrization approach is restricted to using a gradient series expression for $T_s[n]$, since the exact form of $T_s[n]$ is not known, except via the solution of equations (1). The second approach, varying the potential in equation (1a),[74] eliminates this restriction, but it is in general more desirable to parametrize only part of the potential [namely, $\delta v(\mathbf{r})$] as Monnier and Perdew do, rather than all of it.

This work (and that of Paasch and Hietschold) shows a quite noticeable variation of the electron density profile among the different crystal faces; an example is shown in Fig. 12. The use of this nonperturbative scheme, relative to the perturbative treatment described earlier, leads to a correction of 0%–20% to the surface energy (see Table 2) except in the case of Pb, which has a particularly strong pseudopotential. This indicates that first-order pertur-

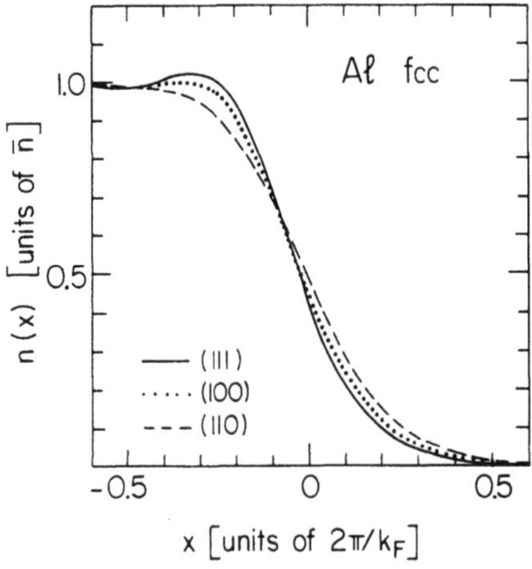

FIGURE 12. Electron density distribution for the surface of fcc Al (plotted along the surface normal direction) obtained by a variational procedure in which the density profile is constrained to be one generated by a particular class of one-dimensional potentials (see text). The profile for the ⟨100⟩ face is very close to that of the uniform-background model. (From Ref. 101.)

bation theory is generally a reasonable approximation for semiquantitative use in this context. Nonlocal exchange-correlation effects were included using the "wave-vector analysis" form described in Sec. 2.3.3; they represent a 10%–20% correction (Table 2). In the case of the work function, the values found using first-order perturbation theory and those obtained in the procedure of Monnier and Perdew generally agree to within ~0.1 eV.

3.2. Nonperturbative Solutions

We discuss here density-functional studies of metal surfaces in which the full three-dimensional lattice potential is taken into account in obtaining the electron distribution (and hence other properties). These analyses have generally proceeded by solving self-consistently the equations of Kohn and Sham [equation (1)]* in the local-density approximation, with the external potential taken to be either the full nuclear potential (the calculations of Gay *et al.*[161,162] and of Appelbaum and Hamann[102] for Cu), or a superposition of ionic pseudopotentials (in which case the core states of the metal are not obtained). We do not discuss here studies in which a small cluster of metal atoms is used to simulate a semi-infinite metal (or is studied in its own right),[163–165] but consider only those in which the metal is taken to be infinite parallel to the surface. In two of these calculations (those of Appelbaum and Hamann[166] for Na and Bohnen and Ying[167] for the alkalis), the metal is semi-infinite in the surface normal direction; in the others, it is a thin slab.

A variety of techniques has been used to solve equation (1) for this problem. The translational symmetry parallel to the surface represents of course an important simplification. In the work of Appelbaum and Hamann[166] for Na⟨100⟩, the wave function was expanded parallel to the surface in plane waves; along the surface normal, it was obtained by numerical integration in the region of the vacuum and the initial several layers of the crystal, after which it was matched to the wave function in the remainder of the crystal (taken to exhibit bulk behavior). In the study of Alldredge and Kleinman[168] for a Li⟨100⟩ slab (13 layers), the wave function was also expanded in plane waves parallel to the surface, and, normal to the surface, in sines and cosines which were chosen to vanish at a location several interplanar spacings into the vacuum. The treatment of Chelikowsky *et al.*[169] for Al⟨111⟩ uses a (12-layer) slab with a vacuum region of several interlayer distances for each surface, which is periodically repeated in order to make the system periodic in all three directions. The wave function is then expanded in plane waves in all directions.

*We also mention here an approach which proceeds via an approximation for the surface dielectric response function.[160]

The analysis of Gay, Smith, and Arlinghaus[161] for a (9-layer) Cu⟨100⟩ slab uses as basis functions Cu atomic orbitals (the occupied core and valence orbitals, augmented by $4p$, $4d$, and $5s$ orbitals) fit to Gaussian functions. Appelbaum and Hamann[102] have treated 3- and 5-layer Cu⟨111⟩ slabs, also using an atomic orbital basis, and have studied in addition a 11-layer slab with the Hamiltonian obtained by combining tight-binding matrix elements from the bulk and 5-layer slab calculations. This same procedure has been used by Feibelman, Appelbaum, and Hamann[170] to study Ti⟨0001⟩ and Sc⟨0001⟩. Wang and Freeman[171] have considered both paramagnetic and ferromagnetic Ni⟨100⟩ slabs (5 layers and 9 layers, respectively) using an LCAO basis set orthogonalized to frozen-core wave functions, with the crystal Coulomb potential constructed on each iteration as a superposition of overlapping spherically symmetric atomic potentials and the exchange-correlation potential determined from a superposition of atomic charge densities.

Louie *et al.* have studied Nb⟨100⟩[172] (9-layer slab) using the procedure of Chelikowsky *et al.*,[169] and Pd⟨111⟩ (7-layer slab)[173] using a similar procedure but with the wave function expanded in a basis of both plane waves and (Bloch sums of) Gaussian orbitals. For Nb, ~1000 plane waves were used, with an additional 1000 treated using perturbation theory; for Pd, ~300 planes waves plus 35 localized functions were used. A calculation for Mo⟨100⟩ similar to that for Nb has been performed by Kerker *et al.*[176]

Finally we mention the study of the alkalis by Bohnen and Ying.[167] Here the metal is taken to be semi-infinite and the analysis proceeds not by solving the equations of Kohn and Sham but by directly minimizing the total energy expressed as a functional of the density matrix. The basis functions are written using three Gaussian orbitals located at each atomic site.*,†

Figure 13 shows the charge density and effective potential v_{eff} for Al⟨111⟩, as well as contours of total charge density, computed by Chelikowsky *et al.*[169] (Since it has been computed for a lattice of ionic pseudopotentials, this charge density is often referred to as a pseudo-charge-density. It does not contain the core charge density or the core-region oscillations of the valence charge density present in the actual metal, but outside the core it should be quite similar to the actual density.[184]) It is seen in the figure that the total charge density differs significantly from the bulk charge distribution only outside the second layer of ions. The Friedel oscillations are visible in the averaged charge density, combined with the oscillations associated with the lattice.

*It should be noted that the kinetic-energy functional $T_s[n]$, whose explicit form as a functional of $n(\mathbf{r})$ is not known, can be expressed as an explicit functional of the density matrix.

†We also take note here of the extensive work on semiconductor surfaces using the density-functional approach.[177–183]

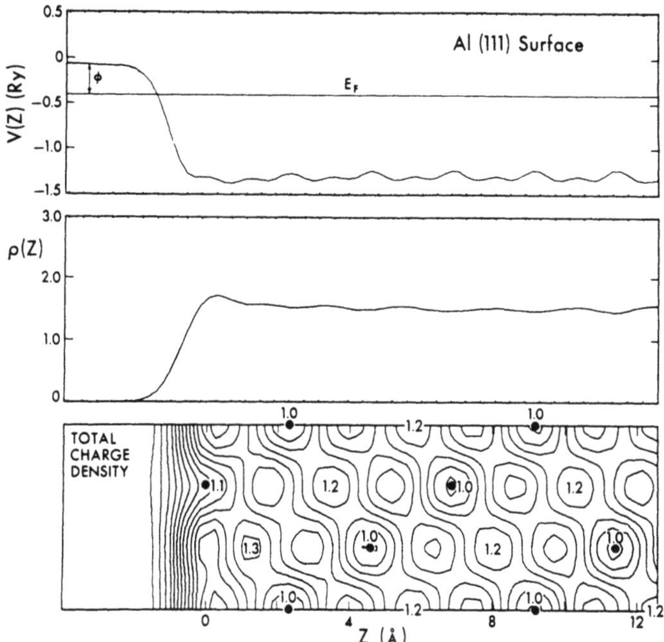

FIGURE 13. Top section: Effective potential v_{eff} for Al⟨111⟩ averaged parallel to the surface and plotted as a function of distance along the surface normal direction. Middle section: pseudo-charge-density averaged parallel to the surface (relative units). Bottom section: contour maps of pseudo-charge-density in ⟨110⟩ plane (relative units). Contour spacing is 0.15; only minima are labeled; ionic positions indicated by black dots. (From Ref. 169.)

In the calculation of Appelbaum and Hamann[166] for Na⟨100⟩, the first peak in the density averaged parallel to the surface is 14% above the average bulk density. These authors note that in the bulk there would be a peak due to the lattice potential at this location of 6%, and that in the uniform-background calculation there is an 8% peak at this position. Thus the two effects appear to superpose linearly. In the calculation of Alldredge and Kleinman,[168] on the other hand, the first peak in the planar average density (Fig. 14) is 15% above the average while the oscillations in the bulk (due to the lattice) and the first Friedel peak in the uniform-background model for Li are each ~5%, so that the calculated result is not simply the superposition of the two. Alldredge and Kleinman note that the fact that the Friedel oscillations are not commensurate with the lattice spacing gives rise to perceptible structure in the charge density as deep as the fourth atomic plane, which is marked by a noticeable minimum where a uniform-background-model minimum and a bulk-crystal minimum closely coincide (Fig. 14).

Figures 15 and 16 show the charge densities for Cu⟨100⟩ (Gay, Smith, and Arlinghaus)[161] and Cu⟨111⟩ (Appelbaum and Hamann).[102] The charge density contours below the surface plane are very similar to those of the bulk. Note the way in which the charge smooths parallel to the surface as it spreads into the vacuum, as originally suggested by Smoluchowski.[185]

We now consider briefly results for the planar density of states, which is the integral of $n(\epsilon,\mathbf{r})$ [defined in equation (4)] over a given layer of the crystal. Even though this quantity is called a density of states in the literature, it is strictly speaking only the density of eigenstates associated with the Kohn–Sham equations [equation (1a)], and not the density of one-electron excitation energies. There is a distinction because the eigenvalue ϵ_i of equation (1a) is in principle simply an auxiliary quantity having no particular connection with excitation energies (but see Ref. 31). It is, however, found to be the case that for sufficiently delocalized states, the eigenstate density associated with equation (1a) and the measured density of excitation energies are rather close. This topic has been discussed in detail in Chapter 4. In our subsequent discussions, whenever we present results for $n(\epsilon)$ or $n(\epsilon,\mathbf{r})$, we will for convenience use the phrase "density of states" to refer to it.

Figure 17 shows the planar density of states from the surface plane,

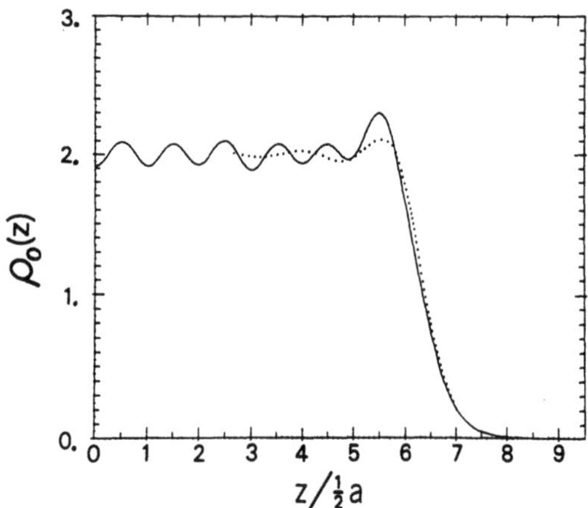

FIGURE 14. Pseudo-charge density for Li⟨100⟩ averaged parallel to the surface and plotted as a function of distance along the surface normal direction, in units of electrons per a^3 (with a the lattice constant). The density in the uniform-background model for Li is shown as a dotted line. (From Ref. 168.)

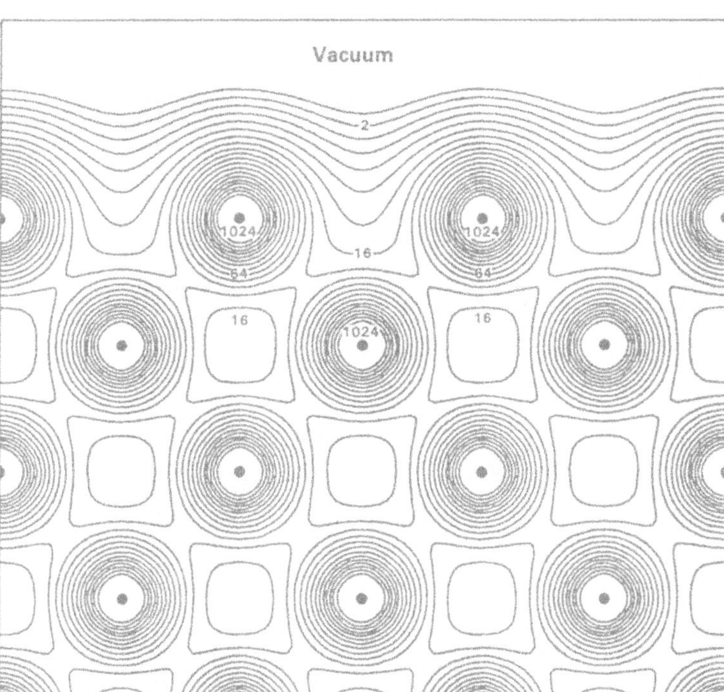

FIGURE 15. Conduction-band charge density at a Cu⟨100⟩ surface plotted on a plane normal to the surface (passing through a line connecting a surface atom with one of its near neighbors in the second plane of atoms). The units of charge density are 2.44×10^{-3} electrons/bohr^3. Charge densities on successive contours differ by a factor $\sqrt{2}$. (From Ref. 161.)

second plane and central plane of a 9-layer Cu⟨100⟩ slab (as well as the total density of states for the slab).(161) While the state density of the second plane is already very similar to that of the central plane, the surface-plane state density shows a definite narrowing of the *d* band (the second moment is reduced to two-thirds of its value for the other planes). The large enhancement of the surface-plane state density in the ~1 eV region at and above the top of the *d*-band is due principally to the presence of surface states and resonances in this energy range.(161) Narrowing of the *d*-band at the surface is also seen in the results for Cu⟨111⟩(102) shown in Fig. 18.

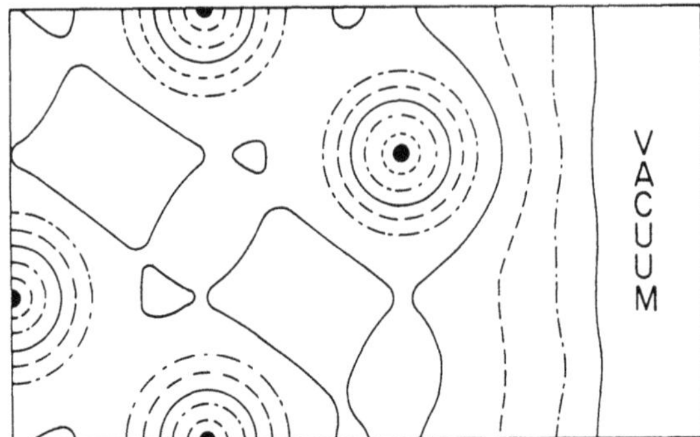

FIGURE 16. Charge density for Cu⟨111⟩ plotted on a plane normal to the surface. The contour nearest to the vacuum region (solid line) is 10^{-3} electrons/bohr^3 and subsequent contours increase geometrically by $\sqrt{10}$. The contour closest to the Cu nuclei is 10 electrons/bohr^3. (From Ref. 102.)

Now true surface states (as opposed to surface resonances) can occur only in gaps in the projection of the bulk band structure onto the surface of interest (see Ref. 172, e.g., for details). In such a gap, a state localized in the surface region will have no bulk states at the same energy or same symmetry into which it can decay. The bands of surface states (and strong surface resonances) calculated by Louie *et al.*[(172)] for Nb⟨100⟩ are shown in Fig. 19, together with the projected bulk band structure. As an example of the density distribution associated with a surface state, we show in Fig. 20 the distribution corresponding to a T1 (see Fig. 19) state at a general point of the two-dimensional Brillouin zone (this state is above the Fermi level). This particular case exhibits primarily $d_{3z^2-r^2}$ character and is quite localized on the outermost plane of atoms. Figure 21 shows the charge density associated with a moderately well localized surface state on Al⟨111⟩ calculated by Chelikowsky *et al.*[(169)] (this state is below the Fermi level).

Surface states below the Fermi level can be observed experimentally using angle-resolved photoemission. Calculations of the type described in this section, using the local-density approximation for exchange-correlation, are able to yield results in quite reasonable agreement with these experiments. For example, a surface state is observed at the center of the two-dimensional Brillouin zone on Cu⟨111⟩ with an energy of -0.4 eV relative to the Fermi level; the calculation of Appelbaum and Hamann places this state at -0.5 eV.[(102)] The full self-consistency of the calculation is found to be important to the proper positioning of these states.

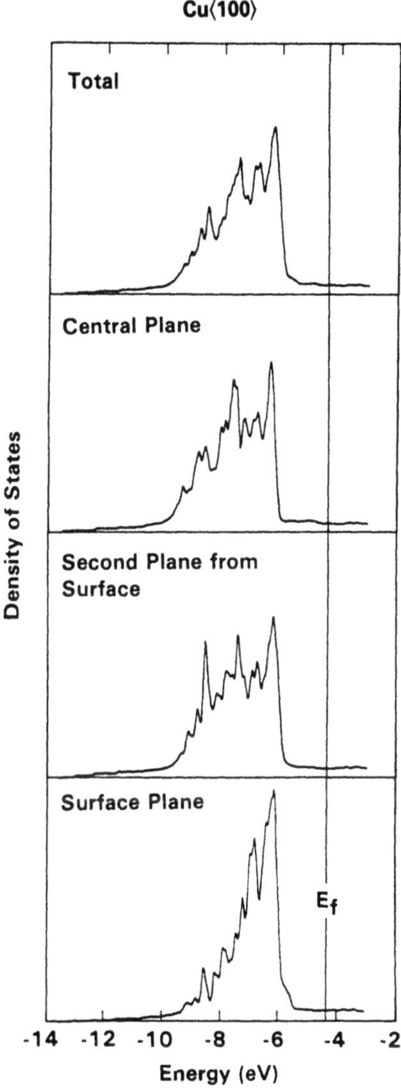

FIGURE 17. Curves of density of states for (9-layer) Cu⟨100⟩ slab. (After Ref. 161.)

The value of the work function is a particularly sensitive test of the self-consistency of calculations of the type described here. The most recent of these give values in reasonable agreement with measured work functions; for example, 5.8 eV (calc.)[173] vs. 5.6 eV (expt.)[174] for Pd⟨111⟩, 5.0 ± 0.1 eV (calc.)[102] vs. 4.85 eV (expt.)[175] for Cu⟨111⟩.

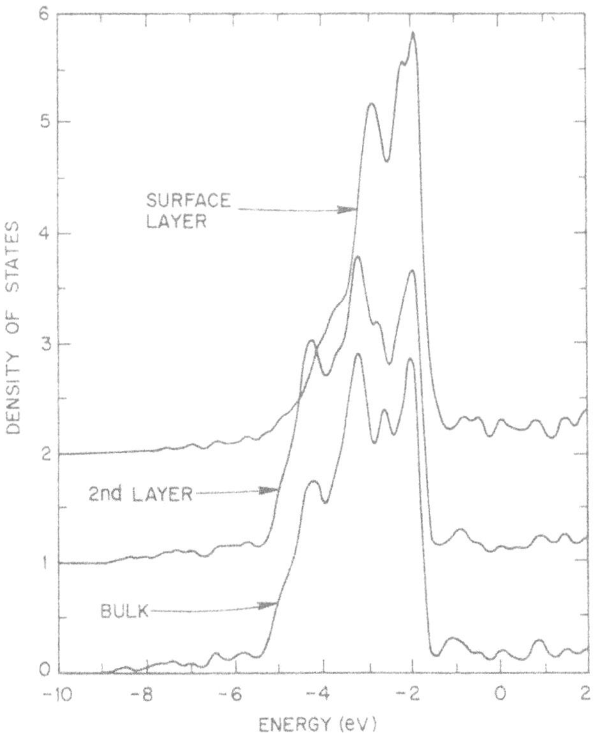

FIGURE 18. Curves of density of states for (11-layer) Cu⟨111⟩ slab. "Bulk" denotes average of central three layers. Curves are displaced vertically from each other by 1 unit (= 1.47 states/eV/atom). (From Ref. 102.)

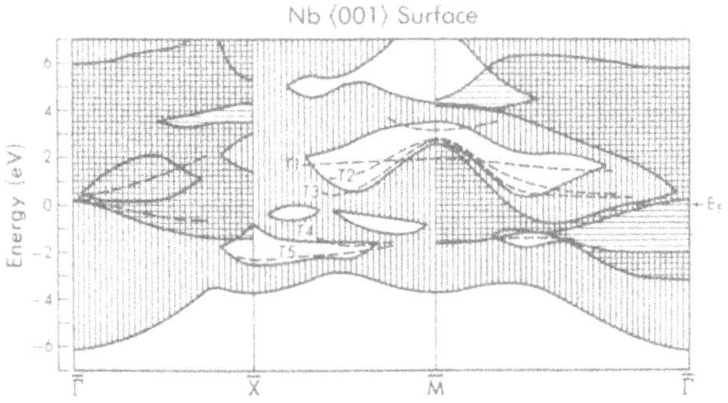

FIGURE 19. Surface bands (dashed curves) and projected bulk band structure (cross-hatched areas) for Nb⟨001⟩. (Type of cross-hatching is related to symmetry of the bulk states.) (From Ref. 172.)

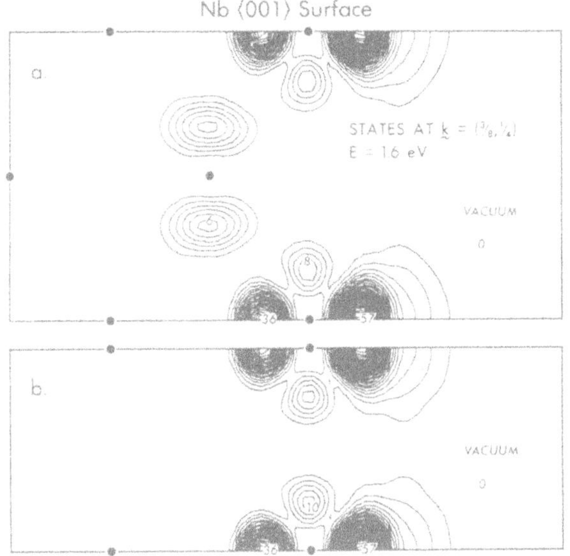

FIGURE 20. Charge density distribution of a T1 (see Fig. 19) surface state for Nb⟨001⟩ at $\mathbf{k} = (3/8, 1/4)2\pi/a$ (with a the bulk crystal lattice constant) plotted on (a) the ⟨110⟩ plane and (b) the ⟨100⟩ plane. Charge density is given in relative units. (From Ref. 172.)

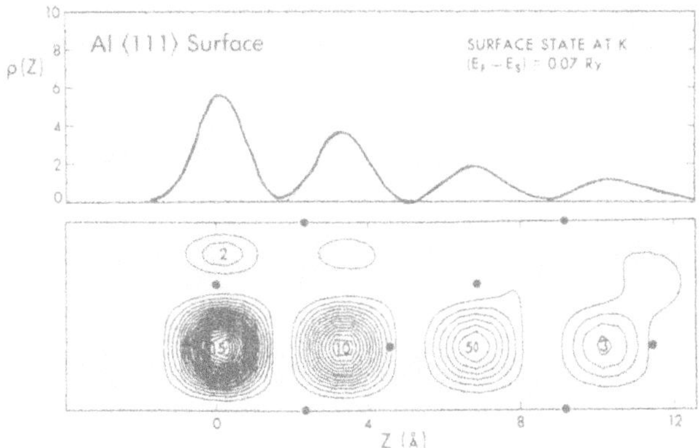

FIGURE 21. Charge density for surface state at $\bar{K}$ in the two-dimensional Brillouin zone of Al⟨111⟩ (arbitrary units). Top section: charge density averaged parallel to the surface and plotted as a function of distance along the surface normal direction. Bottom section: Contour map of the charge density for this state in the ⟨110⟩ plane. Contours are spaced by units of 0.75. (From Ref. 169).

The calculation of Appelbaum and Hamann[(102)] for Cu⟨111⟩ extracts a surface energy by assuming that this energy is size-insensitive and thus that the total slab energy is the sum of a bulk term proportional to the volume and a surface term proportional to the surface area. The total slab energies for two different numbers of layers then yield a value for the surface energy, which in this case is 2130 ergs/cm^2 (an experimental value[(186)] for polycrystalline Cu is 2016 ergs/cm^2).

A number of authors (see in particular Nieminen and Hodges[(187)]) have discussed obtaining surface barriers on metals by combining experimental values of the work function with bulk chemical potential values obtained in a band calculation. This provides a simple alternative to the actual calculation of this quantity. Usually it is equation (13) which is used for this purpose, which provides D^{WS}. Table 3 gives results for D^{WS} for a variety of metals, obtained in this way from measured polycrystalline work functions and values

TABLE 3. Values for the Dipole Barrier D^{WS} Obtained by Combining Experimental Values of the Work Function with Bulk Chemical Potential Values Obtained in a Band Calculation[(188)]: $D^{WS} = \Phi_{expt} + \mu^{WS}$. (See Section 1.1.)

Metal	Φ_{expt} (eV)	D^{WS} (eV)	Metal	Φ_{expt} (eV)	D^{WS} (eV)
Li	3.1[a]	0.9	Cu	4.65[c]	3.6
Be	4.98[b]	7.2	Zn	4.33[a]	4.0
Na	2.7[a]	0.5	Ga	4.4[a]	3.3
Mg	3.66[a]	2.3	Rb	2.21[a]	0.1
Aℓ	4.19[a]	3.8	Sr	2.6[a]	0.8
K	2.39[a]	0.2	Y	3.1[c]	1.8
Ca	2.87[a]	1.0	Zr	4.05[c]	3.5
Sc	3.5[c]	2.1	Nb	4.3[c]	4.1
Ti	4.33[c]	3.7	Mo	4.6[c]	5.0
V	4.3[c]	4.3	Ru	4.71[d]	4.3
Cr	4.5[c]	5.3	Rh	4.98[d]	3.7
Mn	4.1[c]	4.9	Pd	5.55[c]	3.2
Fe	4.5[c]	4.9	Ag	4.0[c]	2.1
Co	5.0[c]	4.7	Cd	4.22[a]	3.2
Ni	5.15[c]	4.2	In	4.12[a]	2.5

[a]References given in footnote 119 of Ref. 21.
[b]T. Gustafsson, G. Brodén, and P.-O. Nilsson, *J. Phys. F* **4,** 2351–2358 (1974).
[c]D. E. Eastman, *Phys. Rev. B* **2,** 1–2 (1970).
[d]H. B. Michaelson, *IBM J. Res. Develop.* **22,** 72–80 (1978).

of μ_{WS} from band calculations of Moruzzi and Williams.[188] These calculations used the augmented-spherical-wave technique of Williams, Kübler, and Gelatt,[189] which involves approximation of the atomic polyhedra by Wigner–Seitz spheres of equivalent volume. All calculations were done in the bcc structure, were paramagnetic, and used the local-density approximation [equation (3)] in solving equation (1).

4. Chemisorption on Metal Surfaces

The density-functional formalism has proven to be a very useful tool in the study of the chemical bond between an atom or molecule (the adsorbate) and a metal surface (the substrate).* Some of the analyses treat the classic case of a single atom (or molecule) on the surface; the adsorbate in this instance destroys the periodicity of the substrate. Other studies consider the case of a layer of atoms on the surface, usually with the same lattice periodicity as the substrate, in which case the problem retains the symmetry of the bare substrate.

A quantity which is often of particular interest to calculate is the difference between the local density of states, defined in equation (4), for the metal–adsorbate system (MA) and for the bare metal (M):

$$\delta n(\epsilon,\mathbf{r}) = n^{MA}(\epsilon,\mathbf{r}) - n^{M}(\epsilon,\mathbf{r}) \qquad (64)$$

The relation between these state densities and actual densities of one-electron excitations is discussed above in Section 3.2 and also in Chapter 4. In terms of $\delta n(\epsilon,\mathbf{r})$, the state density associated with the presence of the adatom is

$$\delta n(\epsilon) = \int \delta n(\epsilon,\mathbf{r})d\mathbf{r} \qquad (65)$$

and the additional electron density is

$$\delta n(\mathbf{r}) = \int_{-\infty}^{\epsilon_F} \delta n(\epsilon,\mathbf{r})d\epsilon \qquad (66)$$

In the following sections, we consider a progression of analyses of the metal–adsorbate system, starting with the case in which both substrate and adsorbate are represented using the uniform-background model, and ending

*We discuss only chemisorption here, but we note several studies of physisorption based at least in part on this formalism.[190–192]

with the case in which the full nuclear potential (or ionic pseudopotential) of the atoms of both substrate and adsorbate are taken into account.

4.1. *Uniform-Background Model for Both Substrate and Adsorbate*

One of the more extensively studied chemisorption systems has in the past been that consisting of a layer of alkali atoms adsorbed on a high-work-function substrate; the most often measured property is the variation of work function Φ with adatom coverage. As atoms are adsorbed, Φ decreases rapidly, reaches a minimum, and then rises to approximately the bulk alkali value with completion of the first full layer of alkali atoms.

This system is modeled in Ref. 193 using a uniform-background representation of both the substrate and the adsorbate. The positive background density for the problem [cf. equation (22)] is taken to be

$$n_+(z) = \begin{cases} \bar{n}^M, & z \leq 0 \\ \bar{n}^A, & 0 < z \leq d \\ 0, & z > d \end{cases} \tag{67}$$

Since the alkalis are monovalent, $\bar{n}^A d = N$, with N the number of adatoms per unit area. The adsorbate background thickness d is taken to be fixed at the interplanar spacing between the most densely packed planes in the bulk alkali, and $\bar{n}^A$ is varied to simulate changes in coverage N. The reason for keeping d fixed is that it is related to the substrate–adsorbate nuclear separation in the actual system, which is not expected to change substantially with coverage.

The electron density was obtained in Ref. 193 using the Kohn–Sham equations [equation (1)] with the local-density approximation for exchange-correlation. The differential equation is integrated in from the vacuum region to the bulk of the metal, just as in the case of the uniform-background model of the bare metal (there is no explicit matching of wave functions). The only difference is that a different external potential [namely, the electrostatic potential due to the $n_+(z)$ of equation (67)] appears in the equations. The substrate density $\bar{n}^M$ was chosen to correspond to $r_s = 2$; this is broadly representative of high-work-function substrates.* Warner[50] has discussed a similar model (with however $d \propto N$), treated in the extended Thomas–Fermi approximation.

An example of a calculated density distribution is shown in Fig. 22. The work function is computed from equations (23) and (24). Graphs of $\Phi(N)$ are shown in Fig. 23; these exhibit the experimentally observed minimum. Figure

*A treatment using this model is given by Wojciechowski.[194]

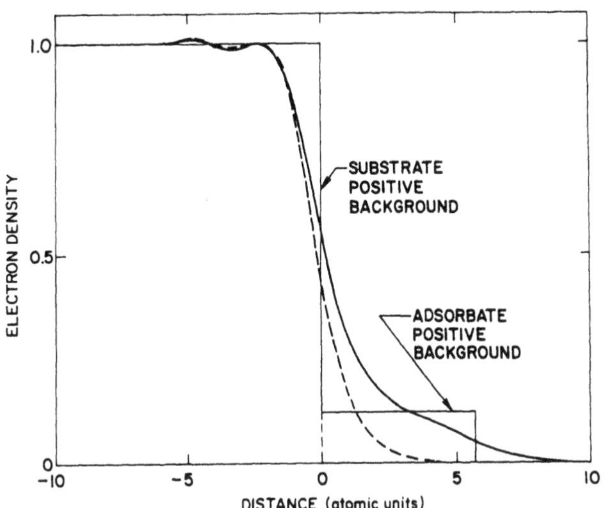

FIGURE 22. Electron density distribution for uniform-background model of alkali adsorption. Solid line gives distribution for case of one full layer of absorbed Na atoms; dashed line gives distribution for bare substrate (r_s = 2). (Atomic unit ≡ bohr.) (After Ref. 193.)

FIGURE 23. Curves of work function vs. coverage for Na and Cs adsorption in the uniform-background model (r_s = 2 substrate). Values of the coverage at the minimum indicated here are smaller than those determined by adsorption on transition-metal substrates, whose work functions are much larger than that of the model substrate. The value of the work function at the minimum, however, is relatively independent of the substrate work function, and so a comparison can be made with experiments on transition metals (see Fig. 24). (After Ref. 193.)

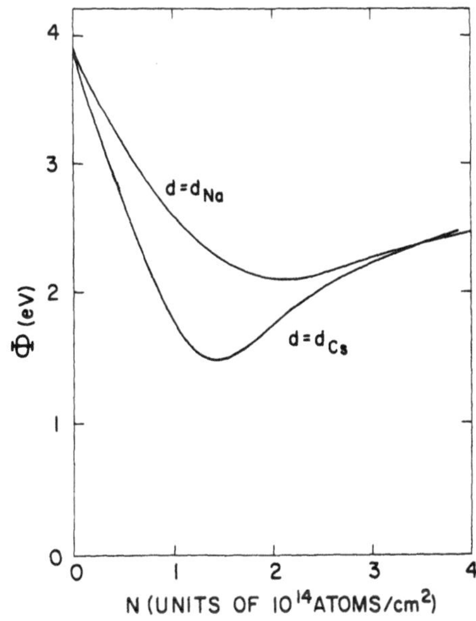

24 compares calculated and measured minimum values. (The experiments were done on transition metals, which are not well represented by a uniform-background model, but the value of the minimum is found experimentally to be quite independent of details of the substrate.)

We can examine several limiting cases of the model in order to elucidate major features of these results. We consider in particular the question of why there is a minimum in the $\Phi(N)$ curve. For these purposes, it is useful to allow

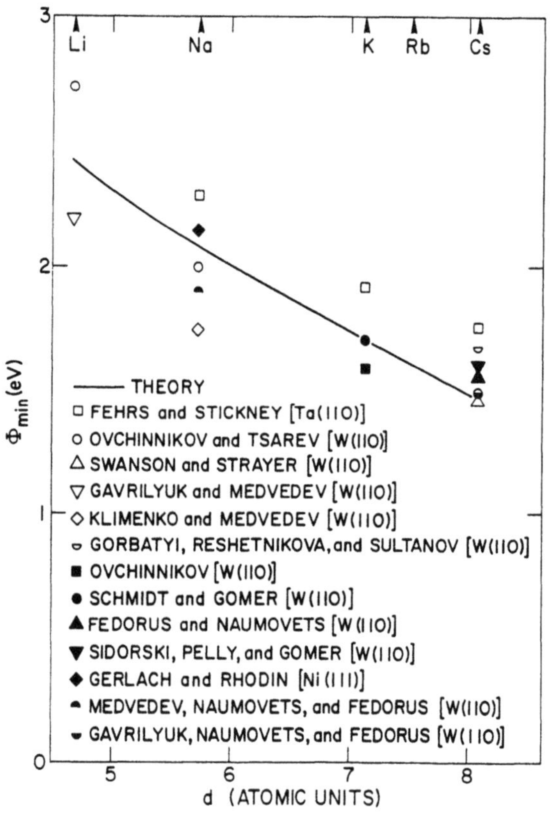

FIGURE 24. Symbols: measured values of the minimum work function for alkali adsorption on close-packed metal substrates of initially high work function. Solid line: minimum value computed using the uniform-background model of alkali adsorption. (Atomic unit ≡ bohr.) (After Ref. 193.) A value of 1.5 eV has been found for Rb/W(100) by T. W. Hall and C. H. B. Mee [*Jpn. J. Appl. Phys. Suppl.* **2** (2), 741–744 (1974)].

FIGURE 25. Schematic representation of $n(z)$ when $\bar{n}^A$ and d are sufficiently large for the substrate–adsorbate system (uniform-background model) to act like a bimetallic junction. (After Ref. 193.)

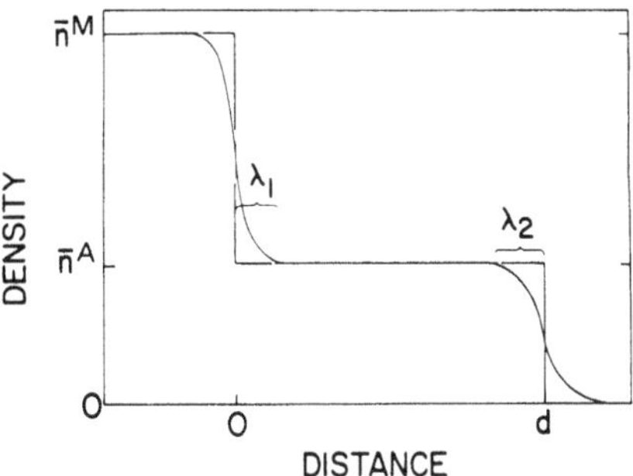

$\bar{n}^A$ to assume any non-negative value, including ones beyond the full-layer density. We also think of Φ as a function of $\bar{n}^A$ rather than of N.

For any value of d, we note that Φ will equal Φ^M when $\bar{n}^A = 0$ and again when $\bar{n}^A = \bar{n}^M$. Thus the presence of a minimum in the $\Phi(\bar{n}^A)$ curve can be demonstrated by showing that $d\Phi/d\bar{n}^A > 0$ at $\bar{n}^A = \bar{n}^M$. (That this minimum occurs below the full-layer density is verified only by the actual calculation.) We consider the case in which d is very large (compared with substrate screening lengths), and examine how the behavior of Φ as a function of $\bar{n}^A$ persists down to smaller values of d.

For $\bar{n}^A \sim \bar{n}^M$, the model is simply that of a bimetallic junction, depicted schematically in Fig. 25, with a semi-infinite left-hand member and a finite- (though large-) thickness right-hand member. Φ is then the work function associated with a bulk sample having mean electron density $\bar{n}^A$—the presence of the left-hand member has no effect on Φ. Since the work function of a bulk metal increases as its mean density is raised,* we see immediately that $d\Phi/d\bar{n}^A > 0$ at $\bar{n}^A = \bar{n}^M$, and hence that there must be a minimum in the $\Phi(\bar{n}^A)$ curve.

Now consider the two double layers in Fig. 25. Note that $\lambda_1 \sim \lambda_2 = O((\bar{n}^A)^{-\gamma})$ in atomic units [$1/6 \lesssim \gamma \lesssim 1/3$]. We assume $\bar{n}^A$ to be small enough so that λ_1 is large relative to any lengths associated with the left-hand member. It seems evident that as $\bar{n}^A$ is decreased through values of $O(d^{-1/\gamma})$, the right-hand member of the junction will cease, with regard to the work function, to behave like a bulk sample. The work function need then no longer decrease monotonically with $\bar{n}^A$: in particular, it becomes possible for the minimum to appear. Since d is large, the work function for the system will exhibit bulk work-function values associated with the density $\bar{n}^A$ until $\bar{n}^A$ is very small, implying that these values, and hence $\Phi_{\min}$, will themselves be quite small. For $\bar{n}^A \rightarrow 0$, Φ must of course return to Φ^M.

*This is true for $r_s \gtrsim 2$ (see Fig. 3).

We show schematically the expected behavior of the $\Phi(\bar{n}^A)$ curve for large d in Fig. 26. The ascending part of the curve (past the minimum region) is simply the bulk work function appropriate to a sample of mean electron density $\bar{n}^A$. The way in which this behavior persists for smaller values of d is illustrated by including the Na and Cs curves of Fig. 23, here with $\bar{n}^A$ instead of N as the abscissa. We thus understand the decrease of $\Phi_{\min}$ along the series of alkalis from Li to Cs, i.e., with increasing d in the model.

4.2. *Uniform-Background Model for Substrate*

4.2.1. Ground State. The calculations that we describe in this section treat the classic chemisorption problem of a single atom bonded to a semi-infinite metal surface. Use of a uniform-background substrate leads to a problem with cylindrical symmetry, which represents an important simplification.

We denote by d the distance between the adatom nucleus and the positive background edge. This distance is fixed by minimizing the total computed energy. The model is completely specified by only two numbers: the density

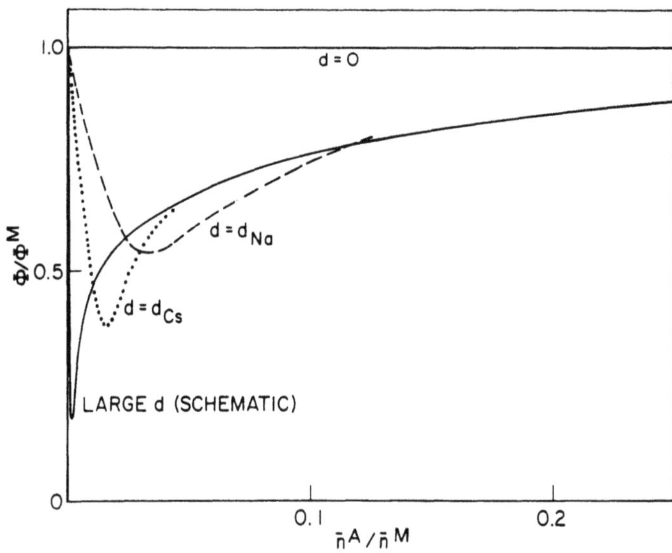

FIGURE 26. This shows the way in which the Φ–versus–$\bar{n}^A$ curve in the uniform-background model depends upon the adsorbate slab thickness d. All curves must meet at $\Phi = \Phi^M$ when $\bar{n}^A = \bar{n}^M$. Note the way in which $\Phi_{\min}$ decreases as d increases. (After Ref. 193.)

parameter r_s of the uniform-background substrate, and the adatom nuclear charge Z.

Two quantities of which we will take particular note are the dipole moment μ and the atomic binding energy ΔE_a. If we take the coordinate origin to be at the adatom nucleus and take $z \rightarrow \infty$ to be in the vacuum, then the dipole moment is defined as

$$\mu = -\int z\delta n(\mathbf{r})d\mathbf{r} \tag{68}$$

The atomic binding energy ΔE_a is the difference in total energy between the metal (M) plus separated atom (A) and the combined metal–adatom system (MA); i.e.,

$$\Delta E_a = E^A + E^M - E^{MA} \tag{69}$$

(a) Extended Thomas–Fermi Treatments. Ying, Smith, and Kohn[134,135] have discussed hydrogen chemisorption on a uniform-background substrate. They have obtained the linear density-potential response function for the substrate, using the extended Thomas–Fermi form for the energy functional $G[n]$ [equation (36)], and have employed this to find the induced density $\delta n(\mathbf{r})$ (and thus other ground-state properties) for a point perturbing charge of unit magnitude ($Z = 1$). The use both of linear response and of the extended Thomas–Fermi energy functional represent significant approximations, as is found in a comparison of the results of this treatment with those of a full wave-mechanical, nonlinear study of the problem [i.e., one based on solving equation (1)—see below].

The calculations were performed for a high-density substrate and a value for the equilibrium position was found that was verified by the subsequent fully wave-mechanical analysis.[144] An ionic desorption energy ΔE_i of 9 eV was obtained. Experimental data are available for H on tungsten, with $\Delta E_i \cong$ 11.3 eV[134] (though the uniform-background model is not particularly intended to represent transition metals). Schrieffer[195] has argued, however, that it is more meaningful to consider the atomic binding energy $\Delta E_a = \Delta E_i + \Phi - I$, with Φ the substrate work function and I the ionization potential of the free atom (for H/W, a measured ΔE_a is 3.0 eV). Ying *et al.*[135] use the measured work function of the most densely packed face of tungsten (5.3 eV) and the exact value of I (13.6 eV) to obtain $\Delta E_a = 0.7$ eV*; they draw the conclusion that the method, which by its nature yields ΔE_i as the primary result, is not precise enough to obtain ΔE_a. The approximations do not appear

*It would, with regard to this evaluation, seem more consistent to use the Φ of the uniform-background substrate actually employed in the calculations (3.9 eV).

to permit giving an adequate account of the dipole moment or of the electron density at the nucleus.*,†

Kahn and Ying[197] have used the response function of Ying *et al.* and a pseudopotential representation for the adatom core to study alkali chemisorption. For large atoms that tend to lose their valence electrons to the substrate, the linear-response screening picture can be expected to be much better than it is for hydrogen.

These authors have also used a different energy functional, in which the coefficient 1/72 of the gradient term in equation (36) is replaced by one that is density dependent. It had been found by Jones and Young[198] that if the susceptibility of a uniform electron gas is calculated using equation (36) (without exchange and correlation) and compared with the known exact result for the noninteracting gas, then in the limit in which the perturbing wave vector $Q \to 0$, the correct coefficient is 1/72, but in the limit $Q \to \infty$ (for which the gradient expansion is not designed to be appropriate), it is 1/8. The coefficient used by Kahn and Ying is made density dependent by identifying Q with the local Thomas–Fermi wave-vector in the inhomogeneous electron gas [$Q_{TF} \propto n(\mathbf{r})^{-1/6}$], and choosing a simple form that interpolates between the $Q \to 0$ and $Q \to \infty$ values. The calculated ionic desorption energies and dipole moments are in reasonable agreement with values found for adsorption on transition metals. The use of the density-dependent coefficient for the gradient term in the energy functional is found to have a significant effect on the value of the dipole moment.

Huntington, Turk, and White[199] have presented an extended Thomas–Fermi treatment for a chemisorbed Na atom, with no linear-response approximation. A variational form for the electron density was employed and the total energy minimized. The adatom core was represented using a pseudopotential; and the substrate density was chosen to be that of Na metal.

(b) Wave-Mechanical Treatments. The problem of an atom chemisorbed on a uniform-background substrate has been treated wave-mechanically [i.e., by solving equation (1)] by Lang and Williams,[144–146] and by Gunnarsson, Hjelmberg, and Lundqvist.[200–205] In the study of Lang and Williams, equation (1a) for the continuum states is cast into Lippmann–Schwinger form

$$\Psi^{MA}(\mathbf{r}) = \Psi^{M}(\mathbf{r}) + \int G^{M}(\mathbf{r},\mathbf{r}';\epsilon)\delta v_{\text{eff}}(\mathbf{r}')\Psi^{MA}(\mathbf{r}')d\mathbf{r}' \tag{70}$$

Here G^M is a Green's function for the bare metal and δv_{eff} is the difference in potential v_{eff} between the metal–adatom system and the bare metal. Now

*See discussion in Ref. 144.

†Wang and Weinberg[196] have discussed another version of this treatment in which the perturbation is taken to be an unpolarized H atom.

the differential equation (1a) can be solved by direct numerical integration outward from the adatom nucleus. Solutions $\Psi_l(\mathbf{r})$ obtained in this way are characterized by their angular behavior near the adatom nucleus and do not in general satisfy the boundary conditions embodied in the Lippmann–Schwinger equation; the desired solution is, however, a linear combination of these fundamental solutions,

$$\Psi^{MA}(\mathbf{r}) = \sum_l C_l \Psi_l(\mathbf{r}) \tag{71}$$

where the coefficients C_l are obtained by substituting (71) into (70). The use of the Lippmann–Schwinger equation provides a natural way of incorporating the fact that $\delta n(\mathbf{r})$ and $\delta v_{\text{eff}}(\mathbf{r})$ are, because of metallic screening, short-ranged about the adatom.

For the adatom core states, the eigenfunctions of equation (1a) satisfy a boundary condition that is simpler to impose than that for the continuum states, and a Lippmann–Schwinger equation is not used. Just as in the case of the continuum states, the direct numerical integration of equation (1a) outward from the nucleus involves the solution of radial equations that couple different l values (thus the core states are allowed to polarize in the full nonspherical potential of the metal–adatom system).

In the treatment of Gunnarsson, Hjelmberg, and Lundqvist, the metal–adatom eigenfunctions Ψ^{MA} are expanded in a basis consisting of the bare-metal eigenfunctions Ψ^{M} and a set of localized functions. The problem of overcompleteness is taken care of by a subsidiary condition on the coefficients of the bare-metal functions. Solving equation (1a) is reformulated as a Green's function problem in terms of matrices in just the localized functions, and $\delta n(\mathbf{r})$ is obtained using this Green's function. It is its short range that justifies expressing $\delta n(\mathbf{r})$ in terms only of the localized functions. Since this formalism was applied just to H chemisorption, it was not necessary to consider the treatment of core states.

Figure 27 shows $\delta n(\epsilon)$ for Li, Si, and Cl chemisorbed on a high-electron-density substrate ($r_s = 2$), as calculated by Lang and Williams.[144] The states constituting the Cl $3p$ resonance are below the Fermi level, and are therefore occupied; those constituting the Li $2s$ resonance are essentially empty. This implies that charge transfer has taken place, toward the Cl and away from the Li, as would be expected from the electronegativities of these atoms. The prohibitively large energy required either to fill or empty the Si $3p$ level forces this resonance to straddle the Fermi level, resulting in the formation of a covalent, rather than ionic bond.

The electron densities associated with the three fundamental bond types are exhibited for comparison in Fig. 28. In the top row of the figure are

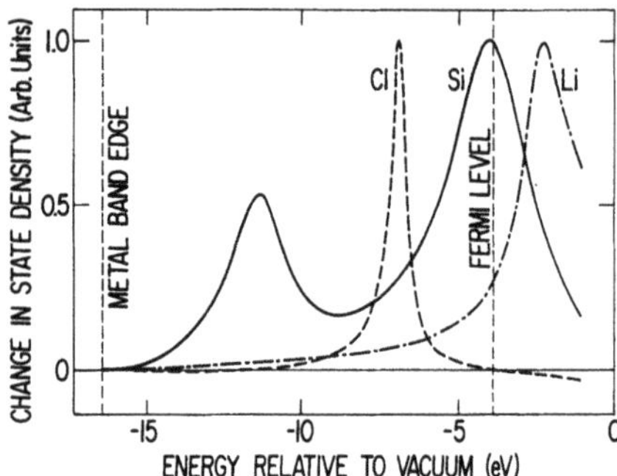

FIGURE 27. Change in state density $\delta n(\epsilon)$ due to chemisorption of one atom on a uniform-background substrate ($r_s = 2$). Curves correspond to metal–adatom distance which minimizes the total energy. Note that the lower Si resonance corresponds to the 3*s* level of the atom; for Cl this is a discrete state below the bottom of the band (band edge). (From Ref. 144.)

contours of constant total electron density. Note first the way in which these contours rapidly regain their bare-metal form away from the immediate region of the atom. This is a graphic illustration of the short range of metallic screening. Note also the way in which the metal contours bend toward the positively charged Li and away from the negatively charged Cl. The very elongated outermost closed contour in the Si plot is characteristic of such plots for covalently bonded diatomic molecules.

The detailed charge rearrangements associated with chemical-bond formation are displayed most clearly by the contour maps of the difference between the electron density in the chemisorption system and the superposition of bare-metal and free-atom densities shown in the second row of Fig. 28. The solid contours indicate regions of charge accumulation; broken contours indicate regions of charge depletion.

The density-difference contours in the case of Li reveal the complexity underlying the notion of charge transfer. Taken alone, this difference plot indicates only that electrons have been displaced from the vacuum side to the metal side of the atom. The plot does not distinguish between the two possible interpretations of the displacement: polarization of the atom, or ionization (followed by metallic screening). The state-density graph in Fig. 27, however, resolves this ambiguity: the fact that virtually none of the states in the Li 2*s* resonance are occupied indicates that the ionization/screening interpretation is the correct one.*

The kidney-shaped depletion contour on the vacuum side of the Li adatom, and even the reverse-dipole contours in the core region, are very similar to those found by Bader and co-workers[206] in difference plots for the LiH

*A further confirmation of this interpretation is provided by the fact that as the Li atom is moved somewhat further away from the surface, the region of charge accumulation in the difference plot tends to stay behind on the surface.

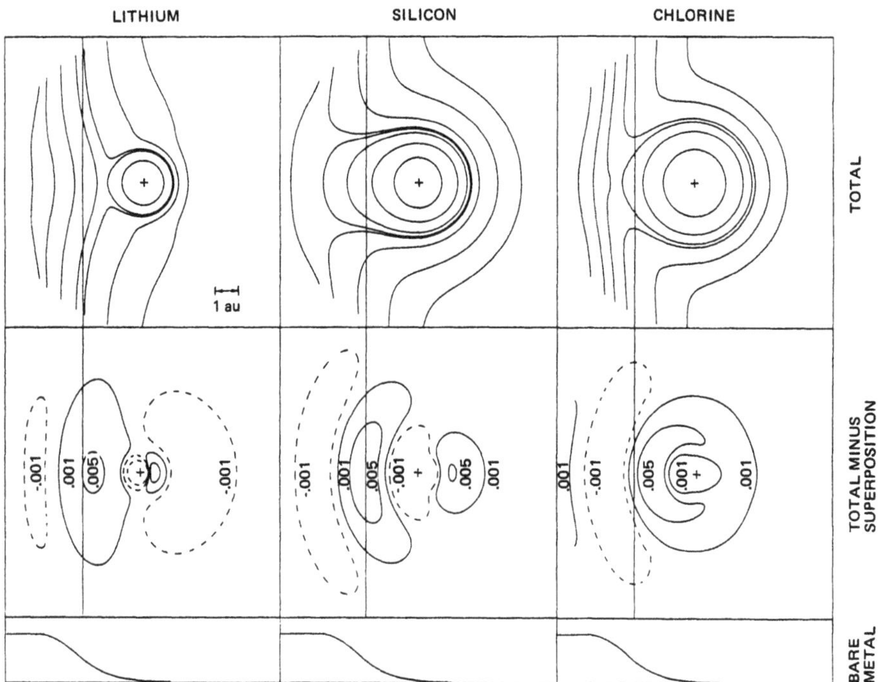

FIGURE 28. Electron-density contours for chemisorption of one atom on a uniform-background substrate ($r_s = 2$). Metal–adatom distances shown minimize the total energy. Upper row: Contours of constant density in (any) plane normal to the metal surface containing the adatom nucleus (indicated by +). Metal is to the left-hand side; positive-background edge indicated by vertical line. Contours are not shown outside the inscribed circle of each square; contour values were selected to be visually informative. Center row: total electron density minus the superposition of atomic and bare-metal electron densities (electrons/bohr^3). The polarization of the core region, shown for Li, was deleted for Si and Cl because of its complexity. Bottom row: bare-metal electron-density profile (shown to establish physical distance scale). (For reference, the bulk metal density is 0.03 electrons/bohr^3.) (From Ref. 144.)

and LiF molecules. Note also that for this case, as well as for Si and Cl, the sequence of contours continues into the metal in the form of Friedel oscillations induced by the perturbing atom.

The difference contours for Cl show clearly that charge has been transferred from the metal to form a polarized negative ion. In the case of Si, there is a central region of charge depletion, and accumulations on both the bond and vacuum sides. The same general configuration of contours seen here for Si is found in difference plots for covalently bonded diatomic molecules in which p orbitals play a significant role.[206]

The connection between the covalent bond charge seen for Si in the contour maps and the partial occupation of the valence resonance is exhibited in Fig. 29. Contour plots of $\delta n(\epsilon,\mathbf{r})$ are shown for four values of ϵ: one value each from the low-energy and high-energy parts of both valence resonances. These two valence resonances are identified as arising from the $3s$ and $3p$ atomic levels. There are two basic interactions of interest in this case: the interaction between each atomic level and the metal, and the intra-atomic mixing of the two levels ($3s$ and $3p_z$) due to the asymmetry of the surface potential. The interaction between each atomic level and the metal broadens each level into a resonance, whose lower part adds charge to the bond region ("bonding") and whose upper part subtracts charge from this region ("antibonding"). The interaction between the levels provides an overall polarization to the charge distributions associated with each level—into the bond region for the "$3s$" charge distribution and into the vacuum for the "$3p$" charge. Note that the two effects tend to work in opposite directions in the cases labeled "antibonding s" and "bonding p" in Fig. 29, particularly in the latter case, as seen in the contour plot. The fact that the upper part of the $3p$ resonance is unoccupied means that the negative contribution it would make to the charge density in the bond region is not present, and that there is a net accumulation of bond charge from the states that are occupied.

Figure 30 shows plots of $-\Delta E_a(d)$ for a hydrogen atom on substrates of three different densities, as calculated by Hjelmberg.[205] Despite the large variation of substrate density, the atomic binding energies ΔE_a [$\equiv$ max. of $\Delta E_a(d)$] vary by only $\sim$25%.[205a] The repulsive barrier of the substrate for the adatom is due essentially to "kinetic-energy repulsion," as pointed out by Gunnarsson.[6] He notes that when the proton is outside the surface, the electrons of the metal–adatom system have the low-potential regions of both the metal and the proton vicinity to move in, whereas when the proton is inside, it no longer contributes a volume of low potential in addition to that of the metal. It is this reduction of volume, i.e., an effective compression of the total electron distribution, that leads to an increase of kinetic energy; this increase is particularly strong at high substrate densities.*

While the calculations of Lang and Williams and of Gunnarsson, Hjelmberg, and Lundqvist predict equilibrium metal–adatom separations, the re-

*This argument must be appropriately qualified in order to discuss inhomogeneous systems, such as a metal *not* treated using the uniform-background model into which hydrogen dissolves interstitially.[207]

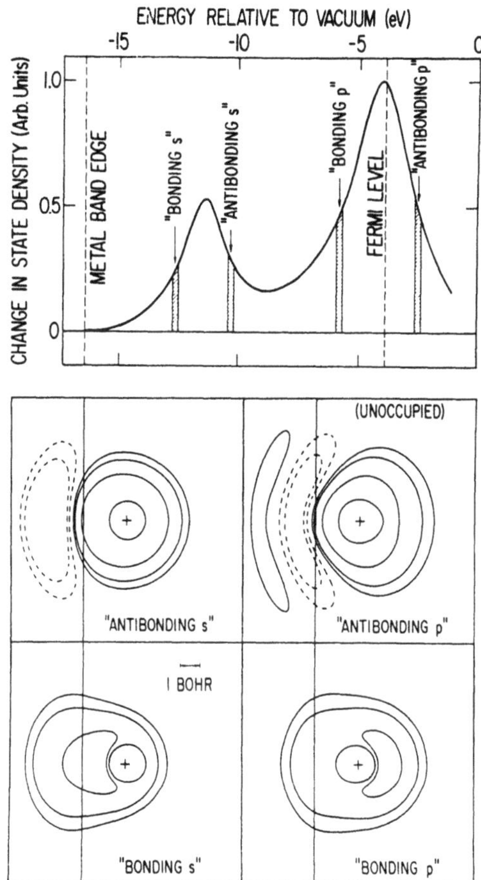

FIGURE 29. Upper part of figure reproduces the state density curve from Fig. 27 for a Si atom chemisorbed at its equilibrium distance on a uniform-background substrate ($r_s = 2$). The two peaks correspond to the $3s$ and $3p$ atomic states. The lower part of the figure shows the density contours [contours of $\delta n(\epsilon,\mathbf{r})$] associated with the four shaded regions in the state density curve. (See caption of Fig. 28 for details of such contour maps.) Solid lines correspond to positive contour values, dashed lines to negative values. The same set of contour values is used for all four cases. The crescent-shaped contours in the two "bonding" maps correspond to maxima in the density. (Contours near the nucleus were deleted for clarity.) (From Ref. 146.)

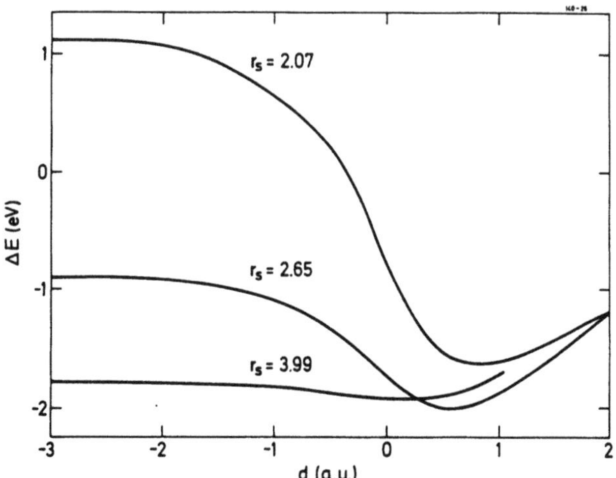

FIGURE 30. Energy change upon chemisorption as a function of distance [negative of the atomic binding energy $\Delta E_a(d)$], for hydrogen atom on uniform-background substrates of three different densities: r_s = 2.07 (Al), 2.65 (Mg), and 3.99 (Na). (A.u. ≡ bohr.) (From Ref. 205.)

sults of calculations performed for other separations yield information important to discussing the dynamics of atoms at surfaces, and provide theoretical insight into the distance dependence of the metal–adatom interaction. Of particular interest are the variations of the state-density difference $\delta n(\epsilon)$[208] and the dipole moment μ.

In Fig. 31, $\delta n(\epsilon)$ is shown for a chemisorbed hydrogen atom at three different distances. At the largest distance, the metal–adatom interaction is not strong, and the resonance that is present below the Fermi level is relatively narrow. When the atom is moved closer, this interaction increases, and the resonance widens considerably. It also moves further below the Fermi level, showing a tendency to follow the bare-metal surface potential. When the atom is moved still closer, the broadening due to increasing metal–adatom interaction is overtaken by narrowing due to the decreasing density of metal states seen by the resonance as it moves down toward the bottom of the metal band, and the resonance narrows again.*,†

*The fact that the resonance position tends to follow the surface potential can be easily demonstrated using first-order perturbation theory (see Ref. 146). A more general analysis of changes in resonances with distance has been given by Gunnarsson, Hjelmberg, and Nørskov.[209]

†Muscat and Newns[210] have interpreted this behavior using a simple model.

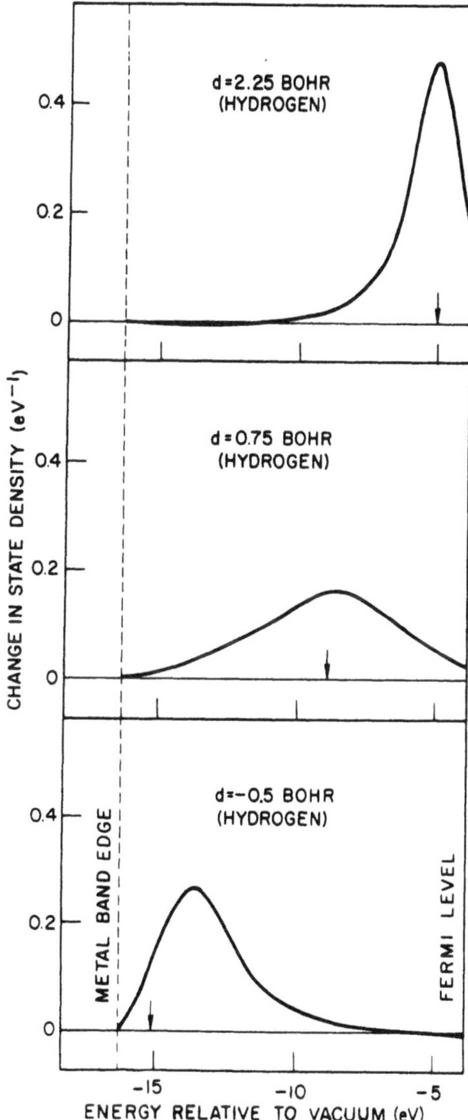

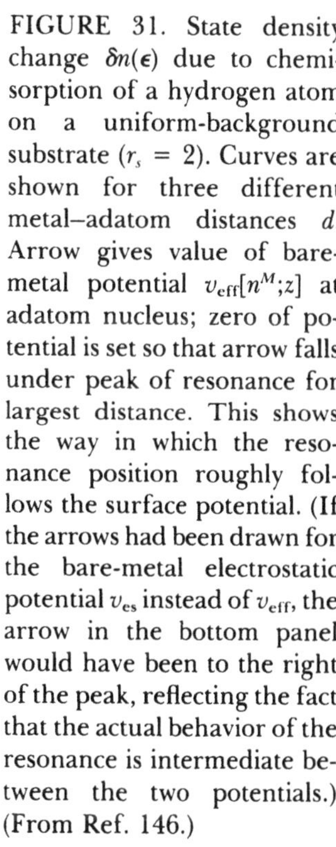
FIGURE 31. State density change $\delta n(\epsilon)$ due to chemisorption of a hydrogen atom on a uniform-background substrate ($r_s = 2$). Curves are shown for three different metal–adatom distances d. Arrow gives value of bare-metal potential $v_{\text{eff}}[n^M;z]$ at adatom nucleus; zero of potential is set so that arrow falls under peak of resonance for largest distance. This shows the way in which the resonance position roughly follows the surface potential. (If the arrows had been drawn for the bare-metal electrostatic potential v_{es} instead of v_{eff}, the arrow in the bottom panel would have been to the right of the peak, reflecting the fact that the actual behavior of the resonance is intermediate between the two potentials.) (From Ref. 146.)

In connection with Fig. 31, we show in Fig. 32 the effective potential v_{eff} for chemisorbed hydrogen (calculated by Hjelmberg[203] for an $r_s = 2.07$ substrate). It is particularly noteworthy that there is only a very small potential barrier between the metal and the adatom. From Fig. 31 we see that the resonance position is above the barriers shown (along the normal direction), implying a strong mixing between metal and adatom states.

Figure 33 shows the changes with distance in the position and width of the valence resonances for Si and Cl chemisorbed on a high-density substrate. The Si–3s resonance shows particularly clearly the narrowing at both small and large distances discussed in connection with hydrogen. The asymmetry of the surface potential splits the p_z and p_{xy} components of the p resonance. This splitting is given in the figure for Si. At shorter distances (not shown) the splitting decreases to zero and changes sign. (Details are discussed in Ref. 146.)

The dipole moment as a function of distance for Na, Si, and Cl (on a

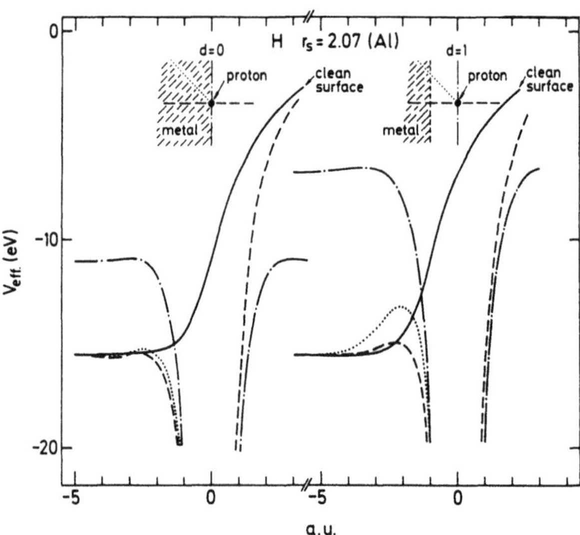

FIGURE 32. The effective potential $v_{eff}[n^{MA};\mathbf{r}]$ for a hydrogen atom chemisorbed on a uniform-background substrate ($r_s = 2.07$, corresponding to Al) at two different metal–adatom separations ($d = 0$ and $d = 1$ bohr). The curves show v_{eff} as a function of distance from the proton along lines normal (– – –), parallel (– · –), and at an angle of 45° (· · ·) to the surface. The solid line shows the effective potential for the bare metal surface. (A.u. ≡ bohr.) (From Ref. 203.)

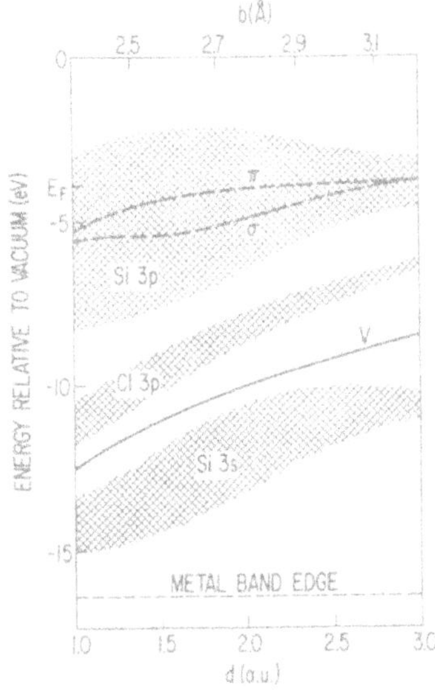

FIGURE 33. Characteristics of Si and Cl resonances (cf. Fig. 27) as a function of metal–adatom separation for adsorption on a uniform-background substrate (r_s = 2). Boundaries of shaded regions indicate half-maximum energies. Lower axis d is separation of adatom nucleus from positive-background edge (a.u. ≡ bohr). The plane through the outermost nuclei of the substrate represented by this background is half an interplanar spacing behind the background edge. Thus d, the crystal structure of the substrate, and the adsorption site determine an adatom to metal-atom bond length b. The upper axis provides b for a threefold site on a (111) surface of Al as an example. Dashed curves show peak positions for p_z (σ) and p_{xy} (π) components of Si–3p resonance. (Not shown for Cl.) Curve V gives effective one-electron potential ($v_{\text{eff}}[n^M;z]$) of bare metal (displaced downward for pictorial reasons). (From Ref. 144.)

high-density substrate) is shown in Fig. 34. (It is useful in thinking about this figure to keep in mind the density difference maps in Fig. 28.) One can define a dynamic charge on the adatom as the slope of such a curve, i.e., as $\mu'(d)$. For distances in the central part of the graph, $\mu'(d)$ is ~ +0.4 for Na, ~ −0.5 for Cl, and ~0 for Si (units are magnitude of the electron charge).

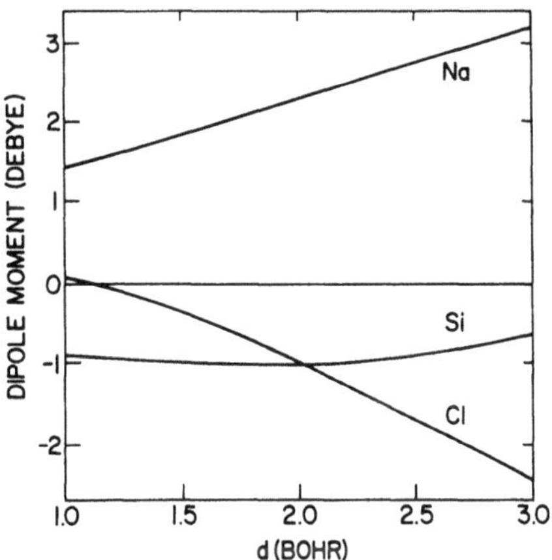

FIGURE 34. Dipole moment as a function of metal–adatom distance d for Na, Si, or Cl atom chemisorbed on a uniform-background substrate (r_s = 2). The sign of the dipole moment is defined so that a negative moment corresponds to an increase in substrate work function [see the definition of the dipole moment in equation (68)]. (From Ref. 146.)

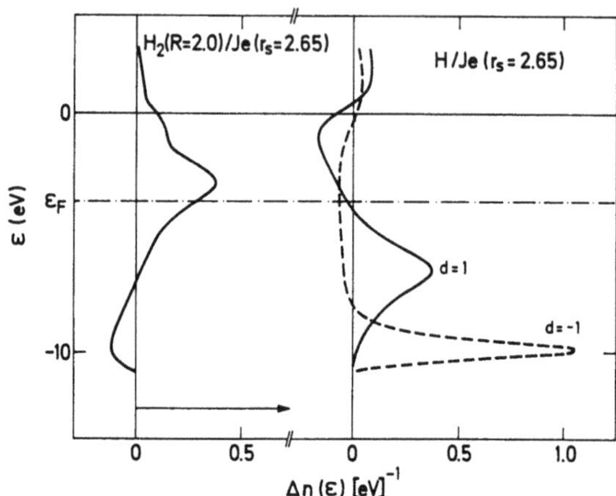

FIGURE 35. This figure discusses the change in state density $\Delta n(\epsilon)$ [$\equiv \delta n(\epsilon)$] induced by two hydrogen atoms at a uniform-background-model surface, one placed 1 bohr inside the background edge and the other 1 bohr outside (r_s = 2.65, corresponding to Mg). In the right half of the figure, the atoms are widely separated laterally (and thus do not interact with one another). In the left half, the atoms have zero lateral separation, thus forming an H_2 molecule that stands upright with internuclear distance R = 2 bohr, centered at the background edge. (From Ref. 212.)

These numbers should not, however, be taken as providing a good measure of static charge transfer because of the importance of other contributions, such as the distance dependence of polarization effects, to the value of $\mu'(d)$.*

Nørskov, Hjelmberg, and Lundqvist[212] have done calculations for H_2 standing upright on a uniform-background substrate, using an extension[213] of the calculational scheme of Gunnarsson, Hjelmberg, and Lundqvist.[200–205] As illustrated in Fig. 35, when the two atomic adsorbates interact, bonding and antibonding states/resonances occur, as in the free molecule. The primary difference is that at the surface, the antibonding resonance may be partly filled, leading to a weakened H–H interaction. In fact, the intramolecular energy E_{intra}, defined as the difference between the chemisorption energy of the molecule and the chemisorption energy of the isolated constituent atoms (E_{extra}), is found to be completely determined by the position of the antibonding resonance relative to the Fermi level. As in the case of atomic adsorbates, the resonance is shifted down and broadened upon approaching

*This point is discussed in the context of diatomic molecules in Ref. 211.

the surface (see Fig. 36). On high-density substrates, the antibonding resonance becomes sufficiently filled to make the H–H interaction repulsive, and the molecule dissociates. If on the other hand the antibonding resonance is not shifted downward appreciably, as in the low-density case (Fig. 36), an attractive H–H interaction remains, and the molecule adsorbs associatively.*

As the molecule approaches the surface from large distances, E_{intra} is found initially to increase more rapidly than E_{extra} (the adsorption energy of the constituent atoms) decreases.[214] This means that the molecular chemisorption energy initially increases, giving rise to an activation energy for adsorption (see Fig. 37). Such an activation barrier arising from the interaction of the molecule with the s–p conduction electrons of the substrate is also expected[215] to be present on transition metals, although it may be diminished (or even outweighed) by the interaction of the molecule with the substrate d electrons.

4.2.2. Adsorbate Core Holes. We discuss here the use of the density-functional formalism, in the form of equation (1), to study core holes in adsorbates. Now this formalism is, strictly speaking, not valid in general for excited states of the system†; it has, however, given results in good agreement with experiment when applied for example to the study of deep core holes in transition metals.[217] The justification for its use for core holes is thus only empirical; and there are particular cases, such as those in which core-hole fluctuations are important, as discussed by Wendin, Ohno, and Lundqvist,[218] where, in an unmodified form, it can be expected to be inappropriate.

Applying equation (1) to this problem involves solving these equations for the case in which the n_i in equation (1b) corresponding to a particular core level is set equal to 0 instead of 1. There will in general be a self-consistent solution for this problem just as there is for the ground state. It is of interest to consider the electron density $n(\mathbf{r},n_i)$, the total energy $E(n_i)$, and the core-state eigenvalue $\epsilon_i(n_i)$, for the cases in which n_i is 0 or 1. These quantities are evaluated both for the metal–adsorbate system (superscript MA) and the free adsorbate, which will be a single atom in the work discussed here (superscript A).

The experimental quantity usually discussed in this problem is the difference in core-level binding energy between the free atom and the solid:

*The filling of the antibonding resonance is also found to control the H_2 vibrational frequency for H_2 standing upright on the surface. As would be expected, the frequency decreases with increased filling of this resonance (i.e., with decreasing intramolecular interaction).[212]

†It is, however, valid for the energetically lowest state of each symmetry.[216]

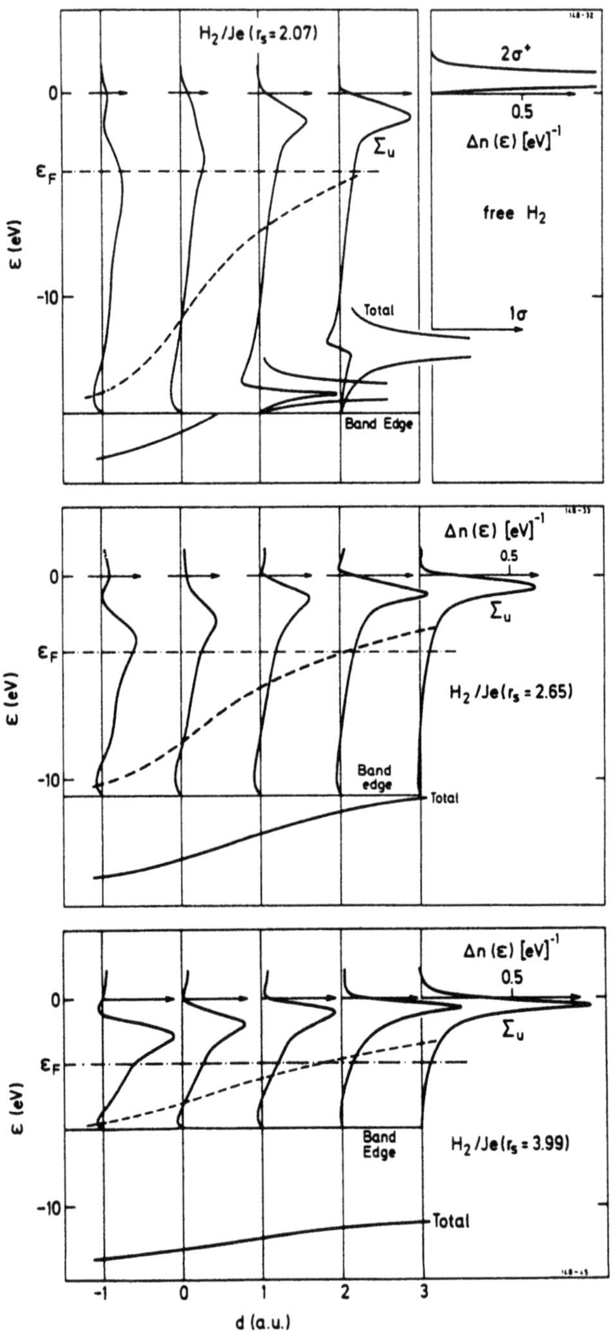

H2/Je (rs = 2.07)
0
εF
ε (eV)
-10
Σu
Total
Band Edge
2σ+
0.5
Δn (ε) [eV]-1
free H2
1σ
Δn (ε) [eV]-1
0.5
0
Σu
εF
ε (eV)
H2/Je(rs = 2.65)
-10
Band edge
Total
Δn (ε) [eV]-1
0.5
0
Σu
εF
ε (eV)
Band Edge
H2/Je(rs = 3.99)
-10
Total
-1
0
1
2
3
d (a.u.)

FIGURE 36. Change in state density $\Delta n(\epsilon)$ [$\equiv \delta n(\epsilon)$] induced by an H_2 molecule standing upright on a uniform-background-model surface, as a function of the distance d between the molecule center and the background edge, for three r_s values: 2.07 (Al), 2.65 (Mg), and 3.99 (Na). In the top section, at the right, the bonding 1σ and antibonding $2\sigma^*$ levels are indicated for free H_2. Immediately to the left of this ($d = 2$) is shown the $\Delta n(\epsilon)$ curve for energies near the bottom of the metal band (labeled "total"); this resonance corresponds to the free-molecule 1σ level. On the same axis is also shown a Σ_u-projection of $\Delta n(\epsilon)$, for energies from the bottom of the band up to a point just above the vacuum level. The resonance in this projection that is situated primarily above ϵ_F corresponds to the free-molecule $2\sigma^*$ level. For $d \lesssim 0.5$ bohr, the resonance in $\Delta n(\epsilon)$ corresponding to the 1σ level becomes a discrete state below the bottom of the band. In the middle and lower sections of the figure, curves are shown for five different distances; in these cases the 1σ-derived state is discrete, as indicated. The dashed curves show the effective potential v_{eff} for the bare metal. Internuclear distance $R = 1.4$ bohr. (A.u. $\equiv$ bohr.) (From Ref. 212.)

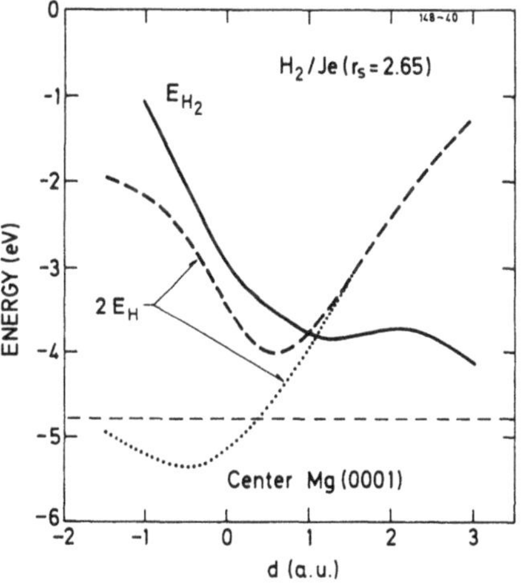

FIGURE 37. Solid curve (E_{H_2}) gives energy change upon chemisorption as a function of distance ("potential energy curve") for H_2 standing upright on a uniform-background substrate ($r_s = 2.65$, corresponding to Mg; internuclear distance $R = 1.4$ bohr.) The activation energy for adsorption is seen to be about 1 eV. The dashed curve gives $2E_H$ for the case in which E_H is the energy change upon chemisorption of a single H atom on the uniform-background substrate. Comparison between the solid and dashed curves shows that dissociative chemisorption is favored. For H_2 parallel to the surface, a similar E_{H_2} curve is expected. In this case, increasing the internuclear distance will give a smooth transition from the E_{H_2} curve to the $2E_H$ curve. The dotted curve gives $2E_H$ for chemisorption at a centered site on Mg⟨0001⟩ (perturbative treatment); this curve suggests that including the lattic correction for H_2 standing upright at the centered site would have a negligible effect on the activation barrier because the correction is only important closer to the surface than the region of the barrier. (A.u. $\equiv$ bohr.) (From Ref. 212.)

$$\Delta = I^A - I^{MA} \tag{72}$$

where

$$I^{MA} = E^{MA}(0) - E^{MA}(1) \tag{73a}$$

$$I^A = E^A(0) - E^A(1) \tag{73b}$$

with the electron removed from the core taken to the vacuum level in both cases. Calculations of Δ for the deep core levels in several adatoms were performed by Lang and Williams[145] by finding the self-consistent solutions of equation (1) and corresponding total energies, for the four cases required: free atom and metal–adatom system, both with and without the core hole.* (These calculations used an $r_s = 2$ uniform-background substrate.) For the case of oxygen chemisorbed on polycrystalline Al, the experimental value of Δ (for the $1s$ level) is 9.9 ± 0.5 eV,[145] while the calculated value is 9.5 eV.[145],† (Experimental data for other atoms chemisorbed on simple metals were not available.)

The change in binding energy Δ can be thought of as having two components: the change which takes place prior to the removal of the electron (chemical or initial-state shift) and the change associated with the screening of the core hole (relaxation or final-state shift). An appropriate definition of the chemical shift (Δ_{chem}) within the framework of the density-functional formalism is the change in eigenvalue that takes place upon chemisorption:

$$\Delta_{\text{chem}} = \epsilon_i^{MA}(1) - \epsilon_i^A(1) \tag{74}$$

This shift occurs because of changes in the environment and chemical state of the atom, such as those due to charge transfer or to partial penetration by the atom of the surface double-layer. The relaxation shift Δ_{relax} is then defined as $\Delta - \Delta_{\text{chem}}$.

*The energy difference for a given system with and without the core hole can be written as $-\epsilon_i(1/2)$ (Slater's transition-state form),[219] thus avoiding the subtraction of total energies. The numerical effect of using this form here instead of subtracting total energies is found to be of no importance. The transition-state form is very useful in the theoretical development, however.

†The evidence in the experiment pointed to the fact that the oxygen, which was dissociated, had penetrated the surface, as readily happens in this system.[220] The calculation therefore was done for an oxygen atom inside the surface. The result given also includes a pseudopotential correction for the substrate lattice (see Section 4.3 below).

It is particularly interesting to examine the internal structure of the charge distribution that screens the core hole and which leads to Δ_{relax}. The extra-atomic screening charge distribution is defined as

$$n(\mathbf{r};\ \text{extra-atomic}) = [n^{MA}(\mathbf{r};0) - n^{MA}(\mathbf{r};1)] - [n^{A}(\mathbf{r};0) - n^{A}(\mathbf{r};1)] \quad (75)$$

In terms of this distribution, a close approximation to Δ_{relax} for deep core levels can be shown to be[2,145]

$$\Delta_{relax} \doteq \frac{1}{2} \int \frac{n(\mathbf{r};\ \text{extra-atomic})}{r}\, d\mathbf{r} \quad (76)$$

i.e., half the electrostatic potential at the adatom nucleus (taken as the origin) due to $n(\mathbf{r};$ extra-atomic).

Figure 38 shows the state-density difference $\delta n(\epsilon)$ for Na chemisorbed on a high-density substrate, both before and after the creation of a deep core

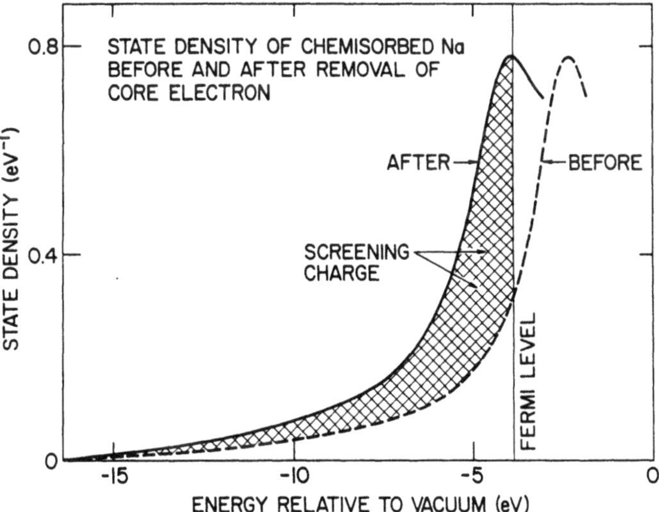

FIGURE 38. Change in state density $\delta n(\epsilon)$ for Na atom chemisorbed on a uniform-background substrate ($r_s = 2$), before and after removal of 2s core electron. Discrete portion of spectrum not shown. [Metal–adatom separation shown is the equilibrum distance for the case in which an Al⟨111⟩ pseudo-potential lattice is included to first order in calculating the total energy.] (From Ref. 145.)

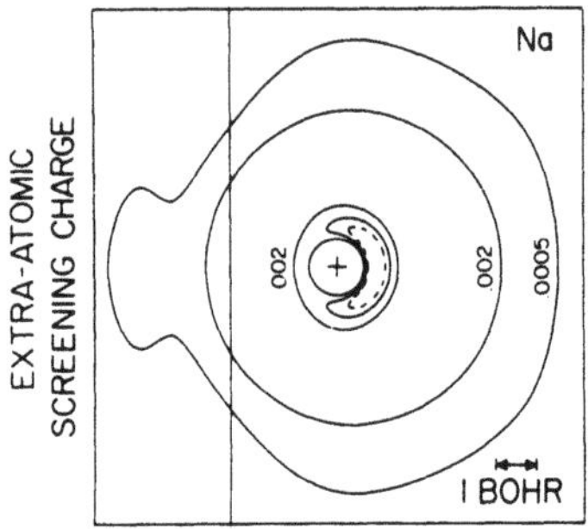

FIGURE 39. Contour map of extra-atomic screening charge density for Na atom chemisorbed on a uniform-background substrate ($r_s = 2$). Contours in core region deleted for clarity. Adatom nucleus denoted by +; see caption of Fig. 28 for other details of map. Metal–adatom separation chosen as in caption of Fig. 38. (2s hole state.) (From Ref. 145.)

hole. It is the additional charge which occupies the resonance when it is pulled down through the Fermi level that screens the hole.

The screening of a core hole is sometimes visualized as the linear response of the bare metal to a point positive charge at the position of the adatom nucleus, i.e., the screening charge is taken to have an imagelike form. That this is not an appropriate picture (in the static limit which we are considering) is seen in the contour plot given in Fig. 39 for the extra-atomic screening charge distribution of the Na case just mentioned. These contours are quite different from those of the bare metal plus point charge case (shown in Fig. 9). Their circularity about the nucleus suggests instead the "excited-atom" screening picture,[145,221] according to which the screening charge resembles the distribution obtained by subtracting the electron density of the free atom with a core hole from that of the excited atom (free atom with a core hole and an extra electron in the lowest unoccupied valence orbital). This is confirmed in Fig. 40, which shows that the radial distribution of $n(\mathbf{r}$; extra-atomic) for the Na case is very similar to that of the excited-atom picture.* This will generally be true so long as there is a partially empty valence level close to the Fermi level in the ground state. (It is true, e.g., for Si, as seen in the figure, but not for Cl, whose valence resonance is filled.)

For the cases of Na, Si and Cl ($r_s = 2$ substrate) considered in Ref. 145, the calculated relaxation shifts were found to be ~5 eV, about twice the magnitude of the chemical shifts obtained in this analysis.

4.3. *Lattice Model for Substrate (Perturbative Treatment)*

A number of the effects of the discrete substrate lattice in chemisorption can be studied via a perturbative reintroduction of the lattice into the uniform-background substrate [The perturbing potential is that given earlier in equa-

*Inserting the actual and the excited-atom screening distributions into equation (76) for Δ_{relax} gives 4.7 and 4.5 eV, respectively.

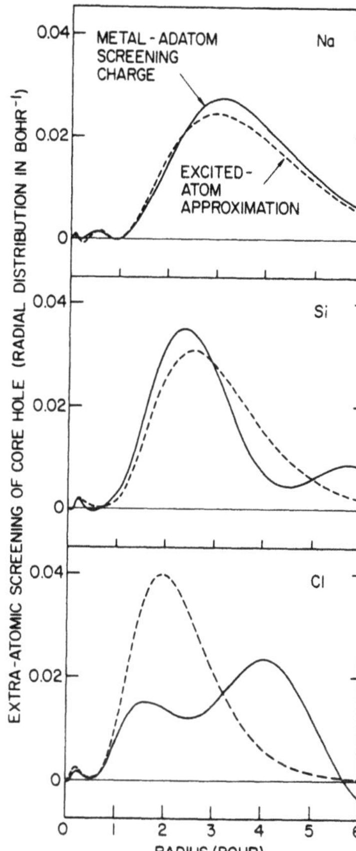

FIGURE 40. Radial distribution of extra-atomic screening charge (charge between r and $r + dr$ divided by 4π) for Na, Si, and Cl atoms chemisorbed on a uniform-background substrate (r_s = 2). Distance is measured from adatom nucleus. See Ref. 145 for metal–adatom separations. ($2s$ hole state.) (From Ref. 145.)

tion (57)]. As in the bare-surface case, this perturbative treatment is suitable only for simple-metal surfaces. The treatments presented have used first-order perturbation theory, and, accordingly, the result they yield is the value of the total system energy as a function of binding site and metal–adsorbate separation. The minimum value of this total-energy function indicates the most favorable binding site and equilibrium adsorbate–substrate bond length, and the adsorbate binding energy (heat of adsorption).

Such an analysis was first presented by Gunnarsson, Hjelmberg, and Lundqvist.[200–205] Curves of (the negative of) the atomic binding energy for a hydrogen atom on Al⟨100⟩ calculated by Hjelmberg[205] are shown in Fig. 41. Analogous curves for Si/Al⟨111⟩, calculated by Lang and Williams,[146] are shown in Fig. 42. The metal–adatom separation at the most favorable adsorption sites is reduced from that appropriate to the uniform-background substrate, and the atomic binding energy at these sites is larger. Estimates of

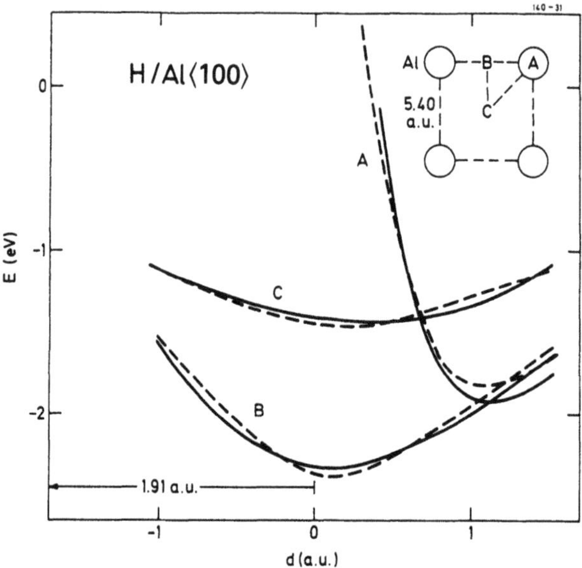

FIGURE 41. Energy change upon chemisorption (negative of the atomic binding energy ΔE_a) as a function of distance d for a hydrogen atom on an Al⟨100⟩ substrate. Dashed lines give results for the case in which the substrate is represented by the uniform-background model ($r_s = 2.07$) with a pseudopotential lattice included using first-order perturbation theory. Solid lines give results for the case in which part of the substrate lattice pseudopotential is included nonperturbatively as discussed in Ref. 101. Separation d is measured from the positive-background edge; note that the outermost plane of substrate nuclei lies half an interplanar spacing behind this edge (corresponding to $d = -1.91$ bohrs in the present case). (A.u. ≡ bohr.) (From Ref. 205.)

quantities such as the dipole moment μ and the state density difference $\delta n(\epsilon)$ in the presence of the pseudopotential lattice can be obtained by evaluating these quantities at the reduced metal-adatom separation.* This of course neglects any direct effect of the lattice.

4.4. *Lattice Model for Substrate (Nonperturbative Treatment)*

We discuss here self-consistent density-functional analyses of chemisorption on metal surfaces in which the full discrete-lattice potential (or pseudopotential) of the metal is taken into account. We omit discussion of studies in which a small cluster of metal atoms is used to simulate a semi-infinite

*Graphs such as Figs. 31, 33, and 34 are appropriate for this evaluation.

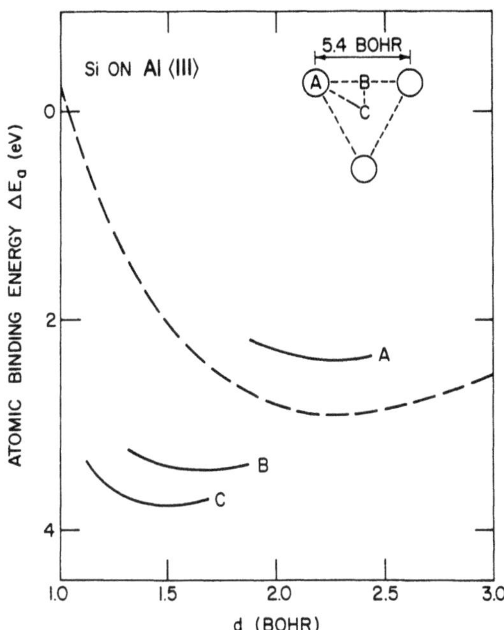

FIGURE 42. Atomic binding energy ΔE_a as a function of distance d for a Si atom on an Al⟨111⟩ substrate. Solid lines give results for the case in which the substrate is represented by the uniform-background model (r_s = 2) with a pseudopotential lattice included using first-order perturbation theory. Dashed line gives result for uniform-background model without lattice correction. Separation d is measured from the positive-background edge; the outermost plane of substrate nuclei lies 2.2 bohrs behind this edge in the present case. The calculated equilibrium position for adsorption in the centered site C (which is over a hole in the second layer of atoms) corresponds to an Al–Si bond length of 2.6 Å. (From Ref. 146.)

substrate,† and consider only those in which the substrate is taken to be infinite parallel to the surface. The reader is referred to the literature[223] for an analysis of chemisorption on clusters. We also note here a number of important density-functional studies of chemisorption on semiconductors.[224–231]

Smith, Arlinghaus, and Gay[232] have studied the chemisorption of nitrogen on Cu ⟨100⟩, using the same general procedure they employed to study the bare Cu surface (see above). Their system consisted of a 3-layer slab of Cu with a monolayer of N on either surface. The full lattice potential was taken into account, and the calculation was done completely self-consistently

†Cf. the discussion of Melius, Upton, and Goddard[222] concerning this simulation.

(no potential constraints such as the muffin-tin approximation were used.) An atomic-orbital basis set was employed for both Cu and N. The adsorption site and adsorbate-substrate separation were obtained from experimental data. The wave-mechanical equations of Kohn and Sham [equations (1)] were solved, using the local-density approximation for exchange-correlation.

The conduction-band charge density obtained is shown in Fig. 43. The bulklike nature of the second (central) Cu plane is fairly evident, and it is found that the central-plane densities in this calculation and that for the bare

FIGURE 43. Conduction-band charge density for nitrogen chemisorbed on Cu⟨100⟩ plotted on a plane normal to the surface (passing through a line connecting a surface Cu atom with one of its near neighbors in the second plane of Cu atoms). (The adatoms occupy fourfold sites directly above the centers of the squares formed by Cu atoms in the surface.) The units of charge density are 1.5×10^{-3} electrons/bohr^3. Charge densities on successive contours differ by a factor of 2. (From Ref. 232.)

Cu slab are virtually identical. This shows, as was seen earlier in Fig. 28, how rapidly the effects of an adsorbate are screened out by the metal. The gross features of the experimentally measured state-density (i.e., whether there is an enhancement or a reduction relative to the bare surface in a given energy region) are reproduced by the calculation. Self-consistency was found to be very important—the total system density of states with the final self-consistent potential is quite different from that for the starting potential constructed using overlapping atomic charge densities.

The calculation of Louie[233] for H/Pd⟨111⟩ proceeds in the same way as that for bare Pd.[173] A 7-layer Pd slab is taken to have one monolayer of H on each surface and a vacuum region equivalent to four layers of Pd; this is repeated along the surface normal to produce a problem periodic in three dimensions. A basis set of plane waves plus d-like Gaussians is used (mixed basis). The Pd ion core is represented using a pseudopotential, and the substrate–adsorbate separation is determined by taking the H–Pd bond length

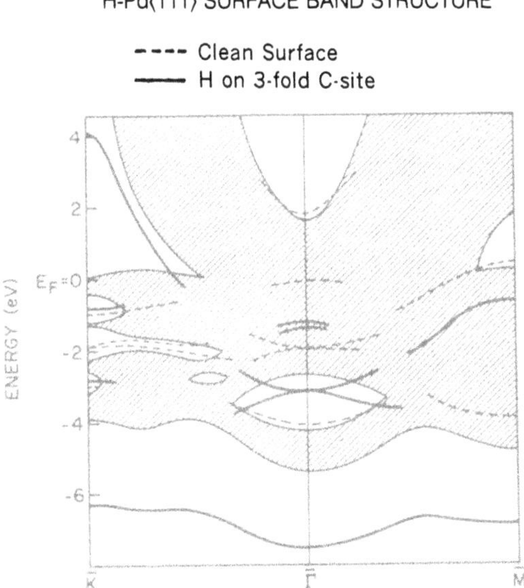

FIGURE 44. Localized states (solid curves) and the projected bulk band structure (shaded areas) in the $\bar{K} = (2\pi/a)(0,-\frac{2}{3},\frac{2}{3})$ and $\bar{M} = (2\pi/a)(-\frac{1}{3},-\frac{1}{3},\frac{2}{3})$ directions for hydrogen chemisorbed at the centered threefold sites on Pd⟨111⟩ (a is the bulk-crystal lattice constant). (The C site is the centered site over an atom in the second Pd layer.) Also indicated are the surface states for the clean surface (dashed curves). (After Ref. 233.)

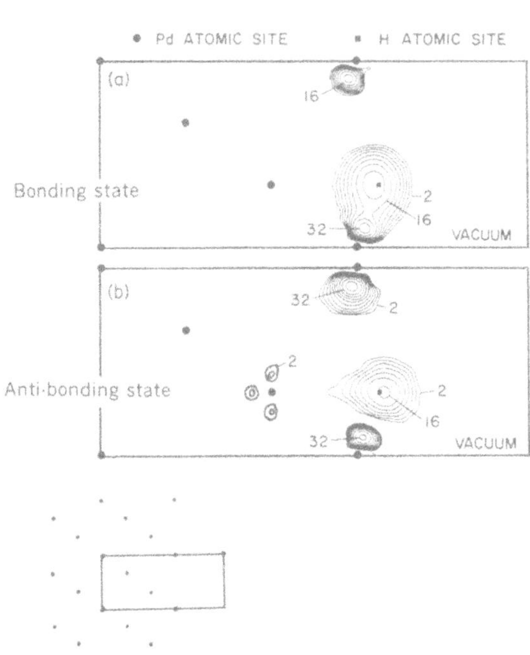

FIGURE 45. Charge density contour plots for adsorbate states at $\bar{K}$ at (a) -6.3 eV and (b) 4.1 eV. [H/Pd⟨111⟩ at C sites.] The charge densities are given in relative units and plotted for a ⟨110⟩ plane cutting the ⟨111⟩ surface. (After Ref. 233.)

to be the sum of the Pd metallic radius and the H covalent radius. The adsorption site is determined by comparing the calculated state density with the experimental photoemission spectrum (see below). Equations (1) are solved, with the local-density approximation for exchange-correlation.

Figure 44 shows the states localized in the surface region for this system, as well as those for bare Pd⟨111⟩. The prominent adsorbate-induced features are an H–Pd bonding adsorbate band below the Pd bulk d bands and a dispersive band of antibonding H–Pd states just above the Fermi level in a gap in the projected band structure (near $\bar{K}$). (Chemisorption also affects the intrinsic surface states, as is evident in the figure.) The bonding and antibonding character of the H–Pd states is clearly seen in the charge density contours of Fig. 45. The picture suggested is that the interaction between the hydrogen 1*s* states and the *d* states of Pd is stronger than the *d*-band width, leading to the formation of well-defined surface molecular bonding and antibonding states (Newns' picture[234]).

Figure 46 shows the surface-region density of states for bare Pd and for the Pd–H chemisorption system. (The surface region is taken to be the region outward from and including the first Pd layer.) The effects of H chemisorption most noticeable in this graph are the presence of the peak at -6.5 eV and the sharp reduction of state density near the Fermi level. The latter effect is due primarily to the removal of intrinsic surface states and resonances from this energy region[233] (see Fig. 44). The antibonding states described above are not visible in Fig. 46 because of their dispersion and because of the fact that they can exist as true localized states only in a rather limited region of the two-dimensional Brillouin zone.[233]

In comparing the state-density results to experiment,[235] Louie constructs the broadened difference spectrum

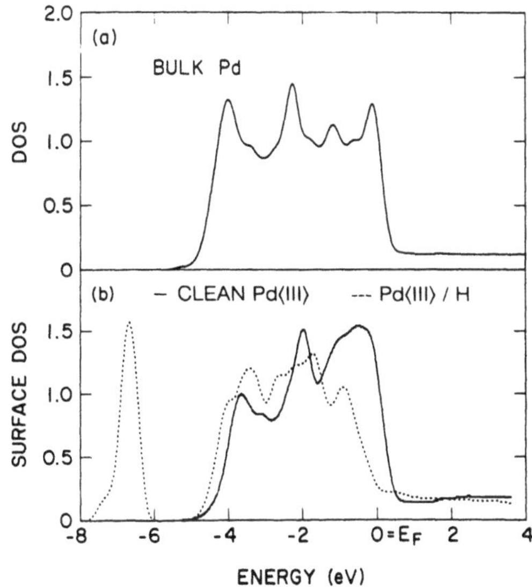

FIGURE 46. Lower half of figure (b) gives surface-region density of states for clean Pd⟨111⟩ and for H/Pd⟨111⟩ (adsorption in *C* sites). (Surface region is the region outward from and including the first Pd layer.) Upper half of figure (a) gives state density of bulk Pd for comparison. (Relative units.) (From Ref. 233.)

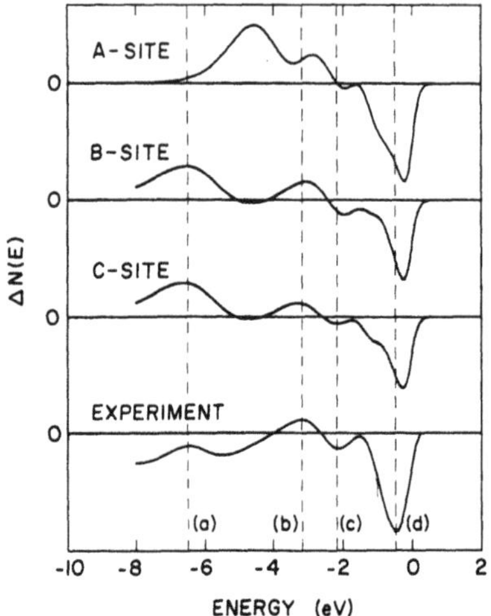

FIGURE 47. Comparison of the calculated hydrogen-induced photoemission difference spectra for H/Pd⟨111⟩ with the experimental (Ref. 235) difference curve. The site over a surface Pd atom is denoted the A-site; the two centered threefold sites are denoted B (over a hole in the second layer of Pd atoms) and C (over an atom in the second Pd layer.) (From Ref. 233.)

$$\delta n_b(\epsilon) \propto \int_{-\infty}^{\epsilon_F} \delta n(\epsilon') \exp\left[-\frac{(\epsilon-\epsilon')^2}{\gamma(\epsilon)}\right] d\epsilon'$$

where $\gamma(\epsilon)$ is an experimentally determined lifetime and instrument broadening factor. Final-state and matrix-element effects are not included. Figure 47 gives the comparison. The case with an H atom on top of each Pd atom (A-site) is clearly ruled out; the agreement is very reasonable for the two possible centered (3-fold) sites (B and C).

The prominent dip (d) in the curve is due as noted above to the removal of intrinsic surface states and resonances from this energy region by chemisorption. The peak (a) corresponds to the bonding band at this energy (Figs. 44 and 46). Peak (b) arises primarily from the appearance of two surface state bands near $\bar{\Gamma}$ in the corresponding energy region (see Fig. 44) and the valley (c) results from the removal of these states, and of clean-Pd surface states near $\bar{K}$, from the energy region of the valley.[233] A rather good description of the experimental spectrum is thus obtained in this calculation.

References

1. W. Kohn and P. Vashista, Chap. 2 in this volume.
2. L. Hedin and S. Lundqvist, in *Solid State Physics*, F. Seitz, D. Turnbull, and H. Ehrenreich, eds., Vol. 23, pp. 1–181 (Academic Press, New York, 1969).
3. N. H. March, in *Orbital Theories of Molecules and Solids*, N. H. March, ed., pp. 95–122 (Clarendon Press, Oxford, 1974).

4. A. K. Rajagopal, in *Advances in Chemical Physics*, I. Prigogine and S. A. Rice, eds., Vol. 41, pp. 59–193 (Wiley, New York, 1980).
5. B. I. Lundqvist, H. Hjelmberg, and O. Gunnarsson, in *Photoemission and the Electronic Properties of Surfaces*, B. Feuerbacher, B. Fitton, and R. F. Willis, eds., pp. 227–278 (Wiley, New York, 1978).
6. O. Gunnarsson, in *Electrons in Disordered Metals and at Metallic Surfaces*, P. Phariseau, B. L. Gyorffy, and L. Scheire, eds., pp. 1–53 (Plenum, New York, 1979).
7. T. B. Grimley, in *Electronic Structure and Reactivity of Metal Surfaces*, E. G. Derouane and A. A. Lucas, eds., pp. 113–162 (Plenum Press, New York, 1976).
8. J. R. Schrieffer, *J. Vac. Sci. Technol.* **9,** 561–568 (1972); **13,** 335–342 (1976).
9. S. Lundqvist, in *Surface Science*, Vol. 1, pp. 331–392 (International Atomic Energy Agency, Vienna, 1975).
10. N. H. March, in *Physics and Contemporary Needs,* Riazuddin, Ed., Vol. 1, pp. 53–95 (Plenum Press, New York, 1977).
11. F. Garcia-Molíner and F. Flores, *Introduction to the Theory of Solid Surfaces* (Cambridge Univ. Press, Cambridge, 1979).
12. J. Hölzl and F. K. Schulte, in *Solid Surface Physics*, Springer Tracts in Modern Physics Vol. 85, G. Höhler, ed., pp. 1–150 (Springer-Verlag, Berlin, 1979).
13. R. C. Brown and N. H. March, *Phys. Rep.* **24C,** 77–169 (1976).
14. F. Forstmann, in: *Photoemission and the Electronic Properties of Surfaces*, B. Feuerbacher, B. Fitton, and R. F. Willis, eds., pp. 193–226 (Wiley, New York, 1978).
15. J. W. Gadzuk, in *Surface Physics of Materials*, J. M. Blakely, ed., Vol. 2, pp. 339–375 (Academic Press, New York, 1975).
16. J. R. Smith, in *Interactions on Metal Surfaces*, R. Gomer, ed., pp. 1–39 (Springer-Verlag, Berlin, 1975).
17. J. P. Muscat and D. M. Newns, *Progr. Surf. Sci.* **9,** 1–43 (1978).
18. S. K. Lyo and R. Gomer, in *Interactions on Metal Surfaces*, R. Gomer, ed., pp. 41–62 (Springer-Verlag, Berlin, 1975).
19. D. S. Boudreaux and H. J. Juretschke, in *Structure and Properties of Metal Surfaces*, Honda Memorial Series on Materials Science, Vol. 1 (Maruzen, Tokyo, 1973).
20. J. A. Appelbaum and D. R. Hamann, *Rev. Mod. Phys.* **48,** 479–496 (1976).
21. N. D. Lang, in *Solid State Physics*, F. Seitz, D. Turnbull, and H. Ehrenreich, eds., Vol. 28, pp. 225–300 (Academic Press, New York, 1973).
22. P. Hohenberg and W. Kohn, *Phys. Rev.* **136,** B864–B871 (1964).
23. W. Kohn and L. J. Sham, *Phys. Rev.* **140,** A1133–A1138 (1965).
24. B. Y. Tong, *Phys. Rev. B* **5,** 1436–1439 (1972).
25. C. H. Hodges and M. J. Stott, *Phys. Rev. B* **7,** 73–79 (1973).
26. R. M. Nieminen and C. H. Hodges, *Solid State Commun.* **18,** 1115–1118 (1976).
27. E. Wigner and J. Bardeen, *Phys. Rev.* **48,** 84–87 (1935).
28. N. D. Lang and W. Kohn, *Phys. Rev. B* **3,** 1215–1223 (1971).
29. N. D. Lang and W. Kohn, *Phys. Rev. B* **8,** 6010–6012 (1973).
30. F. K. Schulte, *Z. Phys. B* **27,** 303–307 (1977).
31. Cf. J. F. Janak, *Phys. Rev. B* **18,** 7165–7168 (1978).
32. N. D. Lang, *Solid State Commun.* **7,** 1047–1050 (1969).
33. N. D. Lang and W. Kohn, *Phys. Rev. B* **1,** 4555–4568 (1970).
34. H. F. Budd and J. Vannimenus, *Phys. Rev. Lett.* **31,** 1218–1221 and 1430 (erratum) (1973).
35. J. Vannimenus and H. F. Budd, *Solid State Commun.* **15,** 1739–1743 (1974).
36. J. Heinrichs and N. Kumar, *Phys. Rev. B* **12,** 802–805 (1975).

37. H. F. Budd and J. Vannimenus, *Phys. Rev. B* **14,** 854–855 (1976).
38. B. Raykov, *Solid State Commun.* **25,** 257–259 (1978).
39. J. Vannimenus and H. F. Budd, *Phys. Rev. B* **15,** 5302–5306 (1977).
40. J. Heinrichs, *Phys. Stat. Sol. (b)* **92,** 185–192 (1979).
41. G. D. Mahan and W. L. Schaich, *Phys. Rev. B* **10,** 2647–2651 (1974); J. P. Perdew and V. Sahni, *Solid State Commun.* **30,** 87–90 (1979).
42. L. H. Thomas, *Proc. Cambridge Phil. Soc.* **23,** 542–547 (1927); E. Fermi, *Z. Phys.* **48,** 73–79 (1928).
43. J. Frenkel, *Z. Phys.* **51,** 232–238 (1928); A. Samoilovich, *Acta Pyhsicochim. URSS* **20,** 97–120 (1945) (notes certain errors in the Frenkel paper); B. Mrowka and A. Recknagel, *Phys. Z.* **38,** 758–765 (1937).
44. P. Gombas, *Z. Phys.* **121,** 523–542 (1943).
45. J. R. Smith, Ph. D. thesis, Ohio State University, Columbus, 1968 (unpublished).
46. J. Vannimenus, Thése de Doctorat d' État, Université Pierre et Marie Curie, Paris, 1976 (unpublished).
47. J. Goodisman, *J. Chem. Phys.* **63,** 4437–4441 (1975).
48. P. Gombas, in *Handbuch der Physik*, S. Flügge, ed., Vol. 36, pp. 109–231 (Springer-Verlag, Berlin, 1956).
49. J. R. Smith, *Phys. Rev.* **181,** 522–529 (1969).
50. C. Warner, in *Thermionic Conversion Specialists Conference, San Diego, 1971*, pp. 170–180 (Institute of Electrical and Electronics Engineers, New York, 1972).
51. W. A. Tiller, S. Ciraci, and I. P. Batra, *Surf. Sci.* **65,** 173–188 (1977).
52. C. Q. Ma and V. Sahni, *Phys. Rev. B* **19,** 1290–1294 (1979).
53. M. M. Pant and A. K. Rajagopal, *Solid State Commun.* **10,** 1157–1160 (1972).
54. J. Ferrante and J. R. Smith, *Surf. Sci.* **38,** 77–92 (1973).
55. J. Vannimenus and H. F. Budd, *Solid State Commun.* **17,** 1291–1295 (1975).
56. R. M. Kobeleva, O. M. Rozental', and A. V. Kobelev, *Fiz. Metal. Metalloved.* **40,** 652–656 (1975); Engl. trans. in *Physics of Metals and Metallography* **40,** (3)180–184 (1975).
57. R. Mehrotra, M. M. Pant, and M. P. Das, *Solid State Commun.* **18,** 199–201 (1976).
58. J. N. Swingler and J. C. Inkson, *J. Phys. C* **10,** 573–579 (1977).
59. I. Gyémánt and G. Solt, *Phys. Stat. Sol. (b)* **82,** 651–655 (1977).
60. L. M. Sander, H. B. Shore, and L. J. Sham, *Phys. Rev. Lett.* **31,** 533–536 (1973); *Solid State Commun.* **27,** 331–333 (1978).
61. H. Büttner and E. Gerlach, *J. Phys. C* **6,** L433–L436 (1973).
62. T. M. Rice, *Phys. Rev. B* **9,** 1540–1546 (1974).
63. T. L. Reinecke and S. C. Ying, *Solid State Commun.* **14,** 381–385 (1974).
64. P. Vashishta, R. K. Kalia, and K. S. Singwi, *Solid State Commun.* **19,** 935–938 (1976).
65. D. C. Langreth, *Phys. Rev. B* **5,** 2842–2843 (1972).
66. J. A. Appelbaum and E. I. Blount, *Phys. Rev. B* **8,** 483–491 (1973).
67. G. Paasch and H. Wonn, *Phys. Stat. Sol. (b)* **70,** 555–566 (1975).
68. J. E. Inglesfield, *J. Phys. C* **10,** 4067–4072 (1977).
69. A. Sugiyama, *J. Phys. Soc. Jpn* **15,** 965–982 (1960).
70. V. E. Kenner and R. E. Allen, *Phys. Rev. B* **11,** 2858–2859 (1975).
71. J. Bardeen, *Phys. Rev.* **49,** 653–663 (1936).
72. A. J. Bennett and C. B. Duke, in *The Structure and Chemistry of Solid Surfaces*, G. A. Somorjai, ed., pp. 25-1 to 25-8 (Wiley, New York, 1969).
73. F. K. Schulte, *Surf. Sci.* **55,** 427–444 (1976).
74. J. H. Rose, Jr. and H. B. Shore, *Solid State Commun.* **17,** 327–330 (1975).

75. G. D. Mahan, *Phys. Rev. B* **12,** 5585–5589 (1975).
76. V. Sahni and J. Gruenebaum, *Phys. Rev. B* **15,** 1929–1935 (1977); V. Sahni, J. B. Krieger, and J. Gruenebaum, *Phys. Rev. B* **15,** 1941–1949 (1977); V. Sahni and J. Gruenebaum, *Solid State Commun.* **21,** 463–465 (1977); V. Sahni, C. Q. Ma, and J. S. Flamholz, *Phys. Rev. B* **18,** 3931–3945 (1978); C. Q. Ma and V. Sahni, *Study of the Inhomogeneous Electron Gas at Surfaces*, research report of Brooklyn College and the Research Foundation of the City University of New York, 1979 (unpublished).
77. J. E. van Himbergen and R. Silbey, *Phys. Rev. B* **18,** 2674–2682 (1978).
78. R. L. Kautz and B. B. Schwartz, *Phys. Rev. B* **14,** 2017–2031 (1976); D. Nagy, *Surf. Sci* **90,** 102–108 (1979).
79. A. J. Bennett and C. B. Duke, *Phys. Rev.* **160,** 541–553 (1967); **162,** 578–588 (1967).
80. J. Ferrante and J. R. Smith, *Solid State Commun.* **20,** 393–396 (1976); **23,** 527–529 (1977).
81. R. M. Nieminen, *J. Phys. F* **7,** 375–384 (1977).
82. J. P. Muscat and G. Allan, *J. Phys. F* **7,** 999–1008 (1977).
83. G. G. Robinson and P. F. de Chatel, *J. Phys. F* **5,** 1502–1511 (1975).
84. M. Manninen, R. Nieminen, P. Hautojärvi, and J. Arponen, *Phys. Rev. B* **12,** 4012–4022 (1975).
85. P. Jena, A. K. Gupta, and K. S. Singwi, *Solid State Commun.* **21,** 293–296 (1977).
86. R. Evans and M. W. Finnis, *J. Phys. F* **6,** 483–498 (1976).
87. J. H. Rose and H. B. Shore, *Phys. Rev. B* **17,** 1884–1892 (1978).
88. R. K. Kalia and P. Vashishta, *Phys. Rev. B* **17,** 2655–2672 (1978).
89. A. R. Williams and U. von Barth, Chap. 4 of this volume.
90. O. Gunnarsson and B. I. Lundqvist, *Phys. Rev. B* **13,** 4274–4298 (1976).
91. O. Gunnarsson, J. Harris, and R. O. Jones, *J. Chem. Phys.* **67,** 3970–3979 (1977); J. Harris and R. O. Jones, *J. Chem Phys.* **68,** 1190–1193 (1978); **70,** 830–841 (1979).
92. V. L. Moruzzi, J. F. Janak, and A. R. Williams, *Calculated Electronic Properties of Metals* (Pergamon Press, New York, 1978).
93. A. K. Gupta and K. S. Singwi, *Phys. Rev. B* **15,** 1801–1810 (1977).
94. V. Peuckert, *J. Phys. C* **7,** 2221–2233 (1974); **9,** 809–817 (1976).
95. T. L. Loucks and P. H. Cutler, *J. Phys. Chem. Solids,* **25,** 105–113 (1964).
96. J. Bardeen, *Surf. Sci.* **2,** 381–388 (1964).
97. J. C. Inkson, *J. Phys. C* **10,** 567–572 (1977).
98. J. E. Inglesfield and I. D. Moore, *Solid State Commun.* **26,** 867–871 (1978).
99. H. B. Huntington, *Phys. Rev.* **81,** 1035–1039 (1951).
100. D. C. Langreth, *Comments Solid State Phys.* **8,** 129–134 (1978).
101. R. Monnier and J. P. Perdew, *Phys. Rev. B* **17,** 2595–2611 (1978).
102. J. A. Appelbaum and D. R. Hamann, *Solid State Commun.* **27,** 881–883 (1978).
103. J. Schmit and A. A. Lucas, *Solid State Commun.* **11,** 415–418 (1972).
104. R. A. Craig, *Phys. Rev. B* **6,** 1134–1142 (1972).
105. V. Peuckert, *Z. Phys.* **241,** 191–204 (1971).
106. W. Kohn, *Solid State Commun.* **13,** 323–324 (1973).
107. P. J. Feibelman, *Solid State Commun.* **13,** 319–321 (1973).
108. M. Jonson and G. Srinivasan, *Phys. Lett.* **43A,** 427–428 (1973).
109. R. A. Craig, *Solid State Commun.* **13,** 1517–1519 (1973).
110. J. C. Phillips, *Comments Solid State Phys.* **6,** 91–94 (1975).

111. W. Kohn and N. D. Lang, *Comments Solid State Phys.* **6,** 95–101 (1975).
112. J. Heinrichs, *Solid State Commun.* **13,** 1599–1602 (1973); *Phys. Rev. B* **11,** 3637–3643 (1975).
113. A. Griffin, H. Kranz, and J. Harris, *J. Phys. F* **4,** 1744–1754 (1974); A. Griffin and H. Kranz, *Phys. Rev. B* **15,** 5068–5072 (1977).
114. G. Paasch, *Phys. Stat. Sol. (b)* **65,** 221–229 (1974).
115. R. G. Barrera and E. Gerlach, *Solid State Commun.* **14,** 979–981 (1974).
116. J. Harris and R. O. Jones, *J. Phys. F* **4,** 1170–1186 (1974).
117. M. Jonson and G. Srinivasan, *Phys. Scripta* **10,** 262–272 (1974).
118. E. Wikborg and J. E. Inglesfield, *Solid State Commun.* **16,** 335–339 (1975); *Physica Scripta* **15,** 37–55 (1977).
119. D. C. Langreth and J. P. Perdew, *Solid State Commun.* **17,** 1425–1429 (1975); *Phys. Rev. B* **15,** 2884–2901 (1977).
120. N. D. Lang and L. J. Sham, *Solid State Commun.* **17,** 581–584 (1975).
121. J. S.-Y. Wang and M. Rasolt, *Phys. Rev. B* **13,** 5330–5337 (1976).
122. K. H. Lau and W. Kohn, *J. Phys. Chem. Solids* **37,** 99–104 (1976).
123. J. H. Rose, Jr., H. B. Shore, D. J. W. Geldart, and M. Rasolt, *Solid State Commun.* **19,** 619–622 (1976).
124. M Rasolt and D. J. W. Geldart, *Phys. Rev. Lett.* **35,** 1234–1237 (1975); D. J. W. Geldart and M. Rasolt, *Phys. Rev. B* **13,** 1477–1488 (1976).
125. J. P. Perdew, D. C. Langreth, and V. Sahni, *Phys. Rev. Lett.* **38,** 1030–1033 (1977).
126. D. C. Langreth and J. P. Perdew, *Phys. Rev. B* **21,** 5469–5493 (1980).
127. M. Rasolt, G. Malmström, and D. J. W. Geldart, *Phys. Rev. B* **20,** 3012–3019 (1979).
128. V. Peuckert, *J. Phys. C* **9,** 4173–4184 (1976).
129. L. I. Schiff, *Phys. Rev. B* **1,** 4649–4654 (1970).
130. A. K. Theophilou and A. Modinos, *Phys. Rev. B* **6,** 801–812 (1972); A. K. Theophilou, *J. Phys. F* **2,** 1124–1136 (1972).
131. J. A. Appelbaum and D. R. Hamann, *Phys. Rev. B* **6,** 1122–1130 (1972).
132. N. D. Lang and W. Kohn, *Phys. Rev. B* **7,** 3541–3550 (1973).
133. V. E. Kenner, R. E. Allen, and W. M. Saslow, *Phys. Rev. B* **8,** 576–581 (1973).
134. J. R. Smith, S. C. Ying, and W. Kohn, *Phys. Rev. Lett.* **30,** 610–613 (1973).
135. S. C. Ying, J. R. Smith, and W. Kohn, *Phys. Rev. B* **11,** 1483–1496 (1975).
136. P. W. Lert and J. H. Weare, *J. Chem. Phys.* **68,** 5010–5019 (1978).
137. H. F. Budd and J. Vannimenus, *Phys. Rev. B* **12,** 509–513 (1975).
138. R. L. Kautz, A. J. Freeman, and B. B. Schwartz, *Phys. Lett.* **57A,** 473–474 (1976).
139. S. C. Ying, L. M. Kahn, and M. T. Beal-Monod, *Solid State Commun.* **18,** 359–362 (1976); S. C. Ying and L. M. Kahn, *Surf. Sci.* **67,** 278–284 (1977).
140. J. P. Perdew, *Phys. Rev. B* **16,** 1525–1535 (1977).
141. V. Peuckert, *J. Phys. C* **11,** 4945–4956 (1978).
142. S. C. Ying, *Nuovo Cim. B* **23,** 270–281 (1974).
143. N. D. Lang, in *Electronic Structure and Reactivity of Metal Surfaces*, E. G. Derouane and A. A. Lucas, eds., pp. 81–111 (Plenum Press, New York, 1976).
144. N. D. Lang and A. R. Williams, *Phys. Rev. Lett.* **34,** 531–534 (1975); **37,** 212–215 (1976).
145. N. D. Lang and A. R. Williams, *Phys. Rev. B* **16,** 2408–2419 (1977).
146. N. D. Lang and A. R. Williams, *Phys. Rev. B* **18,** 616–636 (1978).
147. M. W. Finnis, *J. Phys. F* **5,** 2227–2240 (1975).
148. P. W. Lert and J. H. Weare, *J. Phys. C* **11,** 1865–1875 (1978).

149. N. W. Ashcroft, *Phys. Lett.* **23,** 48–50 (1966); N. W. Ashcroft and D. C. Langreth, *Phys. Rev.* **155,** 682–684 (1967).
150. J. K. Grepstad, P. O. Gartland, and B. J. Slagsvold, *Surf. Sci.* **57,** 348–362 (1976); R. M. Eastment and C. H. B. Mee, *J. Phys. F* **3,** 1738–1745 (1973).
151. J. Ferrante and J. R. Smith, *Phys. Rev. B* **19,** 3911–3920 (1979).
152. J. W. Allen and S. A. Rice, *J. Chem. Phys.* **68,** 5053–5061 (1978).
153. G. Paasch and M. Hietschold, *Phys. Stat. Sol. (b)* **83,** 209–222 (1977).
154. M. Manninen and R. M. Nieminen, *J. Phys. F* **8,** 2243–2260 (1978).
155. R. Monnier, J. P. Perdew, D. C. Langreth, and J. W. Wilkins, *Phys. Rev. B* **18,** 656–666 (1978).
156. G. Paasch and M. Hietschold, *Phys. Stat. Sol. (b)* **67,** 743–754 (1975).
157. M. Hietschold, G. Paasch, and P. Ziesche, *Phys. Stat. Sol. (b)* **70,** 653–662 (1975).
158. M. M. Pant and M. P. Das, *J. Phys. F* **5,** 1301–1306 (1975).
159. V. Sahni and J. Gruenebaum, *Phys. Rev. B* **19,** 1840–1854 (1979).
160. C. Tejedor and F. Flores, *J. Phys. F* **6,** 1647–1659 (1976); C. Tejedor, *J. Phys. F* **7,** 991–997 (1977).
161. J. G. Gay, J. R. Smith, and F. J. Arlinghaus, *Phys. Rev. Lett.* **38,** 561–564 (1977); **42,** 332–335 (1979).
162. F. J. Arlinghaus, J. G. Gay, and J. R. Smith, *Phys. Rev. B* **21,** 2055–2059 (1980).
163. J. R. Smith and J. G. Gay, *Phys. Rev. B* **12,** 4238–4246 (1975).
164. R. O. Jones, P. J. Jennings, and G. S. Painter, *Surf. Sci.* **53,** 409–428 (1975).
165. D. R. Salahub and R. P. Messmer, *Phys. Rev. B* **16,** 2526–2536 (1977).
166. J. A. Appelbaum and D. R. Hamann, *Phys. Rev. B* **6,** 2166–2177 (1972).
167. K. -P. Bohnen and S. C. Ying, *Phys. Rev. B* **22,** 1806–1817 (1980).
168. G. P. Alldredge and L. Kleinman, *Phys. Rev. B* **10,** 559–573 (1974).
169. J. R. Chelikowsky, M. Schlüter, S. G. Louie, and M. L. Cohen, *Solid State Commun.* **17,** 1103–1106 (1975).
170. P. J. Feibelman, J. A. Appelbaum, and D. R. Hamann, *Phys. Rev. B* **31,** 413–417 (1979); *Solid State Commun.* **20,** 1433–1443 (1979).
171. C. S. Wang and A. J. Freeman, *Phys. Rev. B* **19,** 793–805 (1979); *J. Appl. Phys.* **50,** 1940–1943 (1979); *Phys. Rev. B* **21,** 4585–4591 (1980).
172. S. G. Louie, K. -M. Ho, J. R. Chelikowsky, and M. L. Cohen, *Phys. Rev. B* **15,** 5627–5635 (1977).
173. S. G. Louie, *Phys. Rev. Lett.* **40,** 1525–1528 (1978).
174. J. E. Demuth, *Surf. Sci.* **65,** 369–375 (1977).
175. P. O. Gartland and B. J. Slagsvold, *Phys. Rev. B* **12,** 4047–4058 (1975).
176. G. P. Kerker, K. M. Ho, and M. L. Cohen, *Phys. Rev. Lett.* **40,** 1593–1596 (1978).
177. J. A. Appelbaum and D. R. Hamann, *Phys. Rev. Lett.* **32,** 225–228 (1974); *Phys. Rev. B* **12,** 1410–1417 (1975).
178. J. A. Appelbaum, G. A. Baraff, and D. R. Hamann, *Phys. Rev. B* **11,** 3822–3831 (1975); **12,** 1410–1417 and 5749–5757 (1975); **14,** 588–601 and 1623–1632 (1976); **15,** 2408–2412 (1977).
179. M. Schlüter, J. R. Chelikowsky, S. G. Louie, and M. L. Cohen, *Phys. Rev. Lett.* **34,** 1385–1388 (1975); *Phys. Rev. B* **12,** 4200–4214 (1975).
180. J. R. Chelikowsky and M. L. Cohen, *Phys. Rev. B* **13,** 826–834 (1976).
181. J. R. Chelikowsky, *Phys. Rev. B* **15,** 3236–3242 (1977).
182. G. P. Kerker, S. G. Louie, and M. L. Cohen, *Phys. Rev. B* **17,** 706–715 (1978).
183. S. Ciraci and I. P. Batra, *Phys. Rev. B* **15,** 3254–3259 (1977).
184. D. R. Hamann, *Phys. Rev. Lett.* **42,** 662–665 (1979).

185. R. Smoluchowski, *Phys. Rev.* **60,** 661–674 (1941).
186. H. Wawra, *Z. Metallkunde* **66,** 395–401 (1975).
187. R. M. Nieminen and C. H. Hodges, *J. Phys. F* **6,** 573–585 (1976).
188. V. L. Moruzzi and A. R. Williams, private communication. Lattice constant used minimized calculated total energy.
189. A. R. Williams, J. Kubler, and C. D. Gelatt, Jr., *Phys. Rev. B* **19,** 6094–6118 (1979).
190. G. G. Kleiman and U. Landman, *Phys. Rev. B* **8,** 5484–5495 (1973).
191. Cf. E. Zaremba and W. Kohn, *Phys. Rev. B* **15,** 1769–1781 (1977).
192. J. E. van Himbergen and R. Silbey, *Solid State Commun.* **23,** 623–627 (1977).
193. N. D. Lang, *Phys. Rev. B* **4,** 4234–4245 (1971).
194. K. F. Wojciechowski, *Surf. Sci.* **55,** 246–258 (1976); **80,** 253–260 (1979).
195. J. R. Schrieffer, in: *Collective Properties of Physical Systems*, B. Lundqvist and S. Lundqvist, eds., pp. 159–160 (Nobel Foundation, Stockholm, 1973.
196. S. -W. Wang and W. H. Weinberg, *Surf. Sci.* **77,** 14–28 (1978).
197. L. M. Kahn and S. C. Ying, *Surf. Sci.* **59,** 333–360 (1976).
198. W. Jones and W. H. Young, *J. Phys. C* **4,** 1322–1330 (1971).
199. H. B. Huntington, L. A. Turk, and W. W. White III, *Surf. Sci.* **48,** 187–203 (1975).
200. O. Gunnarsson and H. Hjelmberg, *Phys. Scripta* **11,** 97–103 (1975); O. Gunnarsson, H. Hjelmberg, and B. I. Lundqvist, *Phys. Rev. Lett.* **37,** 292–295 (1976).
201. O. Gunnarsson, H. Hjelmberg, and B. I. Lundqvist, *Surf. Sci.* **63,** 348–357 (1977).
202. H. Hjelmberg, O. Gunnarsson, and B. I. Lundqvist, *Surf. Sci.* **68,** 158–166 (1977).
203. H. Hjelmberg, *Phys. Scripta* **18,** 481–493 (1978).
204. P. Johansson and H. Hjelmberg, *Surf. Sci.* **80,** 171–178 (1979).
205. H. Hjelmberg, *Surf. Sci.* **81,** 539–561 (1979).
205a. J. K. Nørskov and N. D. Lang, *Phys. Rev. B* **21,** 2131–2136 (1980).
206. R. F. W. Bader, W. H. Henneker, and P. E. Cade, *J. Chem. Phys.* **46,** 3341–3363 (1967); R. F. W. Bader, I. Keaveny, and P. E. Cade, *J. Chem. Phys.* **47,** 3381–3402 (1967).
207. C. D. Gelatt, Jr., H. Ehrenreich, and J. A. Weiss, *Phys. Rev. B* **17,** 1940–1957 (1978).
208. B. Kasemo, E. Törnqvist, J. K. Nørskov, and B. I. Lundqvist, *Surf. Sci.* **89,** 554–565 (1979); J. K. Nørskov and B. I. Lundqvist, *Phys. Rev. B* **19,** 5661–5665 (1979); *Surf. Sci.* **89,** 251–261 (1979).
209. O. Gunnarsson, H. Hjelmberg, and J. K. Nørskov, *Phys. Scripta* **22,** 165–170 (1980).
210. J. P. Muscat and D. M. Newns, *Phys. Rev. B* **19,** 1270–1282 (1979).
211. R. L. Matcha and S. C. King, Jr., *J. Am. Chem. Soc.* **98,** 3415–3420 and 3420–3432 (1976).
212. J. K. Nørskov, H. Hjelmberg, and B. I. Lundqvist, *Solid State Commun.* **28,** 899–902 (1978); H. Hjelmberg, B. I. Lundqvist, and J. K. Nørskov, *Phys. Scripta* **20,** 192–201 (1979).
213. J. K. Nørskov, *Solid State Commun.* **25,** 995–998 (1978).
214. B. I. Lundqvist, J. K. Nørskov, and H. Hjelmberg, *Surf. Sci.* **80,** 441–449 (1979)
215. B. I. Lundqvist, O. Gunnarsson, H. Hjelmberg, and J. K. Nørskov, *Surf. Sci.* **89,** 196–225 (1979).
216. O. Gunnarsson and B. I. Lundqvist, *Phys. Rev. B* **13,** 4274–4298 (1976).
217. N. D. Lang and A. R. Williams, *Phys. Rev. Lett.* **40,** 954–957 (1978).
218. G. Wendin, M. Ohno, and S. Lundqvist, *Solid State Commun.* **19,** 165–170 (1976).
219. J. C. Slater, *The Self-Consistent Field Method for Molecules and Solids: Quantum Theory of Molecules and Solids*, Vol. 4 (McGraw-Hill, New York, 1974).

220. K. Y. Yu, J. N. Miller, P. Chye, W. E. Spicer, N. D. Lang, and A. R. Williams, *Phys. Rev. B* **14,** 1446–1449 (1976).
221. L. Ley, S. P. Kowalczyk, F. R. McFeely, R. A. Pollak, and D. A. Shirley, *Phys. Rev. B* **8,** 2392–2402 (1973).
222. C. F. Melius, T. H. Upton, and W. A. Goddard III, *Solid State Commun.* **28,** 501–504 (1978).
223. J. Harris and G. S. Painter, *Phys. Rev. Lett.* **36,** 151–154 (1976); R. P. Messmer, C. W. Tucker, Jr., and K. H. Johnson, *Surf. Sci.* **42,** 341–354 (1974); I. P. Batra and P. S. Bagus, *Solid State Commun.* **16,** 1097–1100 (1975); I. P. Batra and O. Robaux, *Surf. Sci.* **49,** 653–656 (1975); N. Rösch and D. Menzel, *Chem. Phys.* **13,** 243–256 (1976); C. H. Li and J. W. D. Connolly, *Surf. Sci.* **65,** 700–705 (1977); R. P. Messmer and D. R. Salahub, *Phys. Rev. B* **16,** 3415–3427 (1977); D. R. Salahub, M. Roche, and R. P. Messmer, *Phys. Rev. B* **18,** 6495–6505 (1978).
224. J. A. Appelbaum and D.R. Hamann, *Phys. Rev. Lett.* **34,** 806–809 (1975); *Phys. Rev. B* **15,** 2006–2011 (1977).
225. J. A. Appelbaum, H. D. Hagstrum, D. R. Hamann, and T. Sakurai, *Surf. Sci.* **58,** 479–484 (1976).
226. J. A. Appelbaum, D. R. Hamann, and K. H. Tasso, *Phys. Rev. Lett.* **39,** 1487–1490 (1977).
227. J. A. Appelbaum, G. A. Baraff, D. R. Hamann, H. D. Hagstrum, and T. Sakurai, *Surf. Sci.* **70,** 654–673 (1978).
228. M. Schlüter, J. E. Rowe, G. Margaritondo, K. M. Ho, and M. L. Cohen, *Phys. Rev. Lett.* **37,** 1632–1635 (1976).
229. K. M. Ho, M. L. Cohen, and M. Schlüter, *Phys. Rev. B* **15,** 3888–3897 (1977).
230. M. Schlüter and M. L. Cohen, *Phys. Rev. B* **17,** 716–725 (1978).
231. Cf. I. P. Batra and S. Ciraci, *Phys. Rev. Lett.* **34,** 1337–1340 (1975); **36,** 170–173 (1976); I. P. Batra, S. Ciraci, and I. B. Ortenburger, *Solid State Commun.* **18,** 563–565 (1978).
232. J. R. Smith, F. J. Arlinghaus, and J. G. Gay, *Solid State Commun.* **24,** 279–282 (1977).
233. S. G. Louie, *Phys. Rev. Lett,* **42,** 476–479 (1979).
234. D. M. Newns, *Phys. Rev.* **178,** 1123–1135 (1969).
235. J. E. Demuth, *Surf. Sci.* **65,** 369–375 (1977).

NOTE: This paper was submitted in 1979.

INDEX

GPSR Compliance
The European Union's (EU) General Product Safety Regulation (GPSR) is a set of rules that requires consumer products to be safe and our obligations to ensure this.

If you have any concerns about our products, you can contact us on

ProductSafety@springernature.com

In case Publisher is established outside the EU, the EU authorized representative is:

Springer Nature Customer Service Center GmbH
Europaplatz 3
69115 Heidelberg, Germany

www.ingramcontent.com/pod-product-compliance
Ingram Content Group UK Ltd.
Pitfield, Milton Keynes, MK11 3LW, UK
UKHW012149240726
13966UKWH00001B/222

* 9 7 8 1 4 8 9 9 0 4 1 6 4 *